Sustainable Coloration Techniques in Textiles

Saptarshi Maiti · Mohammad Shahid ·
Ravindra V. Adivarekar
Editors

Sustainable Coloration Techniques in Textiles

Springer

Editors
Saptarshi Maiti
Department of Fibres and Textile
Processing Technology
Institute of Chemical Technology
Mumbai, Maharashtra, India

Ravindra V. Adivarekar
Department of Fibres and Textile
Processing Technology
Institute of Chemical Technology
Mumbai, Maharashtra, India

Mohammad Shahid
Department of Chemistry
Koneru Lakshmaiah Education Foundation
(Deemed to be University)
Guntur, Andhra Pradesh, India

ISBN 978-981-96-4974-7 ISBN 978-981-96-4975-4 (eBook)
https://doi.org/10.1007/978-981-96-4975-4

This Springer imprint is published by the registered company Springer Nature Singapore Pte Ltd.
The registered company address is: 152 Beach Road, #21-01/04 Gateway East, Singapore 189721, Singapore

Preface

The textile industry has long been a vital part of the global economy, contributing to employment, trade, and the production of goods. However, the environmental challenges associated with textile manufacturing, particularly in dyeing processes, have raised significant concerns. As the world shifts towards more sustainable practices, this book, *Sustainable Coloration Techniques in Textiles*, explores innovative approaches to address the environmental impact of textile dyeing and coloration. The textile dyeing process has traditionally been water- and chemical-intensive, with many methods resulting in harmful effluents, excessive water consumption, and the release of toxic substances into the environment. As a result, sustainable solutions are urgently needed to reduce these adverse effects while maintaining the quality and performance of the dyed fabrics. This book highlights some of the most promising sustainable techniques in textile coloration that can help mitigate these environmental concerns. Chapter 1 sets the stage with an overview of the sustainability challenges within the textile dyeing industry, discussing the environmental impact of synthetic dyes, water consumption, and the importance of considering the entire life cycle of textile production. It also emphasizes the need for environmentally responsible practices and worker well-being. Chapter 2 reviews advancements in machinery and process optimizations that are crucial in reducing the environmental footprint throughout textile coloration processes. The following chapters discuss specific innovations and alternatives to conventional dyeing methods. From waterless dyeing (Chap. 3) to solvent-based dyeing (Chap. 4) and ultrasound-assisted techniques (Chap. 5), each chapter presents methods that aim to reduce environmental impact by minimizing water use, energy consumption, and the need for harmful chemicals. Electrochemical dyeing, discussed in Chaps. 6 and 7, offers another promising avenue: replacing traditional chemical methods with more sustainable electrochemical processes. Supercritical fluid dyeing (Chap. 8) and salt-free dyeing (Chap. 9) further reduce environmental pollution by eliminating wastewater generation and harmful salts. Green chemistry is key in sustainable textile coloration, as illustrated in Chaps. 10 and 11, where micelle dyeing and bio-derived dyes are explored. These methods prioritize environmentally friendly chemicals and materials, helping the industry transition to greener practices. Finally, the book addresses sustainable alternatives in textile wet processing with biosurfactants (Chap. 12) and advancements in textile printing (Chap. 13), providing an in-depth look at how the industry can move towards a more sustainable future. We hope this book will serve as a valuable resource for researchers, industry professionals, and policymakers, guiding the transition towards more sustainable and eco-friendly methods in textile production. We thank all the contributors for their valuable insights and dedication to advancing sustainable practices in textile coloration.

Mumbai, India — Dr. Saptarshi Maiti
Guntur, India — Dr. Mohammad Shahid
Mumbai, India — Prof. Ravindra V. Adivarekar

Contents

About the Editors

Dr. Saptarshi Maiti is an Assistant Professor in the Department of Fibres and Textile Processing Technology at the Institute of Chemical Technology (ICT), Mumbai, India. He completed his post-doctoral research at ICT-Mumbai in 2022. Dr. Maiti holds a Ph.D. (Tech.) and M.Tech. in Fibres and Textile Processing Technology from the same department. He has also worked as a Visiting Research Fellow at the School of Materials, University of Manchester, UK, under the UAA-ICT Dhirubhai Ambani Lifetime Achievement Award. His research interests include graphene and its applications in textiles, dendritic polymers, protein extraction from waste resources, natural dyeing, green processing of textiles, and statistical modeling in textiles. Dr. Maiti has authored more than 50 international publications in reputed journals, contributed to around 22 book chapters in textile chemistry and materials science, and holds one patent. He is a life member of The Indian Natural Fibre Society (INFS).

Dr. Mohammad Shahid Ph.D., is an Assistant Professor in the Department of Chemistry, Koneru Lakshmaiah Education Foundation, India. Before joining KLEF, he worked at various esteemed institutions, including Jamia Millia Islamia, New Delhi; Soochow University, China; University of Glasgow, UK; and the Institute of Chemical Technology, Mumbai. He earned his Master's degree in Chemistry from Shibli National College, Azamgarh, in 2006 and his Ph.D. in Chemistry from Jamia Millia Islamia, New Delhi, in 2014. Dr. Shahid has received several prestigious fellowships to support his research pursuits, including grants from UGC, India, China Postdoctoral Fellowship, EU Marie Curie Fellowship, and Dr. D. S. Kothari Fellowship. He has contributed to over 50 scientific publications with ~4000 citations and h-index of 24. In addition, has also edited and authored several books with renowned international publishing houses.

Dr. Ravindra V. Adivarekar Ph.D., is a distinguished professor in Fibre Chemistry at the Institute of Chemical Technology (ICT), Mumbai, where he has dedicated over 25 years of service. He holds multiple degrees, including an honorary Fellowship of the Textile Association (FTA). Dr. Adivarekar is renowned for his expertise in textile processing and served as the Head of the Department of Fibres and Textile Processing

Technology from 2008 to 2021. He currently serves as the Dean of Human Resource Development (HRD) at ICT Mumbai and as the Director of ICT-IOC Bhubaneswar. His extensive research interests span eco-friendly coloration, natural fiber composites, medical textiles, nanotechnology in textile processing, natural dyes, microbial dyes, and the application of enzymes. A committed mentor, Dr. Adivarekar has guided numerous students and managed research projects with total funding exceeding 18 crores. He has contributed significantly to the academic community with over 200 research papers published in national and international journals, more than 70 conference proceedings, and three patents. Beyond his academic achievements, Dr. Adivarekar is an esteemed member of several textile-related organizations and has received significant recognition for his contributions to the industry. He actively engages in advisory roles and government committees, addressing environmental issues in textile manufacturing.

Chapter 1
Dyes and Dyeing of Textiles: An Overview of Sustainability Issues

Carlos Rafael Silva de Oliveira, Patrícia Viera de Oliveira, Grazyella Cristina Oliveira de Aguiar, Maria Elisa Philippsen Missner, Robert Kimutai Tewo, Afonso Henrique da Silva Júnior, António Benjamim Mapossa, and Catia Rosana Lange de Aguiar

Abstract The textile industry plays a fundamental role in the global economy, contributing significantly to employment and trade, especially in emerging countries. However, the environmental impact of textile production, particularly those arising from dyeing processes, has raised critical concerns. Although synthetic dyes have revolutionized the industry by offering a more comprehensive color range and improved color stability, their production and application have established substantial environmental challenges. In this context, the chemical composition of dyes, emphasizing the persistent use of dangerous substances, such as heavy metals and aromatic amines, which have harmful effects on ecosystems and human health, are

C. R. S. de Oliveira (✉) · G. C. O. de Aguiar · M. E. P. Missner · C. R. L. de Aguiar
Textile Engineering Department, Federal University of Santa Catarina-UFSC, Blumenau, Santa Catarina, Brazil
e-mail: carlos.oliveira@ufsc.br

G. C. O. de Aguiar
e-mail: grazyella.oliveira@ufsc.br

M. E. P. Missner
e-mail: elisa.p.missner@ufsc.br

C. R. L. de Aguiar
e-mail: catia.lange@ufsc.br

C. R. S. de Oliveira · P. V. de Oliveira · A. H. da Silva Júnior
Chemical Engineering Department, Federal University of Santa Catarina-UFSC, Florianópolis, Santa Catarina, Brazil
e-mail: afonso.silva@ufsc.br

R. K. Tewo
Chemical Engineering Department, Dedan Kimathi University of Technology, Kiganjo/Mathari, Kenya

A. B. Mapossa
Chemical and Petroleum Engineering Department, University of Calgary, Calgary, AB, Canada

Chemical Engineering Department, University of Pretoria, Pretoria, South Africa

S. Maiti et al. (eds.), *Sustainable Coloration Techniques in Textiles*,
https://doi.org/10.1007/978-981-96-4975-4_1

Fig. 1.2 Acid Red 37 dye molecule

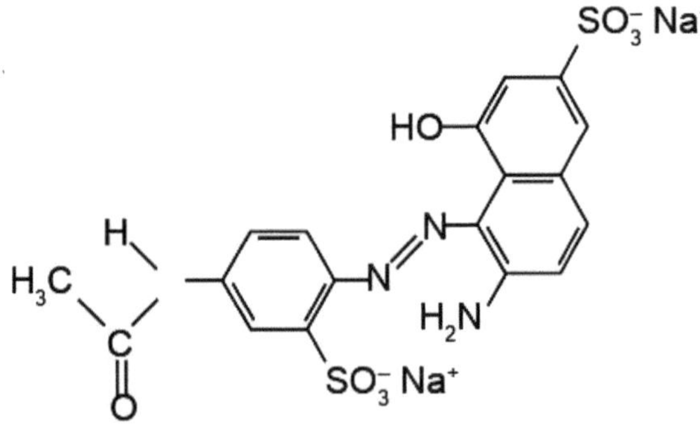

C.I. Acid Red 37, shown in Fig. 1.2 (Mahapatra 2016). The C.I. number is assigned only when the chemical structure of the dye is defined and known. The information provided for each dye also includes chemical structure, first known manufacturer, hue, CAS registry number, and EU number (Burkinshaw 2016).

Dyes must exhibit properties such as intense color and affinity, which can be understood as substantivity or reactivity, solubility, diffusibility, and fastness (Salem 2010). Intense color is an essential factor in dye molecules as it ensures that substrates can be dyed in a range of colors, from light to dark shades, with a sufficient amount of dye that does not cause fiber saturation. Affinity, or substantivity, is achieved by adding functional groups called auxochromes, which allow chemical bonding between the dye and the fiber. The groups responsible for substantivity and solubility are the same in many dyes. Solubility is crucial in cases where the dye needs to be soluble to be adsorbed and diffused into the fiber, such as in dyes for cellulosic fibers. Fastness is a fundamental property for selecting a dye for each type of fiber, its dyeing method, finishing, and its final use. Examples of fastness properties include the sublimation fastness of disperse dyes and the chlorine fastness of acid dyes.

1.1.1 Chemical Composition of Dyes

Dyes are organic compounds with three essential groups in their molecules: the chromophore, the auxochrome, and the matrix. The chromophore group is primarily responsible for the color of textile substrates (Sharma et al. 2021a). The chromophore consists of groups of atoms, with the most common being: nitro ($-NO_2$), azo (–N=N–), nitroso (–N=O), thiocarbonyl (–C=S), carbonyl (–C=O), and alkenes (–C=C–). The absorption of electromagnetic waves by the chromophore is due to the excitation of the electrons in a molecule (Benkhaya et al. 2020). Depending on the chromophore, dyes can be classified as acridine, anthraquinone, triphenylmethane, azo (monoazo, diazo, triazo, polyazo), diphenylmethane, oxazine, xanthene, etc. (Benkhaya et al. 2020; Herath et al. 2024; Rauf et al. 2011; Moyo et al. 2022).

Textile dyes have a unique chemical structure for each color (Benkhaya et al. 2020). The three most common groups of dyes according to their chromophore group are azo, anthraquinone, and phthalocyanine dyes (Herath et al. 2024).

1.1.1.1 Azo Dyes

Azo dyes constitute the largest chemical class, containing at least 70% of all dyes produced in the world (Kamenická 2024). The characteristic feature is the presence of one or more azo groups in their structures, along with hydroxyl, amine, and substituted amine groups as auxochromes (Mahapatra 2016). The main characteristic of azo dyes is their nitrogen-nitrogen double bond (–N = N–), as shown in Fig. 1.3-a, which is attached to at least one aromatic group, either benzene or naphthalene (Herath et al. 2024). The general structure of an azo dye molecule can be represented by the following formula: Ar-N = N-R, where R represents an aryl group, heteroaryl group, or –CH = C(OH)– alkyl derivative, and Ar represents the aromatic ring. The side groups (Ar and R) present in the molecule's structure are responsible for the various colors, with different intensities, ranging from yellowish-green to orange, red, violet, and brown (Abrahart 1968). Azo dyes are produced almost exclusively by diazotization of a primary aromatic amine to give rise to a diazonium salt or diazo compound. The diazo compound is coupled to a second substance, usually a phenol, a ketone, or an aromatic amine, forming an azo compound (Abrahart 1968; Langhals 2004). The dominant colors in this class of dyes are yellow, orange, red, pink, and brown. These dyes are commonly used in various applications, but their primary purpose is the dyeing of natural and synthetic fibers such as wool, silk, polyamide, or cotton (Kamenická 2024). Most azo dyes are carcinogenic and toxic by nature, and due to their resistance to enzymes, they are not easily degradable (Dutta et al. 2024). Therefore, the discharge of azo dyes into water bodies can lead to the accumulation of aromatic amines such as benzidine and 3,3'-dimethylbenzidine, which are highly carcinogenic in the environment (Dutta et al. 2024).

1.1.1.2 Anthraquinone Dyes

Anthraquinone chromophores are responsible for approximately 15% of all dyes (Melati et al. 2023), and anthraquinone dyes can be identified by a standard structure where a p-quinonoid group is fused with two other benzene rings, as seen in Fig. 1.3b (Iqbal 2008). Anthraquinones are reported to be among the oldest known dyes and today represent the second most used group of dyes in the textile industry after the azo dye group. They are chemically stable molecules derived from anthracene containing a quinone structure as the chromophore to which hydroxyl or amino groups can be attached (Berradi et al. 2019a). Two examples of anthraquinones are the dyes C.I. Disperse Red 60 and C.I. Disperse Blue 60 (Mahapatra 2016), as shown in Fig. 1.3c. Anthraquinone dyes are used to dye wool, silk, nylon, cotton, leather, and paper. The most important dye in this group is alizarin, 1,2-dihydroxyanthraquinone. Alizarin

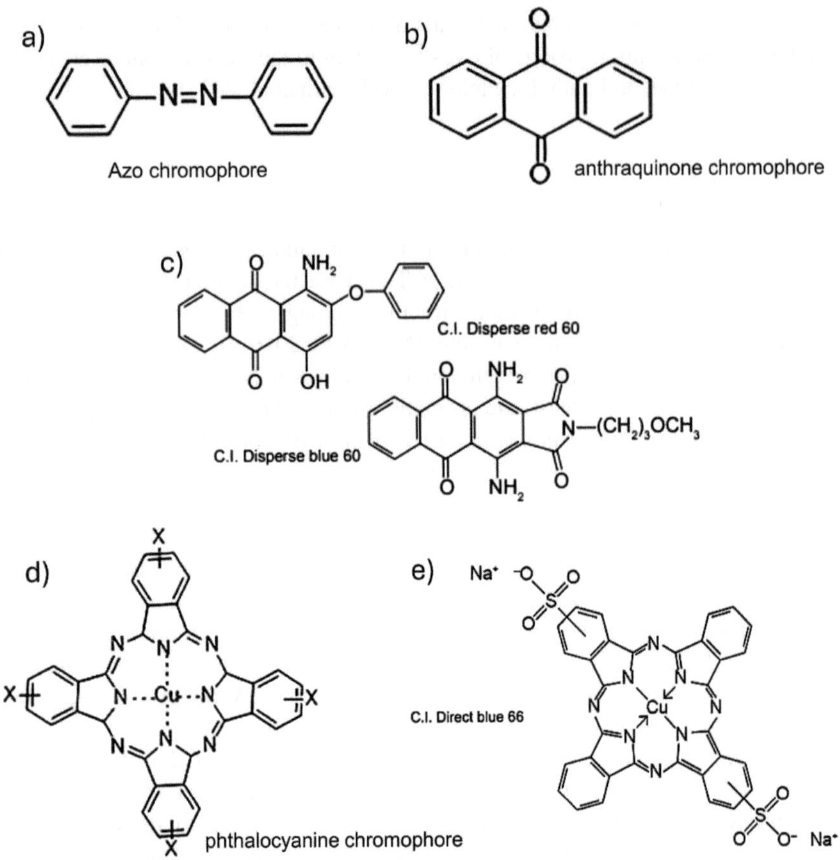

Fig. 1.3 Molecular structure of some dyes. **a** Molecular structure of the azo chromophore; **b** Molecular structure of the anthraquinone chromophore; **c** Disperse dyes containing the anthraquinone group; **d** Molecular structure of the phthalocyanine chromophore; **e** C.I. Direct blue 66. [Adapted from Mahapatra (2016)]

derives its name from the fact that it was first obtained from the roots of the madder plant (Iqbal 2008).

1.1.1.3 Phthalocyanines Dyes

Phthalocyanines coordinate with metal atoms. The most important, contributing about 2% of all dyes, are copper phthalocyanines, used for their bright blue and green colors (Mahapatra 2016). An example is C.I. Direct Blue 66, as shown in Fig. 1.3d and e. A dye molecule presents dyeing capabilities only by adding other groups of atoms called auxochromes (Iqbal 2008). These auxochromic groups allow the fixation of dyes and can modify the dye's color. They can be acidic (COOH, SO_3, and OH) or basic (NH_2, NH-R, and N-R_2). The available auxochromes are (–CO_2H), (–NH_3), (–SO_3H), and (–OH) (Benkhaya et al. 2020). Sulfonic acid groups help in the solubility of the dye in water (Mahapatra 2016). The other atoms in the molecule correspond to the matrix, the third part of the dye (Benkhaya et al. 2020). Regarding the reactivity of dyes, the reactive group is specific to dyes, also known as reactive dyes. Some dyes, such as Reactive Red 3, are derived from inorganic metals like Cu, Cr, and Ni, which do not bind strongly to textile fibers, and during washing, they can be released and mixed with the aquatic environment. The decomposition of these dyes releases toxic heavy metals into the aquatic body, which eventually enter the food chains and damage the entire ecosystem (Dutta et al. 2024). Textile dyes have a range of harmful effects at the cellular level, contributing to the development of potentially fatal diseases, including cancer (Dutta et al. 2024).

1.1.2 Types of Textile Dyes: Classification and Applications

Dyes can be divided into soluble and insoluble dyes. The soluble dye group includes acid, basic, reactive, and direct dyes. In the insoluble dye group are sulfur, disperse, and vat dyes (Dutta et al. 2024). Table 1.1 presents these classes of dyes, their main chemical class, and their applications.

1.1.2.1 Disperse Dyes

Disperse dyes are widely used in the textile industry, primarily for dyeing polyester substrates. These dyes are slightly soluble in water, non-ionic in nature, and applied to hydrophobic fibers in aqueous dispersion (Benkhaya et al. 2020). Most of these dyes are based on azo structures, as can be seen in Fig. 1.4a. Monoazo, disazo, and anthraquinone constitute about 85% of the disperse dyes (Clarck 2011a).

Table 1.1 Some types of dyes, main chemical class, and application example

Dye class	Main chemical class	Main application
Direct	Azo, azine, oxazine, phthalocyanine, thiazole	Cellulosic fibers
Reactive	Azo, anthraquinone, formazan, oxazine, phthalocyanine	Cellulosic fibers, wool, silk and polyamide
Vat	Anthraquinone, indigoid	Cellulosic fibers, wool
Sulphur	Often imprecise structure	Cellulosic fibers
Acid	Azo, anthraquinone, triphenylmethane	Wool, silk, and polyamide
Disperse	Azo, anthraquinone, nitro	Polyester, acetate, and triacetate
Basic	Acridine, azine, methine, oxazine, thiazine, triarylmethane	Acrylic and modacrylic

Source Adapted fromBurkinshaw (2016)

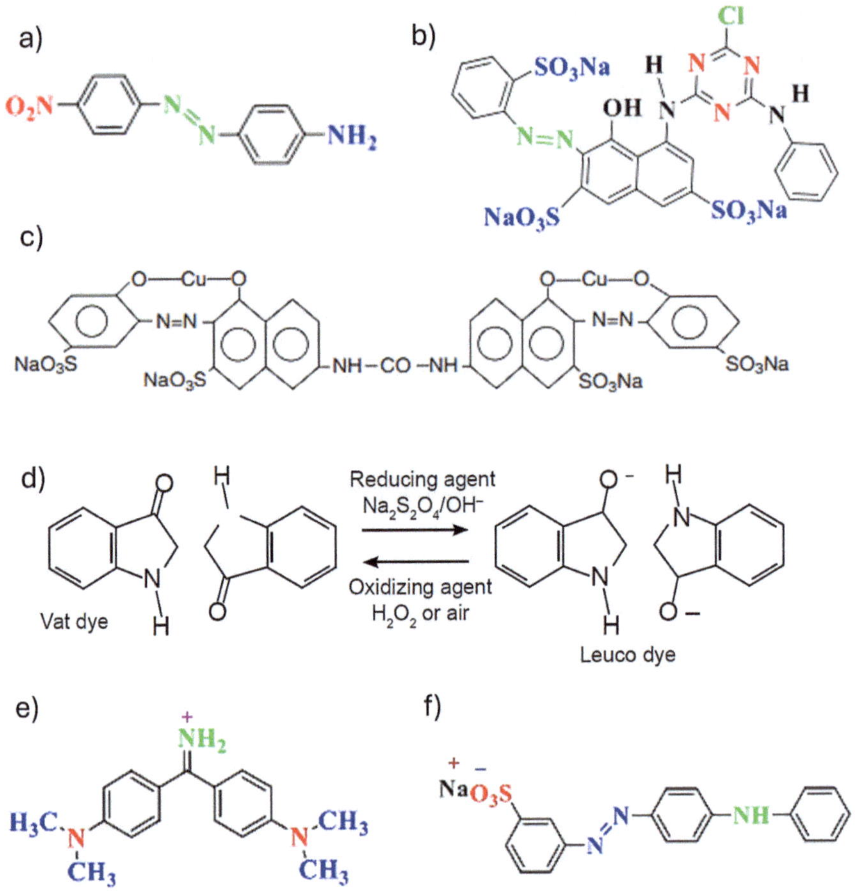

Fig. 1.4 Main types of textile dyes. **a** Disperse dye Orange 3; **b** Reactive dye Red 3; **c** Direct dye Red 83; **d** Reduction and oxidation of a vat dye 93); **e** Basic dye Yellow 2; **f** Acid dye Yellow 36. [Adapted from Iqbal (2008) and Benkhaya et al. (2020)]

1.1.2.2 Reactive Dyes

Reactive dyes have become very popular due to their characteristics of fastness, brightness, and a wide range of shades. They are widely used because of the covalent bond they form with cellulose fibers and the simple dyeing conditions they require. These dyes are anionic and soluble, forming a covalent bond with cellulose. The structure of a reactive dye contains three types of functional groups: chromophore group, responsible for color; solubilizing group, sulfonic groups responsible for solubility, migration degree, substantivity, and washability; and reactive group, which promotes the covalent bond with the fiber and characterizes this class of dyes (Salem 2010). The main chromogens used are azo, metal-azo, anthraquinone systems, and phthalocyanine (Iqbal 2008). Figure 1.4b represents a molecule of a reactive dye with an azo chromophore group, a monochlorotriazine reactive group, and the sulfonic groups for substantivity and solubility. Monochlorotriazine dyes require a higher temperature and a higher pH obtained with a strong alkali compared to diclorotriazine-based dyes (Clarck 2011a).

1.1.2.3 Direct Dyes

The name "direct cotton color" (DCC), or simply direct dye, was derived from the fact that it was the first synthetic dye to have inherent substantivity to cotton (Mahapatra 2016). Direct dyes contain one or more sulfonic acid groups to confer solubility. Direct dyes, also known as substantive colors, differ from basic and acid dyes because cellulosic fibers have a strong affinity for them. Most of the colors belonging to this class are composed of sulfonated azo compounds (Iqbal 2008). Figure 1.4c shows the molecular structure of the dye C.I. Direct Red 83.

1.1.2.4 Vat Dyes

The term vat dye refers to dyes of any chemical class that are applied using the vat dyeing process. Vat dye is so named because indigo was dyed on textiles in wooden vats in ancient times. The old method of fermenting and macerating indigo in a vat gave these dyes their name (Mahapatra 2016). The dyes are insoluble in water and cannot be used directly for dyeing, but when reduced to the leuco form, they become soluble in the presence of an alkali and acquire affinity for cellulosic fibers. A solution of a leuco compound can be applied by dyeing and, upon reoxidation, the original insoluble dye is formed within the fiber structure, as shown in Fig. 1.4d. All vat dyes can be applied to cellulosic fibers, and some of them can also be applied to wool, silk, nylon, and acetate fibers (Iqbal 2008). According to their chemical constitution, vat dyes can be classified as indigoids, which are generally derived from indigotin, or as anthraquinonoids, derived from anthraquinone, which have superior lightfastness properties and are the most widely used.

1.1.2.5 Basic Dyes

Basic dyes belong to different chemical classes, such as Diphenylmethane, Triphenylmethane, Oxazine, Azine, Thiazine, and Xanthene (Mahapatra 2016; Sharma et al. 2021a). Basic dyes, known as cationic dyes (Fig. 1.4e), gained prominence once polyacrylonitrile fibers were introduced in the early 1950s. Initially, acrylic fibers were tough to dye until acidic groups were incorporated as dye sites in the fiber. Cationic dyes are now used in large quantities to dye acrylic fibers (Iqbal 2008).

1.1.2.6 Acid Dyes

Acid dyes are commercially applied mainly to natural proteins (wool and silk), synthetic polyamide, and, to a lesser extent, acrylic and blends of these materials. These dyes are so named because they are applied in the presence of organic and inorganic acids in dye bath solutions (Mahapatra 2016). The electronic arrangement in the chromophores of acid dye molecules is such that they can withstand the degrading effects of UV radiation from the sun (Clarck 2011a). Acid dyes are organic sulfonic acids (Fig. 1.4f), and the commercially available structures are generally their sodium salts, which have good water solubility. Most have one or two $-SO_3Na$ groups that are water-soluble and capable of binding to fibers with cationic sites.

1.1.2.7 Sulfur Dyes

Sulfur dyes are used to dye cellulosic fibers. They are low-cost dyes with good moisture fastness. These dyes are widely used for dyeing black, navy, brown, olive green, and intense, dark colors (Mahapatra 2016). Sulfur dyes do not have a well-defined structure and are better classified by their application methods. The dyes retain sulfur bonds as an integral part of their chromophore. They contain a large proportion of sulfur in the form of sulfide (–S–), disulfide (–S–S–), and polysulfide (–Sn–) bonds in heterocyclic rings. The chromophore systems are mainly based on thiazole, thiazone, and thianthrene, while some dyes retain phenothiazonethianthrone as their chromophores.

1.1.2.8 Azoic Dyes (Naphthol Dyes)

Naphthols are insoluble in water. They are phenols, soluble in alkaline solutions, and substantive to cotton, particularly in the presence of salt. In the presence of a strong alkali, they are converted into naphtholate ions and then become soluble in water. The naphtholate ions are always coplanar and preferably have elongated molecular structures (Clarck 2011a).

1.1.3 Risks and Hazards in the Dye Manufacturing Process

The dye industry began with natural dyes derived from plants and insects, but it quickly shifted to the exploitation of synthetic dyes. Unfortunately, many synthetic dyes, particularly azo dyes, are toxic and mutagenic, leading to their ban in many countries. Despite this, due to their cost-effectiveness and other desirable properties, these dyes continue to be produced and used today (Bafana et al. 2011). Although several synthetic dyes pose significant challenges, there are other synthetic dyes known as imine organic compounds, primarily constituted by Schiff bases and their metal complexes, which are environmentally non-threatening (Yadav et al. 2023; Malik et al. 2022; Ashfaq et al. 2022; Kargar et al. 2022).

The large-scale production and broad application of synthetic dyes have turned them into sources of industrial health and environmental hazards during their synthesis and fiber dyeing (Chavan 2013). Studies corroborate this (Lellis et al. 2019; Afena et al. 2021), which report several health and environmental hazards associated with using and handling dyes and their auxiliary chemicals in various textile companies worldwide. For example, during the textile dyeing process, the concentration of dye loss in wastewater can vary significantly, from 2% for basic dyes to as much as 50% for reactive dyes, leading to high levels of surface and groundwater contamination around textile dyeing industries (Chavan 2013).

During the process of manufacturing dyes, failure to use personal protective equipment can pose significant hazards to both the environment and human health. Several news reports in the media denounce the unhealthy conditions of many workers, especially in underdeveloped countries, which resemble slavery, without any concern for health (Fig. 1.5). It is important to emphasize that there is a high risk of contamination for these workers due to handling chemical products that are precursors of dyes, such as heavy metals and sulfur and nitrogen compounds. However, the risk is not limited to the production of dyes; it extends to dyers. These workers are highly exposed to equipment for processing dyes. Consequently, workers who handle dyes in the dyeing process can develop various health issues such as dermatitis, allergic conjunctivitis, rhinitis, occupational asthma, and other allergic reactions, as can dye producers (Chavan 2013).

Recent studies have discussed the potential health effects of skin absorption of dye and finishing compounds from clothing over both short and long-term periods. These studies have revealed that children often have high levels of these organic compounds in their bloodstream and skin (Chavan 2013). Since dyed clothes remain in prolonged contact with the skin, toxic compounds are frequently absorbed, leading to various diseases through chemical sensitivity in some individuals. For instance, symptoms in adults can range from allergies, skin rashes, headaches, difficulty concentrating, nausea, diarrhea, fatigue, muscle and joint pain, dizziness, difficulty breathing, and irregular heartbeat to seizures. In children, symptoms can include red cheeks and ears, dark circles under the eyes, hyperactivity, and behavioral or learning problems. Some studies even suggest that certain dyes can pose serious health hazards such as cancer (Chavan 2013). The gravity of these potential health issues should not be ignored, and the need for safer alternatives is evident.

Textile dyes' toxicity in humans can manifest as acute toxicity through oral ingestion and inhalation or as chronic toxicity. The primary concerns include skin irritation and sensitization caused by reactions between reactive dyes used for cotton and viscose, acid dyes for polyamide fibers, and disperse dyes for polyester and its compositions with other fibers. The following sections discuss the two main classes of textile dyes used for dyeing all-natural and artificial cellulosic fibers, namely reactive dyes, and the primary class used for dyeing synthetic fibers, such as polyester or disperse dyes.

Fig. 1.5 Workers producing dyes and dyeing fabrics in unhealthy conditions. [Free images from the internet and frames from public videos available on YouTube]

1.1.3.1 Reactive Dyes

The most common hazard associated with reactive dyes is respiratory problems caused by inhaling dye particles. This can sometimes affect a person's immune system, and in extreme cases, repeated inhalation of the dye can trigger severe adverse reactions. This condition, known as respiratory sensitization, may include itching, watery eyes, sneezing, and asthma symptoms like coughing and wheezing (Pizzicato et al. 2023; Jnr and Acquaye 2022).

It has been established that the most prevalent health problems associated with dyeing and finishing processes arise from exposure to chemicals acting as irritants. This exposure can cause skin irritation, itchy or blocked noses, sneezing, and eye sores. Workers may be exposed to heavy metals, formaldehyde-based resins, ammonia, acetic acid, certain shrink-resistant chemicals, some optical whiteners,

Table 1.2 Reactive dyes divided as respiratory and skin sensitizers [Adapted from Chavan (2013)]

Dyes	C I name
Yellow dyes	Reactive yellow 25, reactive yellow 39, reactive yellow 175
Orange dyes	Reactive orange 4, reactive orange 12, reactive orange 14, reactive orange 16, reactive orange 64, reactive orange 67 reactive orange 86, reactive orange 91
Red dyes	Reactive red 29, reactive red 65, reactive red 66, reactive red 123, reactive red 219, reactive red 225
Violet dye	Reactive violet 33
Blue dyes	Reactive blue 114, reactive blue 205
Black dye	Reactive black 5

soda ash, caustic soda, and bleach. Specific reactive, vat, and disperse dyes have also been recognized as skin sensitizers (Jnr and Acquaye 2022; Nicolai et al. 2021).

To significantly reduce the risk to workers from exposure to these dyes, it is crucial to avoid dye in the form of dust. This can be effectively addressed by using dyes in liquid form, low-dusting formulations, and ensuring the use of appropriate personal protective equipment (PPE). These measures are not just recommended, but necessary for the safety and well-being of the workers.

Chavan (2013) listed in Table 1.2 various groups of reactive dyes that cause respiratory and skin sensitization problems for workers due to their constant exposure to these dyes in their workplaces. However, it's important to note that after dyeing and fixation, reactive dyes undergo changes in their toxicological properties due to the loss of reactive groups, and the high wash fastness of the dyed textiles ensures that no dye is exposed to the wearer's skin. Therefore, there have been no reported cases of allergic reactions in consumers caused by wearing textiles containing reactive dyes, providing a reassuring note about the safety of the final product.

1.1.3.2 Disperse Dyes

Specific disperse dyes have caused allergic reactions, mainly when used in skintight, close-fitting clothes made from synthetic fibers. For example, those with low perspiration fastness can lead to this problem. Products made from polyamide or acetate rayon containing disperse dyes with low perspiration fastness allow these dyes to migrate from the clothes to the skin, causing severe allergies. Conversely, polyester products containing disperse dyes generally do not pose a risk due to their high perspiration fastness. Due to the severe allergic reactions caused by these disperse dyes, the German Federal Institute for Consumer Protection and Veterinary Medicine conducted a study to assess the health risks to workers and consumers. They proposed that the authorities cease using these dyes (See Table 1.3). Although no countries have banned their use, some organizations, such as the International Association for Research and Testing in the Field of Textile Ecology, which confers eco-labels

Table 1.3 Dispersion dyes as sensitizing [Adapted from Chavan (2013)]

Dyes	C I name	Dye Type
Blue dyes	Disperse blue 1, disperse blue 3, disperse blue 7, Disperse blue 35, disperse blue 106, disperse blue 124	Aq
Orange dyes	Disperse orange 1, disperse orange 3, disperse orange 37	Azo
Red dyes	Disperse red1, disperse red 17, disperse red 15	Azo and AQ
Yellow dyes	Disperse yellow 1, disperse yellow 9, disperse yellow 3, disperse yellow 39, disperse yellow 49	Nitro, Azo, and Methine

on ecologically and toxicologically safe textiles, do not certify some disperse dyes (Chavan 2013).

1.2 Introduction to Textile Dyeing

Textile dyeing is the process of adding color to fiber, yarn, and fabric by soaking them in a dye solution made up of water, colorant, and other essential ingredients (Ado et al. 2014a; Drumond Chequer et al. 2013). It is a critical aspect in the effective production of textile products, with a history that dates back more than 4,000 years, when natural materials like plants were the primary source of dyes (Wanyama et al. 2011; Samanta and Agarwal 2009). An example of ancient textile dyeing is the scarlet-dyed materials worn exclusively by prominent clergy members in the Middle Ages (Chungkrang and Bhuyan 2020; Kant 2012). Natural dyes were used until about 150 years ago when synthetic dyes were introduced to replace natural dyes (Chungkrang and Bhuyan 2020; Basutkar 2019; Hall 1992).

Synthetic dye was accidentally discovered by William Henry Perkin in 1856. It gained popularity after the royal family began to wear the color (Basutkar 2019; Saremi et al. 2022). Nowadays, synthetic dyes and pigments are so readily available that it is difficult to imagine a time when delicate colors were as precious as gold or silver (Ben Slama et al. 2021). Globally, the process of dying textile materials is constantly changing, mostly to meet energy and environmental sustainability goals (Islam et al. 2022). Innovative dyeing techniques and technologies are being developed, and there are now over 100,000 commercially available dyes. The scale of the industry is staggering, with over 7,107 tons of dyestuff produced annually worldwide. The textile sector, with its 35 million workers, generates $1 trillion in income and contributes 7% of all global exports, making it a significant economic contributor (Hall 1992; Islam et al. 2022). In contemporary dyeing technology, there are various types of dyes, each distinguished by their chemical composition (Saremi et al. 2022; Sharma et al. 2021b).

Some, like reactive dyes, can only make weak chemical interactions with fibers at the molecular level (Berradi et al. 2019b; Lu et al. 2010). Other types, like basic and acid dyes, can form strong bonds. However, direct dyes can also produce weaker connections, such as hydrogen bonds and Van der Waals forces. Table 1.4 summarizes synthetic dyes used in textile dyeing and their specific properties (Saremi et al. 2022). Several steps in the dyeing process are chosen based on the characteristics of the fiber and the dyes and pigments to be used in fabrics (Yaseen and Scholz 2019a; Carmen and Daniel 2012). These factors include chemical structure, classification, commercial availability, and economic considerations (Rapo and Tonk 2021). Traditionally, textile dyeing is done in either a continuous, semi-continuous (pad-batch), or batch processing (exhaust dyeing) system. While both systems are used to apply dyes to fabrics, pad dyeing knitted fabrics is particularly challenging. Therefore, batch-wise processes, where the dye is gradually transmitted to the textiles from the dye bath over a relatively long period, are often preferred (Drumond Chequer et al. 2013; Öner and Sahinbaskan 2011). In the batch-wise process, the textiles to be dyed are placed in enclosed equipment along with auxiliaries. However, varying the duration of contact can affect the dyeing outcome, making control crucial to ensure the correct color is achieved, both in tone and depth, according to the client's specifications. This control also minimizes fabric waste and potential dyeing flaws (Nair 2011; Khattab et al. 2020).

Due to its extensive use of chemicals and large volumes of water, the textile industry is regarded as one of the primary pollutants of surface and groundwater reservoirs. According to recent research by Islam and Mostafa (Islam and Mostafa 2019), the textile industry needs around 1.6 million liters of water daily to produce 8,000 kg of fabric. Recent studies on the environmental sustainability, resource efficiency, and

Table 1.4 Types of synthetic dyes and its properties [Adapted from Basutkar (2019)]

Type of dye	Substrate on which suitable	Nature of bond	Level of fastness
Direct	Cellulosic fibers (cotton, linen, jute, hemp, ramie, etc.)	Van der Waals' forces, H-bonds	Poor
Acid	Cellulosic fibres (feebly), protein fibres (mainly wool, silk, etc.)	Electrovalent bonds	Average with protein fibers, poor with cellulosic fibers
Basic	Protein fibers	Electrovalent bonds	Average to poor
Reactive	Cellulosic fibers	Covalent bonds	Excellent to average
Vat	Cellulosic fibers	Stearic penetration	Excellent to good
Sulphur	Cellulosic fibers	Electrovalent bonds	Average to poor
Pigment	Cellulosic fibers	No bonds, must be kept attached with the help of fixers (chemicals)	Average to poor
Disperse	Synthetic (man-made) fibers (polyester, nylon, acrylic, etc.)	Stearic penetration	Excellent to good

consumer safety of textile dyeing have highlighted significant environmental issues associated with the dyeing process (Ogugbue and Sawidis 2011). Textile dyeing involves using numerous chemicals and a range of non-biodegradable, environmentally hazardous materials (Islam et al. 2022; Islam and Mostafa 2019). During wet processing activities, the textile industry consumes a significant amount of water, resulting in wastewater that contains substantial amounts of dissolved dyestuffs and other contaminants like heavy metals, salts, emulsifiers, dye bath carriers, dispersing and leveling agents, among others (Santos et al. 2007). Additionally, large amounts of chemicals and process water are wasted in the final processing stages (Robinson et al. 2001). Although treating wastewater is necessary to address the variety of hazardous, non-recyclable substances used in different processes, releasing these wastewaters untreated leads to environmental degradation (Drumond Chequer et al. 2013; Rodríguez Couto 2009). Pollution issues, water usage, and chemical exposure along the textile supply chain have been particularly noted using synthetic colors (O'Neill et al. 1999). Therefore, creating environmentally friendly dye formulas and sustainable dyeing techniques is becoming increasingly important in addressing these challenges. The quest for innovative textile dyeing techniques that balance performance with environmental impact is growing. These techniques include waterless dyeing methods, natural dyes sourced from renewable resources, and biodegradable dye additives.

1.2.1 Energy Consumption in Dyeing Processes

The textile sector is one of the most significant energy users; therefore, energy efficiency is critical for the sector's sustainability. In the textile industry, thermal and electrical energy are necessary for dyeing (Hasanbeigi and Price 2015). Textile dyeing techniques, while essential for infusing fabrics with brilliant colors and patterns, are frequently energy-intensive operations (Ado et al. 2014a; Drumond Chequer et al. 2013). Electrical energy is needed to run textile dyeing machinery's motors, pumps, and controls, including printing apparatus, jet dyeing machines, and dyeing vessels. The energy consumption of these machines during dyeing operations can be greatly influenced by their operating settings and efficiency (Koç and Çinçik 2010). Given the pressing sustainability issues in the textile industry, there is a clear and immediate need for methods that can process textiles with significantly less energy, in addition to conventional energy management techniques. Understanding and reducing the energy consumption of dyeing processes is a crucial step in this direction (Branchetti et al. 2019; Palamutcu 2010).

The proportion of fuel and electricity used by the textile industry in each nation's overall final energy consumption is determined by the structure of the country's textile sector (Palamutcu 2010). For example, fuels are the primary energy source for textile wet processing, whereas electricity is the predominant energy source for yarn spinning (Branchetti et al. 2019). Global data from a 2021 study by Khan et al. (Khan et al. 2021) indicated that 39% of the total energy used in the textile sector was electricity and 61% was fuel. Among the 16 major industrial sectors researched the textile industry is the sixth most prominent steam user. Research done by Palamutcu (Palamutcu 2010), has shown that the ultimate energy used in the textile industry worldwide doubled between 1971 and 2004, rising from 47 exajoules (Ej/yr) to 90 Ej/yr. This energy powers textile machinery and controls for heating and cooling, lights, and office supplies (Ado et al. 2014a; Rade et al. 2017). In the conventional textile industry, weaving uses about 23% of energy in process sectors, spinning uses 34%, chemical processing uses 38%, and other operations use an additional 5% (Abu Hasan et al. 2020). Another aspect contributing to energy use is water use, as the textile industry uses much water, especially during wet processing. This generates significant polluted wastewater (Hasanbeigi and Price 2015). The industry's plan to make its production methods more ecologically friendly must also include water conservation and pollution mitigation, especially in regions where water is scarce (Kant 2012; Harane et al. 2014).

In terms of revenue, the European textile sector generated nearly 79,000 million euros in 2015. The sector also employed 61,000. Regarding energy use, the textile sector spent roughly 1,900 million euros on energy (Branchetti et al. 2019). According to Branchetti et al. (Branchetti et al. 2019), the energy cost contribution was more significant than the industrial sector's average energy contribution cost, and the world's top four importers and exporters of textiles in 2016 were China, the European Union, India, and the United States. According to Hasanbeigi and Price (Hasanbeigi and Price 2015), 23% 23% of global textile exports come from the EU, and the textile industry sector in China and the United States contributes 4% and 2%, respectively, to the country's ultimate energy use in manufacturing (Khan et al. 2021). Although the textile sector does not have a high energy intensity, it does entail a huge number of plants that collectively consume a substantial quantity of energy. Also, from the manufacturing census data from 2010 in the US, fuel energy accounted for 53% of the total energy consumed in the textile industry in the country, while electricity accounted for 47% (Bravo and Iturralde 2022).

The overall cost of energy used in the textile industry varies in different countries or regions. Figure 1.6 compares the general cost factor proportions for cotton yarn in China and the United States of America. Usually, a third or a fourth of the product's overall cost goes towards energy costs (Rade et al. 2017; Khude 2017). As seen in Fig. 1.6, spinning uses the most electricity (41%), followed by weaving preparation (18%) and weaving itself. Preparation for wet processing (de-sizing, bleaching, etc.) and finishing together use the most thermal energy (35%). The production and delivery of steam also wastes a substantial amount of thermal energy (35%). Each plant will have a different proportion (Hasanbeigi and Zuberi 2022).

In Turkey, the percentage of energy cost in the textile sector among all the basic components of manufacturing costs for textile goods has been recorded as 6–14% in 1996, 8–10% in 2000, and 5–10% in 2007 of the total production cost. The textile industry's share of Turkey's overall industrial energy consumption (among the top 500 industrial facilities) is estimated to be 7.2% (Palamutcu 2010). Research studies done by Palamutcu (2010) on the actual specific electric energy consumption (SEC) values of production plants show 3.24–3.47 kWh/kg for the yarn spinning

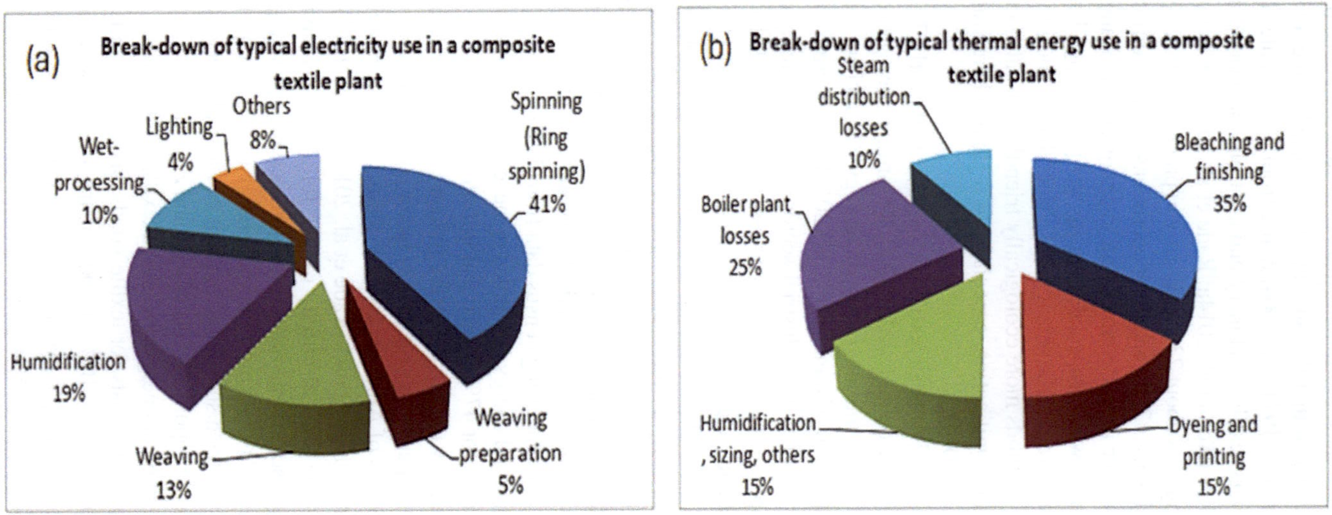

Fig. 1.6 Breakdown of **a** typical electricity and **b** thermal energy use in a composite textile plant. [Reproduced with permission from Hasanbeigi and Price (2015) Copyright 2015 Elsevier Ltd.]

Table 1.5 Important textile processes indicating amount of energy proportions [Adapted from Uddin (2014)]

Process name	Energy requirement	Summary
Fiber production	Electricity in machine operation, water and chemical used, heating in drying	The energy consumed is 21% of the total energy used from fiber to the finished product
Spinning	Mainly electricity and fuel	The energy consumed is 18% of the total energy used from fiber to the finished product
Weaving	Mainly electricity and fuel	The energy consumed is 11.4% of the total energy used from fiber to the finished product
Dyeing and finishing	Electricity, water, chemical, fuel	Heating during the process of dyeing liquor, steaming, soaping, washing, and drying is required. The energy consumed is 25% of the total energy used from fiber to the finished product

plant, 0.0073 kWh/kg for the warping and sizing plant, 1.58–2.24 kWh/kg for the weaving plant, 0.79–1.05 kWh/kg for wet processing, and 0.065–0.195 kWh/kg for the clothing manufacturing plant. This data aligns with the current literature. In 2019, the textile sector in Japan used about 33 PJ of energy. Compared to the Chinese textile sector, which uses 51% of its energy from electricity, the Japanese textile industry uses 42% of its energy from electricity. In Taiwan, in 2019, the textile industry utilized approximately 46 PJ of energy, with electricity making up 47% of the industry's overall energy mix (Hasanbeigi and Price 2015). A summary of the highlights of important textile processes that use much energy is shown in Table 1.5.

In summary, a coordinated effort from industry stakeholders is needed to address the complicated and diverse issue of energy usage in textile dyeing processes. Textile producers can minimize their environmental impact and maintain their competitiveness in the global market by understanding the factors influencing energy use, implementing energy reduction plans, and exchanging best practices. Investing in energy-efficient technology and practices is not just a moral obligation but also a wise business move that can result in long-term profitability and growth as sustainability gains importance in the textile industry.

1.2.1.1 Water Consumption in Dyeing Processes

Water consumption in the textile dyeing process is not just a concern, it is a pressing issue of significant environmental and economic relevance (Marszałek and Żyłła 2021). The textile industry, one of the largest consumers of water in the world, is grappling with this high consumption at several stages of textile processing, from the preparation of fibers to the final dyeing of the fabrics (Chakraborty and Ahmad 2022). A detailed understanding of these steps and the urgent implementation of more sustainable technologies are crucial to mitigating the environmental impact of this sector (Ammayappan et al. 2016).

Textile dyeing involves several phases, each with a significant use of water. Fiber preparation, for example, includes processes such as washing and desizing, which are essential for removing impurities and preparing the fibers to absorb dyes evenly (Chakraborty and Ahmad 2022). This preparation is estimated to consume between 30 and 50 L of water per kilogram of fabric (Ammayappan et al. 2016). Another necessary process is mercerization, used for cotton fibers, which improves the fabric's affinity with the dye and gives the fabric shine, consuming approximately 50 L of water per kilogram of fabric (Uddin et al. 2023).

The dyeing phase is the most intensive in terms of water consumption. Depending on the type of fiber and the method used, dyeing may require 50 to 150 L of water per kilogram of fabric (Ammayappan et al. 2016). Dyeing methods vary widely, the most common being batch and continuous processes. After dyeing, a series of washes are required to remove excess unfixed dye, which is crucial to prevent the dye from fading during use and home washing of the fabric. Water consumption at this stage can reach 100 L per kilogram of fabric (Yaseen and Scholz 2019b).

Dyeing methods can be mainly classified into batch dyeing and continuous dyeing (Chakraborty and Ahmad 2022). Batch dyeing is highly versatile and can be used for small and large batches of fabric, but it is notoriously water-intensive. Processes such as autoclave dyeing and jet dyeing can use between 50 and 150 L of water per kilogram of fabric, depending on the complexity of the process and the type of fiber (Ammayappan et al. 2016). Continuous dyeing is more efficient regarding water and energy consumption, but it is more suitable for the large production of uniform fabrics. In this method, the water used can be significantly less, ranging from 40 to 80 L per kilogram of fabric, thanks to the recycling and reuse of water (Ranganathan et al. 2007).

The type of textile fiber directly influences water consumption during dyeing due to its physical and chemical properties. Cotton, for example, is one of the most common fibers and one that consumes the most water (Ammayappan et al. 2016). From the cultivation phase to dyeing, cotton can consume between 20,000 and 27,000 L of water per kilogram of fiber, considering all stages of the product's life cycle (Varadarajan and Venkatachalam 2016). As a synthetic fiber, polyester consumes significantly less water in its initial production. However, the dyeing process can still use between 40 and 60 L of water per kilogram of fabric, depending on the method used (Chakraborty and Ahmad 2022). Wool, a natural fiber, also requires high water consumption during dyeing, varying between 80 and 120 L per kilogram of fabric due to the need for several washes to remove fats and natural impurities (Ammayappan et al. 2016). Nylon, similar to polyester, is also a synthetic fiber with less water consumption in initial production, but dyeing may require 50–70 L of water per kilogram of fabric (Uddin 2021).

Several technologies and practices are being adopted to reduce water consumption in textile dyeing (Jeong et al. 2023). Dry dyeing, for example, uses supercritical carbon dioxide instead of water to disperse dyes, eliminating water use and significantly reducing the use of chemicals and energy (Khattab et al. 2020). Water reuse and recycling are other sustainable practices. Many industries are implementing water treatment systems that allow the reuse of washing and dyeing water, potentially reducing total water consumption by up to 60% (Jeong et al. 2023). Furthermore, using low-bath ratio dyes, which require less water to disperse and fix on the fabric, can reduce water consumption in batch dyeing processes. Process monitoring and control, using advanced technologies, can optimize water use at all stages of dyeing, reducing waste and increasing efficiency. Education and training of textile workers in more efficient and sustainable dyeing techniques are also crucial strategies for reducing water consumption (Wijannarong et al. 2013).

Water consumption in the textile dyeing process is a significant challenge facing the industry. However, through the adoption of innovative technologies and sustainable practices, it is not just a challenge, but a potential opportunity. Transitioning to more efficient dyeing methods not only preserves a vital resource like water but can also lead to substantial savings for the industry (Chakraborty and Ahmad 2022). This shift, while responding to growing demands for more environmentally friendly production practices, can pave the way for a more sustainable and profitable future for the textile industry.

1.2.1.2 Consumption of Chemicals in Dyeing Processes

In addition to dyes, the textile industry uses more than 8,000 types of chemicals for various purposes in the coloring and finishing of textile substrates in general, including dyeing and printing (Kant 2012). Table 1.6 presents a fraction of these products, with an emphasis on those used in dyeing and printing processes. [NO_PRINTED_FORM] (Oliveira et al. 2023a) mapped the risks and hazards that each of the compounds in Table 1.6 poses to human health in their work. Many of these chemicals are toxic to human health and highly polluting the environment, either directly or indirectly. The auxiliary chemicals used in the dyeing of fibers, yarns, and fabrics are applied together with the dyes using large volumes of water, generating large volumes of effluents to be treated and many tons of contaminated sludge resulting from the treatment of wastewater (Oliveira et al. 2023b). Among the contaminants that cannot be effectively removed from dyeing effluents are the large amounts of salts, such as chlorides, sulfates, and sodium bicarbonates, which are necessary to promote the interaction and fixation of various classes of dyes with the fibers (Oliveira et al. 2021; Chen et al. 2022; Azanaw et al. 2022). Excess salt in the water discharged into rivers can cause salt stress and ionic imbalance in aquatic biomes, leading to the death of more sensitive organisms, complicating the treatment of water for domestic use, and contaminating the soil, making it sterile. The presence of salts, naphthols, nitrates, soaps, auxiliary commercial substances such as emulsifiers, color fixatives, dispersants, and synthetic dyes, whose composition often contains heavy metals such as chromium, copper, lead, cobalt, among others, make textile sludge highly dangerous and polluting (Gregory 2007; Giovanazzi et al. 2024). About 40% of dyes used globally contain organically bound chlorine, a known carcinogen. Other harmful chemicals present in textile wastewater can include formaldehyde-based binders,

Table 1.6 Chemical auxiliaries most used in the pretreatment, dyeing, and printing processes of textile substrates

Pre-treatment and dyeing processes		References	Printing processes		References
Chemical substance	CAS-number		Chemical substance	CAS-number	
Surfactants[a,b,d,f,g,h] (commercial product)	–	Lacasse and Baumann (2004), Clarck (2011b)	Starches (e.g., corn, manioc, rice)	–	Izdebska and Sabu (2016)
Wetting agent[c,d,e,h,i] (commercial product)	–		Modified starches (e.g., carboxymethyl cellulose)	–	
Enzymes[c,e] (commercial product)	–		Gums (e.g., guar, arabica, agar, and carob)	–	
Sodium hydroxide[a,b,d,f,g,h] (Caustic Soda)	1310-73-2	Lacasse and Baumann (2004), Clarck (2011b), Sax (1986), Sweet et al. (1986), OECD (2023)	Sodium alginate	–	
Sodium carbonate[a,b,d,f,g,h]	497-19-8	Lacasse and Baumann (2004), Sweet et al. (1986), OECD (2023)	Sodium bicarbonate	144-55-8	Clarck (2011b), OECD (2023)
Magnesium sulphate[b]	7487-88-9	OECD (2023)	Sodium m-nitrobenzenesulfonate	127-68-4	Sweet et al. (1986), OECD (2023)
Hydrogen peroxide[b,f,i]	7722-84-1	Clarck (2011b), Sax (1986), OECD (2023)	Ammonium sulphate	7783-20-2	Sax (1986), OECD (2023)
Sodium persulphate[b]	7775-27-1	Sweet et al. (1986), OECD (2023)	Sodium dichromate	10,588-01-9	Sweet et al. (1986), OECD (2023)
Sodium silicate[b,i]	1344-09-8	OECD (2023)	Sodium Hydroxymethane Sulfonate	870-72-4	OECD (2023)

(continued)

Table 1.6 (continued)

Pre-treatment and dyeing processes		References	Printing processes		References
Chemical substance	CAS-number		Chemical substance	CAS-number	
Sodium sulphite[b,d]	7757-83-7	OECD (2023)	Dihydroxybutanedioic acid	133-37-9	Sweet et al. (1986), OECD (2023)
Acetic acid[a,b,c,e,f,h]	64-19-7	Lacasse and Baumann (2004), Clarck (2011b), OECD (2023)	Tin chloride	7772-99-8	Sax (1986), OECD (2023)
Formic acid[f]	64-18-6	Sweet et al. (1986), OECD (2023)	Urea	57-13-6	Sax (1986), Sweet et al. (1986), OECD (2023)
Ammonium hydroxide[h]	1336-21-6	Clarck (2011b), Sax (1986), OECD (2023)	Thiourea	62-56-6	Sweet et al. (1986); OECD (2023)
Sulphuric acid[a]	7664-93-9	Lacasse and Baumann (2004, Clarck (2011b); Sax (1986), Sweet et al. (1986)	Sodium formaldehydesulfoxylate	6035-47-8	Sigma-Aldricht (2006)
Hydrochloric acid[a]	7647-01-0	Lacasse and Baumann (2004), Sax (1986), OECD (2023)	Formaldehyde	50-00-0	Sweet et al. (1986), OECD (2023)

(continued)

Table 1.6 (continued)

Pre-treatment and dyeing processes		References	Printing processes		References
Chemical substance	CAS-number		Chemical substance	CAS-number	
Sodium hypochlorite[b]	7681-52-9	Sax (1986), Sweet et al. (1986), OECD (2023)	Pigments (commercial product)	–	Clarck (2011b), Izdebska and Sabu (2016)
Sodium chlorite[b]	7758-19-2	Clarck (2011b), Sweet et al. (1986), OECD (2023)	Synthetic gums (e.g., polyacrylates) (commercial product)	–	Clarck (2011b), Izdebska and Sabu (2016)
Sodium bromide[b]	7647-15-6	OECD (2023)	Softeners (commercial product)	–	
Sodium bisulfite[b,f,g]	7631-90-5	OECD (2023)	Antifoamer agents (commercial product)	–	
Sodium hydrosulfite[b,f,g]	7775-14-6	Lacasse and Baumann (2004; OECD (2023)	Dispersing agents (commercial product)	–	
Sequestering agents[a,b,c,d,e,f,g,h] (commercial product)	–	Lacasse and Baumann (2004), Clarck (2011b)	Binding agents (commercial product)	–	
Magnesium silicate[i]	1343-88-0	OECD (2023)	Emulsifying agents (commercial product)	–	

(continued)

Table 1.6 (continued)

Pre-treatment and dyeing processes		References	Printing processes		References
Chemical substance	CAS-number		Chemical substance	CAS-number	
Sodium nitrate[f]	7631-99-4	Sweet et al. (1986), OECD (2023)	Solvents such as: toluene/ benzene, ethylbenzene, nitrobenzene/xylene/ n-hexane/n-alkanes/ hexamethylene, diisocyanate/cyclohexane, acetone/ cyclohexanone/butanone, butan.1.ol/ propan.2.ol/ isobutanol/2-methylpropan.1-ol/ methanol/ benzyl alcohol Used as diluents for ready-to-use inks, purchased as: Matt ink, Acrylic ink, White spirit, Glue, Cellulose thinner, Plastisol thinner, Plastisol remover, among others	–	Rose et al. (2023)
Sodium phosphate[f]	7558-80-7	OECD (2023)	Plastisol (Vinyl chloride / Chloroethylene)	–	Rose et al. (2023)

Notes[a]Hydrolytic desizing; [b]Oxidative desizing; [c]Enzymatic desizing; [d]Scouring; [e]Bioscouring; [f]Oxidative bleaching; [g]Reductive bleaching; [h]Mercerization, [i]Scouring-bleaching combined hydrocarbon-based softeners, and quaternary ammonium, as well as high loads of optical brighteners, hypochlorites, and peroxides used in fabric bleaching processes (Oliveira et al. 2023a). Another environmental impact factor includes textile effluents with high pH due to the large loads of strong alkaline compounds, such as sodium hydroxides and carbonates, used during dyeing and printing processes (Kant 2012).

1.3 Socio-Environmental Impacts from Dyeing Processes

The textile industry plays a significant role in the global economy (Oliveira et al. 2021), In addition to providing a wide range of essential consumer products, which are marketed in various forms, such as clothing, home textiles, furniture coverings, and technical fabrics, it is one of the sectors that generate the most employment and income, especially in developing countries (Kant 2012; Oliveira et al. 2021, 2023c). Conversely, the dyeing processes associated with textile production have significant socio-environmental impacts, challenging science to develop new materials, technologies, and more sustainable processing methods (Oliveira et al. 2023c; da Sliva Júnior et al. 2022).

The intensive use of water and chemicals in dyeing and printing processes generates high volumes of wastewater, requiring significant energy and chemical inputs for treatment (Hasanbeigi and Price 2012). Despite the water treatment required by environmental agencies in many countries, it is reported that in many places worldwide, textile effluents are discharged without treatment, causing soil contamination, the death of aquatic animals, and water poisoning (Oliveira et al. 2023b). Moreover, even the most common treatment techniques used today, such as biological digestion followed by physicochemical treatment and separation by coagulation/flocculation, are not capable of removing 100% of the contaminants present in textile effluents, such as salts, metals, and other chemical species that can disrupt aquatic biomes in rivers and lakes (Lu et al. 2010; Natarajan et al. 2018; Sher et al. 2020; Li et al. 2020). The improper disposal of textile effluents containing toxic chemicals contaminates the water bodies where they are discharged, degrading water quality and contaminating aquatic animals. If local inhabitants ingest these animals, they can cause various health complications (Gregory 2007; Kumar and Konar 2011; Giorgi et al. 2000).

Excessive water consumption in dyeing processes also contributes to water scarcity in many regions, exacerbating challenges related to water resource management. Another concerning aspect is the emission of greenhouse gases associated with dyeing processes (Oliveira et al. 2023b; Chiango et al. 2022). From the energy required to heat large volumes of water in boilers fueled by wood chips, fossil fuels, and natural gas to the transportation of chemicals, textile production has a significant carbon footprint. These emissions contribute to global climate change, exacerbating existing environmental challenges. China is considered the largest textile producer in the world, accounting for approximately 35% of global production. [NO_PRINTED_FORM] (Lin and Bai 2020) reported that the energy consumption of the Chinese textile industry tends to increase; it recorded an energy consumption of 16.67 Mtoe (million tons of oil equivalent) in 1985, followed by 21.14 Mtoe in 2000 and 51.07 Mtoe in 2016. [NO_PRINTED_FORM] (Hasanbeigi and Price 2015) reported that about 41% of the energy in the textile industry is spent on spinning processes, 18% on fabric manufacturing (weaving and knitting), and about 35% on wet finishing processes (pre-treatment, dyeing, and finishing). Furthermore, approximately 15% of all energy spent on textile finishing is related to dyeing processes (Oliveira et al. 2021).

Dyeing processes also have significant social impacts, especially in communities where textile factories are located. Workers' exposure to toxic chemicals can lead to health problems, including respiratory diseases, dermatitis, and neurological issues. Additionally, poor working conditions and low wages are common in the textile industry in many parts of the world, with reports of labor conditions akin to slavery in countries where labor rights and workplace safety are not clearly established, resulting in exploitation and social injustice (Galatti and Baruque-Ramos 2022; Strand et al. 2024). Besides directly impacting workers, dyeing processes can also affect local communities through air, water, and soil pollution. It can lead to community conflicts, forced displacement of original populations, and the deterioration of the quality of life in local communities.

1.3.1 Water Contamination

With its complex composition and vast volume, textile wastewater stands out as one of the most significant contributors to water pollution among industrial sectors (Roy Choudhury 2014). Given the complex global supply chain of textiles, involving numerous stages involving multiple companies, brand owners play a pivotal role in effecting necessary changes (Roy Choudhury 2014). Throughout various stages of textile production, the industry extensively utilizes potable water and a wide range of synthetic chemicals (Kang et al. 2020; Ağtaş et al. 2020). Consequently, textile manufacturing generates copious amounts of highly pigmented wastewater laden with hazardous persistent coloring pollutants, which infiltrate aquatic ecosystems (Bharagava et al. 2018; Kumar et al. 2020). Also, in the industry, water consumption and polluted water from textile processes rank among the highest anthropogenic activities (Lellis et al. 2019).

A typical textile industry consumes approximately 1.6 million liters of groundwater daily to produce 8000 kg of textile fabric, with around 30–40% of the water used in the dyeing process, 60–70% during the washing, and approximately 10–50% of unused dyes released into aquatic resources alongside generated wastewater (Kumar et al. 2020; Rosa et al. 2020; Cai et al. 2020). With thousands of synthetic dyes used during the dyeing process and global production, ~7×10^7 tons of different dyes are produced annually, and more than 10,000 tons of synthetic dyes are used in the textile industry (Chandanshive et al. 2020). Additionally, textile sludge poses challenges due to its excessive volume and unwanted composition, often containing high levels of organic matter, micronutrients, heavy metal cations, and pathogenic microorganisms (Lellis et al. 2019; Bhatia 2017). Discharging of untreated effluents accounts for a significant portion of environmental damage attributed to the textile industry, constituting about 80% of total emissions (Lellis et al. 2019; Bhatia 2017). Dumping heavy metals, solvents, toxic sludge, and other waste into water bodies by industries, including textiles, raises significant global environmental concerns (Roy Choudhury 2014; Nieminen-Kalliala 2003). The presence of toxic metals like arsenic, lead, chromium, antimony, cadmium, and mercury in textile wastewater underscores its complex matrix of pollutants originating from color pigment production and other industrial processes (Rosa et al. 2020; Hubadillah et al. 2020).

Moreover, groundwater contamination is a pressing concern exacerbated by various industrial sectors, including paper mills, pharmaceuticals, textiles, breweries, soap and detergent manufacturers, soft drink producers, tanneries, and chemical plants. This pollution presents a significant threat to the quality and availability of clean drinking water, with adverse impacts experienced daily. Developing nations, especially, confront heightened obstacles due to the exorbitant costs of establishing and maintaining treatment facilities, coupled with lenient regulatory practices surrounding industrial and domestic discharges. Organizations such as the World Health Organization (WHO), UNESCO, the US Environmental Protection Agency (USEPA), and the Bureau of Indian Standards (BIS) are actively setting permissible limits (PL) for groundwater, which is crucial for ensuring its quality and safety for end user (Jain et al. 2021). Textile wastewater contains polycyclic aromatic hydrocarbons (PAHs), such as benzene (C_6H_6), naphthalene ($C_{10}H_8$), toluene (C_7H_8), and xylenes or dimethylbenzene (C_8H_{10}), among others. Most PAHs display low solubility in water, often adhering to solid particles and settling at the bottoms of water bodies. PAHs tend to tightly bind to particles in soil, with certain compounds capable of migrating through the soil and contaminating underground water sources (Sivaram et al. 2018a).

The immediate consequences of hazardous effluent discharge include groundwater and water body contamination, decreased dissolved oxygen levels, and compromised aquatic ecosystems. Over the past five years, there has been a notable increase in effluent discharge, attributed to the increased fabric dyeing activity, resulting in more wastewater entering natural environments (Hossain et al. 2018). Human health remains susceptible to metal and metalloid exposure, even at trace levels, due to their persistence in environmental media and acute toxicity (Karunanidhi et al. 2021).

Synthetic dyes, including azo, vat, direct, reactive, sulfide, acidic, and basic dyes, are extensively used in textile dyeing (Cao et al. 2019; Asgari et al. 2020). Despite all efforts, many of these dyes fail to adhere to target fibers during the dyeing process (Cao et al. 2019; Asgari et al. 2020). Some of the unfixed reactive dye often remains in the residual dye baths due to competing hydrolysis reactions and contributes to the coloration of the wastewater, leading to their discharge into aquatic ecosystems, including ponds, rivers, streams, and lakes, along with the wastewater (Européenne 2001; Orts et al. 2018). Studies reveal residues of hazardous chemicals from textile manufacturing persisting in clothing items and leaching into water bodies upon washing, further exacerbating pollution (Roy Choudhury 2014;

Patra and Pariti 2022; Kishor et al. 2021). Soluble organic compounds present in textile dyes, particularly those with reactive, direct, basic, and acidic classifications, pose challenges for conventional removal methods due to their high water solubility (Hassan and Carr 2018). The aesthetic damage caused by textile dyes extends beyond water bodies, hindering light penetration and reducing photosynthesis rates and dissolved oxygen levels, adversely affecting aquatic biota (Hassan and Carr 2018). Additionally, other harmful chemicals like formaldehyde-based dye fixatives, hydrocarbon-based softeners, and nonbiodegradable dyeing chemicals, further degrade water quality (Sivaram et al. 2018b).

Despite the deleterious effects of textile dye pollution, bioremediation methods promise to transform or mineralize contaminants through enzymatic action from plant, bacterial, extremophile, and fungal biomass (Lellis et al. 2019). Pre-treatment of textile wastewater is imperative to address high levels of nonbiodegradable organic compounds, especially textile dyes, before discharge into natural effluents (Orts et al. 2018). Various methods, such as advanced oxidation processes, ozonation, enzymatic degradation, and adsorption techniques, have been explored for addressing color and toxicity in textile dyeing effluents (Orts et al. 2018). Therefore, the treatment of wastewater from the textile industry is imperative to safeguard public health and the environment. Despite being a significant challenge, effective treatment methods must be prioritized, given the severe repercussions of untreated discharge, including eutrophication and groundwater pollution. Adequate measures, including stringent regulations, adoption of cleaner production practices, and investment in innovative remediation technologies, are essential to mitigate the adverse impacts of textile industry wastewater on aquatic ecosystems and human well-being.

1.3.2 Soil Contamination

Textile waste constitutes a significant part of the solid waste generated by industry consumers, and its volume has been steadily increasing over recent decades. This textile waste surge brings about many environmental challenges, including a heavy water footprint, soil pollution, and the loss of natural biodiversity. The industry is devising innovative and intricate textiles to keep pace with the rapidly evolving fashion trends, significantly complicating disposal methods (Piribauer and Bartl 2019; Rahman et al. 2022). Soil plays a pivotal role in ecosystems, providing anchorage for roots and serving as a reservoir for water and nutrients. However, soil pollution arises from various sources, such as the leaching of metals and chemicals (monomers, solvents, reactive initiators, dyes/pigments) and chemical industries, accidental spills, and underground leaks, altering the composition of soil (Markandeya and Mohan 2022).

A significant concern is the presence of azo-type textile dyes, of which approximately 15–50% fail to bind to the fabric during the dyeing process, and subsequently enter wastewater. Additionally, complex-colored textile mill effluents significantly influence soil properties, leading to effluents characterized by strong color, fluctuating pH, high COD, and biotoxicity (Manzoor et al. 2006). In developing countries, this wastewater is commonly utilized for agricultural irrigation, adversely affecting soil microbial communities and hindering the germination and growth of plants (Rehman et al. 2018; Imran et al. 2015).

Chemical processes involved in textile production, such as sizing, mercerizing, and finishing, contribute to wastewater characterized by high levels of biochemical oxygen demand (BOD), chemical oxygen demand (COD), and toxicity, posing risks to water and soil habitats. Sizing chemicals like Polyvinyl alcohol (PVA), polycyclic acids, polyacetate waxes, wetting agents, starch, and carboxymethyl cellulose (CMC) are applied to enhance textile fabric potency, resulting in wastewater with high BOD and COD levels (Saxena et al. 2021). Mercerizing chemicals such as Sodium hydroxide and cotton wax are used to improve textile properties, but they lead to wastewater with elevated pH, suspended solids (SS), dissolved solids (DS), and BOD levels, which are toxic to water and soil ecology (Saxena et al. 2021). Similarly, finishing processes involving chemicals like chlorophenols, short-chain chlorinated paraffin, hexabromobiphenyls, formaldehyde, and perfluorinated chemicals aim to provide softening, waterproofing, antimicrobial activity, and UV protection, but the resulting wastewater is highly toxic to public health and contributes to aquatic and soil pollution (Saxena et al. 2021). Disposing nonbiodegradable fibrous waste into landfills perpetuates soil and groundwater pollution, perpetuating environmental degradation (Peña-Pichardo et al. 2018). Textile effluents containing inorganic salts and organic contaminants disrupt the balance of surface water quality, negatively impacting aquatic flora and fauna and increasing the toxicity of chemicals (Madhav et al. 2018a).

The textile industry faces scrutiny for its unsustainable practices, contributing to pollution, water poisoning, habitat destruction, soil salination, reduced fertility, and biodiversity loss (Wanyama et al. 2011). The concentration of pollutants in soil adversely affects soil fertility and quality, with far-reaching implications for the food web and human health (Markandeya and Mohan 2022). Microplastics in agricultural land near textile industries have been extensively documented, revealing a higher prevalence of fibers in textile sludge and found in fragments in soil samples. Remarkably, soil samples exhibit a higher percentage of transparent/white microplastic, while the sludge samples contain a higher frequency of multicolor and others. Hossain et al. (Hossain et al. 2023) conducted recent research in the agricultural soil of Bangladesh identifying nine types of polymers in the soil, including polystyrenes, ethylene vinyl acetate, latex, high-density polyethylene, polyvinyl chloride, acrylonitrile butadiene styrene, low-density polyethylene, and polypropylene (Hossain et al. 2023). Despite knowledge of the detrimental effects of microplastics on soil structure and physicochemical parameters, such as bulk density, porosity, aggregation, electrical conductivity, water holding capacity, and pH value, their impact on soil ecosystem, health and including adverse effects on some soil organisms like plants, bacteria, fungus, and pathogens (Hossain et al. 2023; Zhang et al. 2022). Generally, the presence of microplastic has harmful impacts, including interactions with persistent pollutants, changes in soil structure and water retention, reduced plant growth, decreased photosynthesis, effect on nutrient availability, and impacts on biogeochemical cycles (Oliveira et al. 2023d). Additionally, significant changes in plant root biomass and soil phosphatase levels further indicate the potential impact of microplastics on soil nutrient and geochemical element cycling. Species sensitivity distribution curves were also constructed, revealing that invertebrates exhibit higher sensitivity to microplastics. This heightened sensitivity is attributed to microplastics' ability to pass through the gut wall of soil nematodes, leading to oxidative stress and alterations in gene expression (Qiu et al. 2022). Wan et al. (2019) found that microplastics from plastic films accelerated water evaporation and created pathways for water movement in soil. This effect was more pronounced with smaller microplastics. Consequently, environmental microplastic contamination can disrupt soil water cycles, worsening water scarcity (Wan et al. 2019).

Analysis of the wastewater samples revealed the presence of copper (Cu), zinc (Zn), and chromium (Cr). These heavy metals are frequently utilized in the textile industry to improve color efficiency (Aleem et al. 2020). Nevertheless, vegetation cultivated in areas with environmental pollution can absorb contaminants like heavy metals through root uptake or foliar absorption, leading to their accumulation in different plant components such as roots, stems, leaves, fruits, and grains. The soil contamination with heavy metals poses a critical concern, as it significantly contributes to the pollution of diverse food crops (Kaur et al. 2021). Forts toward reusing and recycling textile waste offer more sustainable alternatives to landfilling, presenting environmental and socio-economic benefits (Chandra and Bharagava 2009). Urgent remediation measures are required, particularly in developing and under-developed nations, to address widespread soil pollution from effluents and sludge applications (Chandra and Bharagava 2009). From soil contamination with heavy metals to the proliferation of microplastics on agricultural land, the consequences of discharging textile waste are manifold. By addressing these challenges, we can mitigate the negative impact of textile waste on soils, plants, and ecosystems and promote a more sustainable future.

1.3.3 Contamination of Living Organisms

The textile industry significantly contributes to water and soil pollution, posing various threats to living organisms due to the discharge of wastewater containing harmful substances (Kishor et al. 2021). The bleaching process in the textile industry uses chemicals like chlorine, dioxins, sodium hydroxide, sodium hypochlorite, hydrogen peroxide, sodium phosphate, acids, surfactants, cotton fiber, sodium silicate, and others. This leads to the generation of wastewater with high alkalinity and suspended solids, and harmful to human health like cardiovascular disease, porphyria, endometriosis, reduced testosterone, and thyroid hormones, altered immunologic response (Saxena et al. 2021).

Similarly, the dyeing process involves using various chemicals like dyes, reducing agents, urea, oxidizing agents, detergents, wetting agents, and acetic acid. The wastewater characteristics are highly dark color, high BOD (Biochemical oxygen demand), and low heavy metals content (Saxena et al. 2021). They are hazardous to the environment due to their carcinogenic, mutagenic, and cytotoxic effects (Saxena et al. 2021; Newman 2009). Introducing textile dyes into aquatic ecosystems threatens aquatic flora, fauna, and human health. Both treated and untreated textile industry effluents contain azo dyes and organic contaminants, with treated wastewater often carrying amino acids and being more noxious than untreated effluents. Untreated effluents adversely affect living organisms and aquatic ecosystems (Mudhoo et al. 2020; Khan et al. 2023). Additionally, inorganic salts in textile effluents harm soil and surface water quality. Organic contaminants reduce dissolved oxygen (DO) levels and increase BOD in surface water. pH-balancing chemicals further disrupt aquatic ecosystems, leading to toxicity and eutrophication (Madhav et al. 2018b). Moreover, the visibility of colored wastewater affects light penetration in water bodies, reducing photosynthesis and impacting aquatic biota. This disruption in the aquatic environment exacerbates oxygen depletion and obstructs sunlight, affecting the biological processes of aquatic organisms (Khan et al. 2023; Ben Mansour et al. 2007; Sharma et al. 2022).

Overall, textile effluents contribute significantly to environmental degradation and human health issues. About 40% of globally used colorants contain organically bound chlorine, and the presence of organically bound chlorine in colorants poses carcinogenic risks. Organic materials in wastewater react with disinfectants, potentially harming human health through allergic reactions and other adverse effects (Sivaram et al. 2018b; Parisi et al. 2015; Ozturk et al. 2015).

Furthermore, heavy metals present in textile operations, whether from incoming fiber, water, dyestuffs, or finishing chemicals, pose serious health risks upon accumulation in internal organs such as the liver or the kidney, with severe effects on health, especially when high concentrations are reached. Heavy metals can also be contained in plant fibers because they are absorbed from the soil in which they are grown ((Yusuf and Sonibare 2004; Lu et al. 2010)). These metals, including lead, cadmium, mercury, chromium, and nickel, have been associated with various health disorders, including neurological damage, cancer, and respiratory issues (Lu et al. 2010). Lead (Pb), for example, can cause damage to the brain, nervous system, and kidneys, and in some cases, it can be classified as carcinogenic (Roy Choudhury 2014). Mercury (Hg) poses significant risks, especially regarding brain damage. Cadmium (Cd) and cadmium oxide are classified as carcinogens and toxic to aquatic life, linked to respiratory, kidney, and lung disorders (Roy Choudhury 2014). Chromium (Cr) is a strong oxidant and skin irritant, causing respiratory issues, including cancer upon inhalation (Roy Choudhury 2014). Arsenic (As) can cause skin cancer, hyperpigmentation, keratosis, and black foot disease. These health risks underscore the importance of rigorous measures to control and minimize exposure to these metals and metalloids in the textile industry, aiming to protect the health of workers and communities near textile factories (Yusuf and Sonibare 2004).

Textile operations also involve the use of volatile organic compounds (VOCs) and chlorinated solvents, which can lead to health disorders, including irritation, headaches, and even cancer. Chlorinated solvents like carbon tetrachloride, chloroform, dichloromethane, tetrachloroethylene, trichloroethane, and trichloroethylene harm human and ecological health (Sivaram et al. 2018b; Karthik and Gopalakrishnan 2014). These solvents can contaminate groundwater and cause long-term environmental damage (Sivaram et al. 2018b). Oxygenated solvent such oxygenated solvents are methanol, ethanol, propanol, butanol ethylene glycol, diethyl ether, ethyl acetate, acetone, butanone, methyl isobutyl ketone, and 2-hexanone; these solvents can produce harmful ground-level ozone and adverse effects on human health, as well as for all flora and fauna; and prolonged exposure to solvents can cause blindness, irregular heartbeat, and damage to the kidneys, liver, lungs, and CNS (Sivaram et al. 2018b).

Chemicals like hexachlorobenzene are also hazardous, with potential impacts on human health and the environment, including hormone disruption and reproductive issues (Roy Choudhury 2014). Phthalates like DEHP are reprotoxic and classified as toxic to reproduction, leading to restrictions on their use (Roy Choudhury 2014). Aliphatic and aromatic hydrocarbons in textile effluents can cause asphyxiation, and neuropathy (Sivaram et al. 2018b; Carreón and Herrick 2012). The hazards caused by grease and oils to living beings are eye damage, eye irritation, and toxic and harmful effects on aquatic life with long-lasting effects (Robinson and Hsu 2017). Furthermore, oils used in textile processing can cause skin irritation upon direct single or repeated and prolonged contact (Robinson and Hsu 2017; Speight and Exall 2014). Moreover, microplastics released into the environment can adversely affect organisms across environmental compartments and trophic levels. They are detected in many organisms throughout aquatic and terrestrial environments and have been linked to internal injuries and decreased organisms. Additionally, the adverse effects of microplastics are extended to cellular and molecular levels (Zhang et al. 2022). Also, several process operations in the textile industry produce sound in the range of 90 dB (A), leading to noise pollution (Bhatia 2017).

The textile industry has a negative reputation for being unsustainable because it causes environmental problems such as pollution, poisoning, lack of oxygen in the water supply, endangerment of fauna, reduction of soil fertility, and loss of ecological biodiversity. Such problems exist due to textiles' production and post-production phases (Textiles 2009). The textile industry's impact on the environment and human health is considerable. The effluents containing harmful substances threaten ecosystems and individuals. Rigorous measures are essential to mitigate these risks and protect the integrity of the environment and human well-being (Roy Choudhury 2013).

1.4 Other Issues Related to the Sustainability of Textile Dyeing

1.4.1 Life Cycle Assessment of Dyes and Textiles

The Circular Economy is an industrial economy aimed at sustainability. This economy has the main objective of recovering the value of tangible goods through a closed cycle of reuse and restoration, which can both increase environmental and economic performance, as well as recycling and energy, that is, making materials continue to circulate for longer, in closed and ideally continuous cycles. In the circular economy, through new product planning behind the idea that waste could be reduced, manufacturing procedures and supply chains are used to keep resources flowing continuously in a closed cycle (da Pinto 2021). To date, an example is that there is no practical methodology for recycling polyester or PLA clothing to obtain the fiber in its raw state for use in the textile industry.

The circular economy begins with the principle that resources are used rather than consumed. With this, seek to redesign products so they can be reused, which helps reduce the amount of waste and eliminate waste (Weetman 2019). According to Griebler et al. (2022), the circular economy goes beyond zero waste in landfills or materials recycling; it is focused on expanding the value chain by covering the entire product life cycle, from supply, manufacturing, distribution, and sales. This means that the circular economy can include product redesign, the creation of new eco-products, and the recovery in the production process of the value of leftover materials used.

Environmental education and technical training are not just tools, but crucial pillars for the development of practices aimed at a more sustainable textile production process in local clothing industries. They integrate the sectors and people involved in this process, emphasizing the importance of recognizing textile waste as reusable and recyclable materials. This recognition is a key step in reducing waste and minimizing environmental impacts through correct environmental disposal. Therefore, an economic model focused on the circular economy is a potential and fundamental instrument for increasingly sustainable economic growth and development. During the cultivation and production of natural and artificial fibers, this model can significantly contribute to minimizing chemical waste in water tributaries, reducing CO_2 emissions, and considerably reducing the use of natural resources (Griebler et al. 2022). Textile waste must be separated on a large scale globally to reduce carbon footprint, improve socioeconomic benefits, and mitigate environmental pollution (Abagnato et al. 2024; Zhou et al. 2022; Galeti Narimatsu et al. 2021).

There are two points of view regarding low recycling: the management point of view and the consumer point of view. From a management point of view, we can attribute the low efficiency of recycling clothing waste to poor organization. Therefore, from the consumer's point of view, the existing clothing recycling system cannot be fully utilized due to differences in understanding among consumers about recycling waste clothing (Xie et al. 2021). Figure 1.7 presents a promising trend in the

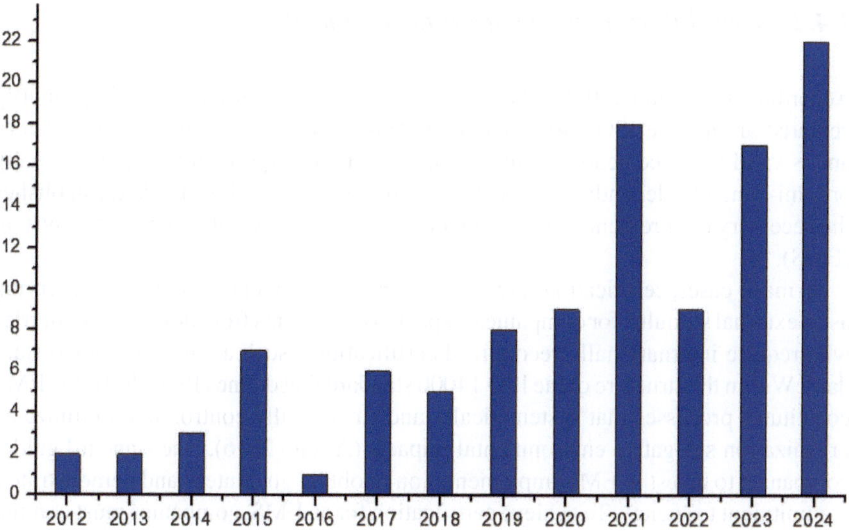

Fig. 1.7 Distribution of the publication of 109 papers on clothing recycling and reuse by publishing year

literature related to the recycling of clothing waste. The time range was 2012–2024, and the deadline was May 5, 2024. The research type was clothing waste recycling, the literature type was articles, and the recovery topic was clothing waste recycling. Driven by environmental responsibility and the circular economy, the number of articles published between 2012 and 2024 fluctuated upward, with an average of 8 articles published yearly. This data reflects a significant increase in research into recycling and reusing clothing waste, with the number of articles published in 2024 reaching 22 in just 5 months. This clearly indicates the growing attention and optimism in the field of recycling and reusing clothing waste.

In 2022, the Global Fashion Agenda (GFA) highlighted that more than 92 million tons of textile waste have been discarded in recent years. A survey conducted in Brazil shows that over 8 billion pieces of clothing are produced annually, corresponding to 42 new units per person each year. This textile waste can contain various substances harmful to the environment, including dyes, plastics (synthetic fibers), and chemicals. According to the UN, 85% of textiles are in landfills yearly (UNECE 2018). Washing certain types of clothes also impacts the amount of microplastics in the ocean, as around 60% of clothes are made of polyester, a plastic material. It is estimated that washing clothes can release an equivalent of 50 billion plastic bottles into the ocean annually. In the last 15 years, global clothing production has nearly doubled; eighty-five percent of raw materials used in clothing production are cotton and polyester, both of which are not environmentally friendly (Ikram 2022; Bailey et al. 2022; Lin et al. 2018).

1.4.2 Regulatory Framework and Standards

According to Araujo (2016), the family of environmental standards is primarily centered around the ISO 14001 standard. This certificate, which is not mandatory and is valid for three years, requires companies to undergo internal audits annually or semi-annually, depending on the organization's contract. The standard establishes the necessary requirements for implementing an Environmental Management System (EMS).

In many cases, certifications are validated by environmental quality seals, serving as an external stimulus for companies, especially for clients from developed countries who require internationally recognized certifications, such as the ISO 14000 standard. Within the structure of the ISO 14000 standard that defines ISO 14001, the EMS constitutes processes that systematically and strategically control and minimize an organization's negative environmental impacts (Araujo 2016). The standard guides companies to invest in EMS implementation to obtain guarantees and demonstrate a commitment to society. To achieve certification for an EMS, companies must commit to continuous improvement, maintaining, enhancing, and demonstrating compliance with their policy. Initially adopted by the largest companies, the EMS has shown beneficial results such as cost/benefit minimization and the reduction of environmental impacts. Environmental instructions within the EMS serve as operational documents that outline responsible environmental practices.

The United Nations (UN), addressing sustainability and ESG (Environment Social Governance) demands, has outlined the Sustainable Development Goals as a global call to action to end poverty, protect the environment and climate, and ensure that people everywhere can enjoy peace and prosperity. The American standard for textile effluents focuses on reducing pollutants such as suspended solids, intense color, odor, pH, COD, TDS, EC, and heavy metals in wastewater. Various treatment methods, including filtration, biosorption, aeration, solar energy-assisted electrolytic precipitation, pH balance, and germicidal UV-C treatment, are explored (Bapat et al. 2021). While conventional methods like aerated lagoons and activated sludge processes are efficient for BOD and SS removal, they struggle with color removal (Judkins and Hornsby 1978). Textile wastewater is characterized by high color content, suspended solids, salts, nutrients, and toxic substances like heavy metals and chlorinated compounds. Effluent degradation methods involve a combination of biochemical, chemical, and physical processes, with adsorption being a promising approach for dye removal.

Various countries, including the United States, Canada, and those in the European Union, seek to control their effluents through specific regulations. The United States has regulations listing effluent values in a mass of constituent per mass of manufactured product, which is not directly comparable to other countries' regulations that present wastewater values in mg/L. These regulations include USEPA Code of Federal Regulations (CFR) Title 40 Part 410—Textile Mills Point Source Category (USEPA, 2012a) and USEPA CFR Title 40 Part 425—Leather Tanning and Finishing Point Source Category (USEPA, 2012b). Canada has instituted SOR/

2012–139, Wastewater Systems Effluent Regulations 2015. This regulation applies when a prescribed harmful substance is deposited into a wastewater system.

The European Union (EU) has Directive 91/271/EEC, concerning Urban Wastewater Treatment (European Commission—Environment 1991), applicable to discharges from urban wastewater treatment plants. Spain has the Industrial Liquid Dumping Law to Integrated System of Sanitation systems (The Statute of Autonomy of the Community of Madrid 2015). Germany's Ordinance on Requirements for the Discharge of Wastewater into Waters applies to waterbody discharges (Germany Federal Ministry for the Environment, Nature Conservation, Building and Nuclear Safety, 2004). Italy's Legislative Decree 152 applies to discharges to surface waters and the public sewer (Zdohcp 2016).

Cambodia's Sub-Decree on Water Pollution Control applies to discharges into public water or sewage areas (Royal Government of Cambodia, 1999). China's GB 4287–2012 + XG1-2015 determines Water Pollutant Discharge Standards for Dyeing and Finishing Textile Industry (Ministry of Environmental Protection of China, 2015). These standards apply to discharges into the environment and public wastewater treatment systems. Honduras does not have a legal standard for wastewater effluent limits, but industrial complexes commonly adopt standards from other countries (Zdohcp 2016). India's standard IS: 2490 stipulates tolerance limits for industrial effluents discharged into inland surface waters (Bureau of Indian Standards, 1981), with specific annexes for textile and cotton dyeing industries (Indian Ministry of Environment, Forest and Environmental Change, 1986).

Indonesia has specified regulations for the quality of raw wastewater for industries (Republic of Indonesia, 2014). In Malaysia, the Department of Environment, Ministry of Natural Resources and Environment Malaysia, established the Environmental Quality (Industrial Effluents) Regulations in 2009. This comprehensive regulation includes Annex 7, which is tailor-made for the textile industry, and Annex 5, which specifically refers to industrial effluents. It applies to the discharge of industrial effluents or mixed effluents into any land, inland waters, or waters of Malaysia (Zdohcp 2016). In South Korea, the Ministry of Environment established the Water Quality and Ecosystem Conservation Law [Law No. 12519] and Sewerage Law [Law No. 12466] in 2008. These laws, with their stringent parameters, apply to discharges into surface waters and include additional parameters such as anionic surfactant, trichloroethylene, and vinyl chloride (Zdohcp 2016).

For Taiwan, the Taiwan Effluent Standards were created by the Taiwan Environmental Protection Administration in 2003. These standards apply to effluents discharged into water bodies from industries, sewage treatment plants attached to buildings, and sewage networks. The document includes parameters for specific insecticides such as endosulfan, toxaphene, and DDT (Zdohcp 2016). In Thailand, the Factory Law and the Industrial Effluent Standard were instituted by the Ministry of Natural Resources and Environment, Pollution Control Department, in 1996. This regulation applies to the discharge of industrial wastewater into public receiving waters or the environment.

In 2015, Vietnam established the National Technical Regulation on Textile Industry Effluent QCVN 13-MT: 2015/BTNMT. This regulation applies to wastewater from the textile industry discharged into receiving waters. The Industrial Wastewater Discharge Regulations 2005, TCVN 5945-2005, apply before industrial wastewater is discharged into a body of water. Sri Lanka instituted the National Environmental Regulations No. 1 of 2008 for protection and quality, covering textile wastewater discharged into inland surface waters (Zdohcp 2016).

In 1981, Brazil established its Environmental Policy through Law 6,938 of 08/31/1981, under the generic name of National Environmental Policy (PNMA). The PNMA is the set of decisions relating to environmental objectives. Subsequently, standards for the classification of solid waste emerged. According to NBR 10,004/2004 (ABNT 2004), solid waste can be class I, which is considered dangerous and poses a risk to public health. Its characteristics are dangerous, such as flammability, reactivity, corrosiveness, toxicity, and pathogenicity; class II A are those considered non-inert, which differ from class I (hazardous) waste because they have properties such as combustibility, biodegradability, or solubility in water.

1.4.3 Social Sustainability in Dyeing Processes

In the mid-70 s, with the oil crisis, the need arose to create a new strategic plan for large industries in the textile sector to survive the economic crisis. In this scenario, a system was born that later became known as fast fashion, a system of producing fashion items initially created for market survival. This system became popular in the 1990s, based on the cheapening of raw materials and labor and the agility in the production and distribution of the product (do Nascimento et al. 2017). The concept of fast fashion comes from the analogy to fast food and, as the name suggests, "fast." It is a dynamic that uses design and emerging fashion trends to promote "fast" consumption, like what is employed in the fast-food industry. The ability to interpret consumption needs, the speed of market response, and low product costs boosted the textile industry and the mass production of fashion items. In this context, the rapid consumption of pieces made by companies using this system produced six times more textile waste compared to the market before fast fashion. The push for consumption has had significant environmental impacts, making the textile industry the second most polluting industry, second only to the oil industry (Zonatti 2016).

The Spanish group Inditex (*Industria de Deseño Têxtil AS*), with its Zara stores, was the first company to draw attention to implementing the fast fashion model in its business dynamics concerning the production and distribution of new models–pre-conception, conception, and post-conception in the shortest possible time. New models are already in stores in a short period, creating a fast cycle to keep up with market demand and differentiate from competitors (do Nascimento et al. 2017). This speed of product turnover, the demand for new items, and ease of payment drive consumption, making it possible for consumers to always be in fashion at an affordable price. This concept, adopted by many retailers, popularizes the latest trends. What has just been paraded at significant fashion events by luxury brands is soon adapted by major stores to be produced and sold at a much lower price.

In the case of fast fashion, since the pieces are cheaper, more accessible, and have an aesthetic appeal that reflects current fashion trends, a rapid exchange system is created. It is suitable for the consumer, who acquires a fashionable product at an affordable price and good for the company, which increases its revenue. However, this dynamic is not synonymous with quality, and the economic factor often outweighs other determinants for purchase. The lifespan of these pieces is usually shorter precisely because of the cheapening of inputs, production, and labor. Thus, unrestrained and unsustainable consumption, the low price of the product, the planned obsolescence of products created to have a shorter life cycle, and often undervalued labor end up generating a high environmental and social impact. Among the notable consequences is the Atacama Desert in Chile, which houses one of the largest textile dumps on the planet. Textile waste is still a solid waste difficult to recycle due to the high costs involved in its reprocessing and because it is often composed of different fibers, especially synthetics, causing environmental damage such as toxic smoke, microplastics, and groundwater pollution, as well as social damage such as poor waste disposal and poor working conditions (Zonatti 2016; Costa and Bacellar Zaneti 2022).

The movement opposed to the fast fashion market system is known as slow fashion, a term coined from the analogy to the slow food movement created in Italy by Carlo Petrini. The slow food movement advocated the preservation, management, and care of the soil, the value of rural producers, local culture, and local production (Santos 2017). Slow fashion, as the name implies, promotes a slower fashion, favoring a system that values cleaner and more sustainable production of products, considering their entire life cycle: from the extraction of raw materials to disposal at the end of their usefulness and reprocessing of the raw material to become a new item. Slow fashion arises from the saturation of exaggerated consumption driven by the fast fashion system. In this perspective, the products created are not necessarily tied to the culture of instant fashion, tend to have a more timeless aesthetic, prioritize creating new products, and focus on the quality of raw materials, production, and finishes. By incorporating these aspects, products tend to be more expensive economically but more consumer and environmentally friendly.

In Brazil, some brands adopt this system, such as Flávia Aranha, which works with organic fabrics, natural dyeing, small collections (capsule collections), and modeling and shapes, emphasizing a more identifiable aesthetic language. Her pieces prioritize exclusivity and originality. The footwear and accessories brand Insecta Shoes initially became known for making its products from thrift store clothes with minor defects, but some parts still had the quality to be used in making shoes. The Catarina Mina brand values craftsmanship, local culture, and the people involved in production, bringing greater transparency to the production of each exclusive piece. The brand's pieces have a very unique aesthetic language.

In this sense, when it comes to sustainability, its concept is much broader than the ecological principle, as it encompasses economic, social, cultural, spatial, and political dimensions. Ecological sustainability is usually the best-known and most correlated dimension to the theme. It is briefly linked to the preservation of natural resources, the production of renewable resources, and the limitation of the use of non-renewable resources. Economic sustainability is associated with economic efficiency evaluated in macro-social terms and not just business profitability, balanced economic development between different sectors, and competitiveness in the international economy.

Social sustainability is briefly related to the need for material and non-material resources, aiming for greater equity in income distribution, greater social homogeneity, equal access to resources and social services, higher employability, and quality of life, thus guaranteeing the rights of the population. Spatial or territorial sustainability: it is directed towards the pursuit of balance between rural and urban development, aiming for better territorial distribution and use in balance with economic activities, biodiversity conservation, and eco-development awareness. Cultural sustainability is related to the preservation of the culture of each place, aiming to ensure continuity and balance between tradition and innovation. Political sustainability: it is linked to democracy, directed towards the effectiveness and enforcement of environmental, labor, and human rights legislation, the management of the environment and natural resources, and the management of global heritage.

1.5 Sustainable Dyeing Technologies

The textile industry has long been associated with significant environmental impacts, particularly in dyeing. Conventional dyeing methods often involve toxic chemicals, large amounts of water, and energy, leading to pollution and resource depletion. In response to these challenges, sustainable dyeing techniques have been developed to minimize environmental impact and promote eco-friendly practices. Several methods are gaining traction in the industry, each offering unique benefits and addressing different aspects of sustainability. Among these, Green Chemistry has emerged, proposing a set of principles to minimize the adverse environmental impact of chemicals. Green chemistry involves new studies, resources, methods, and the latest technologies for preventing waste generation based on life cycle assessment (LCA), anticipating and directing product development, and creating strategies for cleaner production (Roy Choudhury 2013). Green chemistry factors favor solutions to many health, environmental, and economic problems characteristic of industrial chemicals, proposing the reduction or elimination of harmful substances, low contamination, and reduced raw material consumption (Choudhury 2017). Below are some sustainable dyeing technologies that have gained attention in recent years, focusing on using natural dyes.

1.5.1 Textile Natural Dyeing

Natural dyeing is one of the oldest methods, utilizing dyes derived from plants, minerals, and sometimes insects. This technique has seen a resurgence due to its eco-friendly attributes. Natural dyes are biodegradable and typically non-toxic, making them a sustainable alternative to synthetic dyes. Examples include indigo from the indigo plant, madder root for red hues, and turmeric for yellow. However, the cultivation of dye plants must be managed sustainably to prevent land and water resource depletion. Natural pigments and dyes are innocuous and non-toxic chemical substances and can be extracted from plant, animal, or mineral origins. The practice of using natural dyes is ancient. There are records of natural dyeing work in Ancient China before 3000 B.C. By around 2500 B.C., India had already mastered dyeing and printing techniques (Ferreira 1997).

In Brazil, indigenous people mastered the technique of painting with natural inputs using annatto seeds (Bixa orellana–Bixaceae), brazilwood (Leguminosae–Caesalpinoideae), and charcoal to make body paintings and dye utensils (Francisco et al. 2024). With Portuguese colonization, brazilwood, a tree once abundant in the Atlantic Forest and whose economic value inspired the country's name, faced the risk of extinction due to excessive extraction. It was used for various purposes, especially for the textile trade, to extract the reddish dye (Araújo 2006). Besides Brazilwood, the country also became known for the extraction of natural indigo (Indigofera tinctoria) during the nineteenth century (Guaratini and Zanoni 2000). Until the mid-nineteenth century, all dyes were extracted from leaves, roots, fruits, flowers, substances of animal origin like carmine red from the Mexican insect cochineal (Dactyloplus Coccus), and mineral pigments like colored earth (Guaratini and Zanoni 2000).

With the emergence of synthetic dyes in 1856, the use of natural dyes for textile dyeing was gradually replaced, and their application diminished as textile dyeing processes advanced (Saravanan and Chandramohan 2011). The wide range of colors, reduced inputs, greater reproducibility and colorfastness, and lower cost made synthetic dyes widely used during the twentieth century (Samantaa and Agarwal 2009). With climate change and the need for environmental preservation, the macro behavioral trend related to sustainability is gaining space and strength, along with a reflection on overconsumption and product production methods. In this context, applying natural dyes on natural fibers has recently gained more space due to global environmental awareness (Samantaa and Agarwal 2009). Scientific research on the use of natural dyes in textile materials is also growing for the following reasons: wide variety and experimental feasibility, minimization of environmental impact, toxic and allergic effects of some synthetic dyes, and the contrary effects of natural dyes.

Natural dyes can be applied to different materials and fibers, but it is essential to understand their properties and affinities to achieve satisfactory results concerning colorfastness and intensity. Those who work with natural dye tend to choose natural fibers because they are biodegradable. Natural fibers are mainly extracted from plant and animal origins. Examples of well-known animal fibers include mulberry silk, sheep wool, and cashmere wool. Natural dyes have a strong affinity with animal fibers due to their protein composition. The most known plant fibers are cotton, linen, ramie, jute, sisal, and hemp (Ado et al. 2014b).

1.5.2 Low-Impact Dyes

Low-impact dyes are synthetic dyes designed to have a reduced environmental footprint. These dyes require less water, produce less waste, and contain fewer toxic substances compared to conventional dyes. They also tend to have a higher absorption rate, meaning that more dye bonds with the fabric, reducing runoff and contamination of water bodies. Low-impact dyes are often used in conjunction with closed-loop systems, where water and chemicals are continuously recycled within the dyeing process.

1.5.3 Digital Printing

Digital printing on textiles is a modern technique that uses inkjet technology to apply dye directly onto fabrics. This method significantly reduces water and chemical use, as it eliminates the need for pre-treatment and washing processes associated with traditional dyeing. Additionally, digital printing allows for precise application, minimizing waste. The ability to print on demand also reduces excess inventory and waste from unsold products (Kumelachew et al. 2023).

1.5.4 CO_2 Dyeing

CO_2 dyeing, also known as supercritical carbon dioxide dyeing, is an innovative technique that replaces water with supercritical CO_2 as the dyeing medium. In this process, CO_2 is pressurized until it becomes supercritical, a state where it exhibits the properties of both a liquid and a gas. This allows the dye to dissolve and penetrate the fabric effectively. CO_2 dyeing is water-free, eliminates the need for auxiliary chemicals, and the CO_2 used can be recycled within the system, making it highly sustainable (Oliveris et al. 2021, 2023b).

1.5.5 Bio-Dyeing

Bio-dyeing involves the use of microorganisms, such as bacteria and algae, to produce color. These organisms can produce pigments as part of their metabolic processes, which can then be used to dye fabrics. This method is still in its nascent stages but holds great promise due to its potential for low energy and resource use. Additionally, bio-dyes are typically biodegradable and non-toxic, aligning well with environmental sustainability goals (Fazal-ur-Rehman et al. 2022; Chimileski 2017).

1.5.6 Waste-Based Dyeing

Another innovative approach to sustainable dyeing is using waste materials as a source of dye. For instance, food waste such as onion skins, avocado pits, and pomegranate rinds can be used to extract natural dyes. This not only provides a use for waste materials but also reduces the reliance on conventional dye sources. Similarly, industrial by-products and waste from other sectors are being explored as potential dye sources, contributing to a circular economy (Mia et al. 2024; Kishor and Singh 2024; Benli and Bahtiyari 2024).

1.6 Future Perspectives

The dyeing processes in the textile industry have generated significant socio-environmental impacts that require urgent attention and coordinated action. However, by adopting integrated approaches that consider both environmental and social aspects, we cannot only minimize the negative impacts of dyeing processes but also pave the way for a more sustainable and responsible textile industry. Commitment to technological innovation, responsible waste management, improved working conditions, and community engagement are fundamental to achieving this goal and ensuring a more sustainable future for the global textile industry, with the potential for improved public perception and increased market competitiveness. Integrating approaches that consider both environmental and social aspects is necessary to mitigate the socio-environmental impacts of dyeing processes. Some promising strategies include:

- Sustainable Dyeing Technologies: Developing and adopting dyeing technologies that reduce water, energy, and chemical consumption can help minimize the environmental impacts of dyeing processes.
- Responsible Waste Management: Implementing responsible waste management practices, including proper treatment of textile effluents and recycling of chemicals, is crucial to reducing environmental pollution.
- Improvement of Working Conditions: Promoting safe and fair working conditions in the textile industry is essential to protecting workers' health and well-being and promoting social justice.

- Community Engagement: Involving local communities affected by dyeing processes in decision-making and ensuring their active participation in seeking solutions is fundamental to adequately addressing their concerns.

1.7 Final Remarks

The ecological footprint of dyeing methods is an urgent and critical concern in the textile industry, given their extensive and detrimental environmental impacts. Traditional dyeing processes, known for their resource-intensive nature, consume vast amounts of water and energy while generating significant chemical waste. These methods often involve the use of synthetic dyes, which are highly toxic and challenging to treat. When these dyes are discharged into water bodies without proper treatment, they can cause severe aquatic pollution, posing risks to aquatic life and human health. For instance, many dyes contain heavy metals and other harmful substances that can bioaccumulate in the food chain, leading to long-term ecological risks. Moreover, the high water usage in dyeing processes worsens water scarcity issues in many regions, as large volumes of fresh water are diverted from other essential uses to support textile manufacturing.

Energy consumption is another major factor contributing to the ecological footprint of dyeing methods. The need to heat large quantities of water for dye baths and the operation of machinery involved in dyeing processes result in significant greenhouse gas emissions. This reliance on fossil fuels for energy contributes to global climate change, further intensifying the environmental footprint of the textile industry. In China, the world's largest textile producer, the industry's energy consumption has steadily increased, highlighting the pressing need for more energy-efficient and sustainable practices.

Innovations in dyeing technology offer some solutions to these challenges. Waterless dyeing techniques, such as supercritical CO_2 dyeing, significantly reduce water consumption and eliminate the need for effluent treatment. Similarly, developing natural dyes and low-impact synthetic dyes can mitigate the harmful effects of conventional dyes. Natural dyes, derived from plants and other renewable resources, present an environmentally friendly alternative, though their scalability and colorfastness remain areas for improvement. Additionally, advancements in dye fixation technologies can enhance the uptake of dyes by fibers, reducing the amount of residual dye that needs to be treated or disposed of.

Furthermore, implementing closed-loop systems in dyeing processes can significantly reduce waste and resource use. These systems recycle water and chemicals within the production cycle, thereby cutting down on fresh water and raw material inputs. Adopting best practices and regulatory frameworks to reduce the environmental impact of dyeing can drive industry-wide improvements. By prioritizing sustainable dyeing methods, the textile industry cannot only reduce its ecological footprint but also contribute to broader environmental conservation efforts, paving the way for a more sustainable future.

References

Abagnato S, Rigamonti L, Grosso M (2024) Life cycle assessment applications to reuse, recycling and circular practices for textiles: a review. Waste Manag 182:74–90. https://doi.org/10.1016/j.wasman.2024.04.016

ABNT-Associação Brasileira de Normas Técnicas, NBR 10004/2004 – Resíduos Sólidos Classificação, Rio de Janeiro-RJ (Brazil), 2004.

Abrahart EN (1968) Dyes and their intermediates, 1st ed., Pergamon Press

Abu Hasan H, Muhammad MH, Ismail NI (2020) A review of biological drinking water treatment technologies for contaminants removal from polluted water resources. J Water Process Eng 33:101035. https://doi.org/10.1016/j.jwpe.2019.101035

Ado A, Yahaya H, Kwalli A, Abdulkarir S (2014a) Dyeing of textiles with natural dyes: a review. Int J Environ Monit Prot 1:76–81. https://doi.org/10.5772/21341

Ado A, Yahaya H, Kwalli AA, Abdulkadir RS (2014) Dyeing of textiles with eco-friendly natural dyes: a review. Int J Environ Monit Prot 76–81

Ağtaş M, Yılmaz Ö, Dilaver M, Alp K, Koyuncu İ (2020) Hot water recovery and reuse in textile sector with pilot scale ceramic ultrafiltration/nanofiltration membrane system. J Clean Prod 256. https://doi.org/10.1016/j.jclepro.2020.120359

Aleem M, Cao J, Li C, Rashid H, Wu Y, Nawaz MI, Abbas M, Akram MW (2020) Coagulation-and adsorption-based environmental impact assessment and textile effluent treatment. Water Air Soil Pollut 231:1–8. https://doi.org/10.1007/S11270-020-4400-X/FIGURES/4

Afena AS, Boateng DK, Darkwah L, Adjaottor AA (2021) Decolourisation of textile wastewater by dye degrading microorganisms isolated from textile effluent. J Environ Prot (Irvine, Calif) 12:767–783. https://doi.org/10.4236/jep.2021.1210046

Ammayappan L, Jose S, Arputha Raj A (2016) Sustainable production processes in textile dyeing. In: Environmental footprints and eco-design of products and processes, pp 185–216. https://doi.org/10.1007/978-981-10-0111-6_8

Araújo MEM (2006) Corantes naturais para têxteis–da antiguidade aos tempos modernos

Araujo GM (2016) Sistema de Gestão Ambiental ISO 14001 comentada, GVC, Rio de Janeiro-RJ (Brazil)

Ashfaq M, Tahir MN, Muhammad S, Munawar KS, Ali S, Ahmed G, Al-Sehemi AG, Alarfaji SS, Ibraheem Khan ME (2022) Shedding light on the synthesis, crystal structure, characterization, and computational study of optoelectronic properties and bioactivity of imine derivatives, ACS Omega 7:5217–5230. https://doi.org/10.1021/acsomega.1c06325

Asgari G, Shabanloo A, M. Salari, F. Eslami, Sonophotocatalytic treatment of AB113 dye and real textile wastewater using ZnO/persulfate: Modeling by response surface methodology and artificial neural network, Environ Res 184 (2020). https://doi.org/10.1016/j.envres.2020.109367.

Azanaw A, Birlie B, Teshome B, Jemberie M (2022) Textile effluent treatment methods and eco-friendly resolution of textile wastewater. Case Stud Chem Environ Eng 6:100230. https://doi.org/10.1016/j.cscee.2022.100230

Bafana A, Devi SS, Chakrabarti T (2011) Azo dyes: past, present and the future. Environ Rev 19:350–371. https://doi.org/10.1139/a11-018

Bailey K, Basu A, Sharma S (2022) The environmental impacts of fast fashion on water quality: a systematic review. Water (Basel) 14:1073. https://doi.org/10.3390/w14071073

Bapat S, Jaspal D, Malviya A (2021) Integrated textile effluent treatment method. Water Environ Res 93:1060–1076. https://doi.org/10.1002/wer.1494

Basutkar MR (2019) Review -a history and introduction of synthetic dye and their usage, vol 7, pp 396–406

Ben Mansour H, Corroler D, Barillier D, Ghedira K, Chekir L, Mosrati R (2007) Evaluation of genotoxicity and pro-oxidant effect of the azo dyes: Acids yellow 17, violet 7 and orange 52, and of their degradation products by Pseudomonas putida mt-2. Food Chem Toxicol 45:1670–1677. https://doi.org/10.1016/j.fct.2007.02.033

Ben Slama H, Bouket AC, Pourhassan Z, Alenezi FN, Silini A, Cherif-Silini H, Oszako T, Luptakova L, Golińska P, Belbahri L (2021) Diversity of synthetic dyes from textile industries, discharge impacts and treatment methods. Appl Sci 11:1–21. https://doi.org/10.3390/app11146255

Benkhaya S, M'rabet S, El Harfi A (2020) A review on classifications, recent synthesis and applications of textile dyes, Inorg Chem Commun 115(2020):107891. https://doi.org/10.1016/j.inoche.2020.107891

Benli H, Bahtiyari Mİ (2024) An approach for the valorization of bio-waste pistachio shells (Pistacia vera L.): dyeing of cellulose-based fabrics. J Clean Prod 445:141213. https://doi.org/10.1016/j.jclepro.2024.141213

Berradi M, Hsissou R, Khudhair M, Assouag M, Cherkaoui O, El Bachiri A, El Harfi A (2019a) Textile finishing dyes and their impact on aquatic environs. Heliyon 5:e02711. https://doi.org/10.1016/j.heliyon.2019.e02711

Berradi M, Hsissou R, Khudhair M, Assouag M, Cherkaoui O, El Bachiri A, El Harfi A (2019) Textile finishing dyes and their impact on aquatic environs. Heliyon 5. https://doi.org/10.1016/j.heliyon.2019.e02711

Bharagava RN, Mani S, Mulla SI, Saratale GD (2018) Degradation and decolourization potential of an ligninolytic enzyme producing Aeromonas hydrophila for crystal violet dye and its phytotoxicity evaluation. Ecotoxicol Environ Saf 156:166–175. https://doi.org/10.1016/J.ECOENV.2018.03.012

Bhatia SC (2017) Pollution control in textile industry. Woodhead Publishing India. https://doi.org/10.1201/9781315148588

Branchetti S, Petrovich C, Ciaccio G, De Sabbata P, Frascella A, Nigliaccio G (2019), Energy consumption characterization based on a self-analysis tool: a case study in yarn manufacturing. In: Smartgreens 2019-Proceedings of the 8th international conference on smart cities and green ICT systems, pp 40–50. https://doi.org/10.5220/0007680300400050

Bravo MVC, Iturralde WMP (2022) Energy efficiency in the competitiveness of the textile industry. Open J Bus Manag 10:1755–1767. https://doi.org/10.4236/ojbm.2022.104090

Burkinshaw SM (2016) Physico-chemical aspects of textile coloration, Wiley. https://doi.org/10.1002/9781118725658

Cai H, Liang J, Ning X, Lai X, Li Y (2020) Algal toxicity induced by effluents from textile-dyeing wastewater treatment plants, J Environ Sci (China) 91:199–208. https://doi.org/10.1016/j.jes.2020.01.004

Cao J, Sanganyado E, Liu W, Zhang W, Liu Y (2019) Decolorization and detoxification of direct blue 2B by indigenous bacterial consortium. J Environ Manag 242:229–237. https://doi.org/10.1016/j.jenvman.2019.04.067

Carmen Z, Daniel S (2012) Textile organic dyes–characteristics, polluting effects and separation/elimination procedures from industrial effluents–A critical overview. Organic pollutants ten years after the stockholm convention-environmental and analytical update. https://doi.org/10.5772/32373

Carreón T, Herrick RL (2012) Aliphatic hydrocarbons, patty's toxicology, pp 1–102. https://doi.org/10.1002/0471435139.tox049.pub2

Chakraborty R, Ahmad F (2022) Economical use of water in cotton knit dyeing industries of Bangladesh. J Clean Prod 340:130825. https://doi.org/10.1016/j.jclepro.2022.130825

Chandanshive V, Kadam S, Rane N, Jeon BH, Jadhav J, Govindwar S (2020) In situ textile wastewater treatment in high rate transpiration system furrows planted with aquatic macrophytes and floating phytobeds. Chemosphere 252. https://doi.org/10.1016/j.chemosphere.2020.126513

Chandra R, Bharagava R, Yadav S undefined (2009) Accumulation and distribution of toxic metals in wheat (Triticum aestivum L.) and Indian mustard (Brassica campestris L.) irrigated with distillery and tannery. J Hazardous Mater. (Chandra R, Bharagava RN, Yadav S, Mohan D (eds) Elsevier (n.d.))

Chavan RB (2013) Health and environmental hazards of synthetic dyes. Garment and Fashion Design

Chen Y, Lin M, Zhuang D (2022) Wastewater treatment and emerging contaminants: bibliometric analysis. Chemosphere 297:133932. https://doi.org/10.1016/j.chemosphere.2022.133932

Chiango LB, Buffo A, Vanni M, Nierstrasz V, Ferri A (2022) Simulation of an industrial scale scCO2 beam dyeing process. J CO2 Util 64:102147. https://doi.org/10.1016/j.jcou.2022.102147

Chimileski S (2017) Bacterial dyes in fashion. https://Asm.Org/Articles/2017/November/Bacterial-Dyes-in-Fashion

Choudhury AKR (2017) Green chemistry and textile industry. J Text Eng Fashion Technol 2. https://doi.org/10.15406/jteft.2017.02.00056

Chungkrang L, Bhuyan S (2020) Natural dye sources and its applications in textiles: a brief review. Int J Curr Microbiol Appl Sci 9:261–269. https://doi.org/10.20546/ijcmas.2020.910.034

Clarck M (2011a) Handbook of textile and industrial dyeing-Part II: dye types and processes, vol 1. Woodhead Publishing Limited, Cambridge, UK

Clarck M (2011b) Handbook of textile and industrial dyeing-part I: general aspects of dyeing, vol 1. Woodhead Publishing Limited, Cambridge, UK

Costa MFBF, Bacellar Zaneti ICB (2022) Impactos ambientais do fast fashion: o lixão têxtil internacional do Atacama, Chile, Revista Tecnologia e Sociedade 18:129. https://doi.org/10.3895/rts.v18n53.15794

Commission Européenne (2001) Integrated pollution prevention and control (IPPC) Reference document on best available techniques in the pulp and paper industry Dec 2001, Paper Industry

da Silva Júnior AH, Mulinari J, de Oliveira PV, de Oliveira CRS, Júnior FWR (2022) Impacts of metallic nanoparticles application on the agricultural soils microbiota. J Hazard Mater Adv 100103. https://doi.org/10.1016/j.hazadv.2022.100103

da Pinto ASRS (2021) Economia Circular e Análise de Ciclo de Vida: Estudo de casos na Indústria Têxtil.

De Giorgi MR, Cadoni E, Maricca D, Piras A (2000) Dyeing polyester fibres with disperse dyes in supercritical CO2. Dyes Pigm 45:75–79. https://doi.org/10.1016/S0143-7208(00)00011-5

dos Santos SDM (2017) Entre Fios e Desafios: Indústria da Moda, Linguagem e Trabalho Escravo na Sociedade Imperialista. Revista Latino-Americana De Estudos Em Cultura e Sociedade 3:1–15

dos Santos AB, Cervantes FJ, van Lier JB (2007) Review paper on current technologies for decolourisation of textile wastewaters: Perspectives for anaerobic biotechnology. Bioresour Technol 98:2369–2385. https://doi.org/10.1016/j.biortech.2006.11.013

Drumond Chequer FM, de Oliveira GAR, Anastacio Ferraz ER, Carvalho J, Boldrin Zanoni MV, de Oliveir DP (2013) Textile dyes: dyeing process and environmental impact. INTECH Open Science. https://doi.org/10.5772/53659.

Dutta S, Adhikary S, Bhattacharya S, Roy D, Chatterjee S, Chakraborty A, Banerjee D, Ganguly A, Nanda S, Rajak P (2024) Contamination of textile dyes in aquatic environment: adverse impacts on aquatic ecosystem and human health, and its management using bioremediation. J Environ Manag 353:120103. https://doi.org/10.1016/j.jenvman.2024.120103

Fazal-ur-Rehman, Adeel S, Liaqat S, Hussaan M, Mia R, Ahmed B, Wafa H (2022) Environmental friendly bio-dyeing of silk using Alkanna tinctoria based Alkannin natural dye. Ind Crops Prod 186:115301. https://doi.org/10.1016/j.indcrop.2022.115301

Ferreira EL (1997) Corantes naturais da flora brasileira: guia prático de tingimento com plantas. Optagraf Editora e Gráfica, São Paulo-SP (Brazil)

Francisco IBM, dos Baraldi RFS, Dos Santos EG, Da Silva Júnior AH, De Oliveira PV, Felipe BHS, De Oliveira CRS, De Aguiar CRL (2024) Acrylic yarns pre-treated with different mordants and their dyeing potential with annatto dye. Braz J Animal Environ Res 7:191–206. https://doi.org/10.34188/bjaerv7n1-015

Galatti LG, Baruque-Ramos J (2022) Circular economy indicators for measuring social innovation in the Brazilian textile and fashion industry. J Clean Prod 363:132485. https://doi.org/10.1016/j.jclepro.2022.132485

Galeti Narimatsu BM, Do Bem NA, Wachholz LA, Linke PP, de los Perez Lizama MA, Soto Herek Rezende LC (2021) Corantes naturais como alternativa sustentável na indústria têxtil. Revista Valore 5. https://doi.org/10.22408/reva502020507e-5030

Giovanazzi LC, Missner MEP, Francisco IBM, Rodrigues NMM, da Silva Júnior AH, Pellenz L, de Aguiar CRL, de Oliveira CRS (2024) Photocatalytic degradation using TiO2 P25: a comparative study for different textile dyes. Braz J Animal Environ Res 7:125–144. https://doi.org/10.34188/bjaerv7n1-011

Gregory P (2007) Toxicology of textile dyes. In: Environmental aspects of textile dyeing. Elsevier, pp 44–73. https://doi.org/10.1533/9781845693091.44

Griebler CB, dos Marques GS, de Faria EO, Parente HT (2022) Economia circular como oportunidade de crescimento para as pequenas empresas têxteis. Concilium 22:922–937. https://doi.org/10.53660/CLM-470-559

Guaratini C, Zanoni MVB (2000) Corantes têxteis. Quim Nova 1:71–78

Hall CW (1992) Emerging technologies for drying

Harane RS, Mehra NR, Tayade PB, Adivarekar RV (2014) A facile energy and water-conserving process for cotton dyeing. Int J Energy Environ Eng 5:1–10. https://doi.org/10.1007/s40095-014-0096-2

Hasanbeigi A, Zuberi MJS (2022) Electrified process heating in textile wet-processing industry: a techno-economic analysis for China, Japan, and Taiwan, Energies (Basel), vol 15. https://doi.org/10.3390/en15238939

Hasanbeigi A, Price L (2012) A review of energy use and energy efficiency technologies for the textile industry. Renew Sustain Energy Rev 16:3648–3665. https://doi.org/10.1016/j.rser.2012.03.029

Hasanbeigi A, Price L (2015) A technical review of emerging technologies for energy and water efficiency and pollution reduction in the textile industry. J Clean Prod 95:30–44. https://doi.org/10.1016/j.jclepro.2015.02.079

Hassan MM, Carr CM (2018) A critical review on recent advancements of the removal of reactive dyes from dyehouse effluent by ion-exchange adsorbents. Chemosphere 209:201–219. https://doi.org/10.1016/j.chemosphere.2018.06.043

Herath IS, Udayanga D, Jayasanka DJ, Hewawasam C (2024) Textile dye decolorization by white rot fungi–a review. Bioresour Technol Rep 25:101687. https://doi.org/10.1016/j.biteb.2023.101687

Hossain L, Sarker SK, Khan MS (2018) Evaluation of present and future wastewater impacts of textile dyeing industries in Bangladesh. Environ Dev 26:23–33. https://doi.org/10.1016/J.ENVDEV.2018.03.005

Hossain MN, Rahman MM, Afrin S, Akbor MA, Siddique MAB, Malafaia G (2023) Identification and quantification of microplastics in agricultural farmland soil and textile sludge in Bangladesh. Sci Total Environ 858:160118. https://doi.org/10.1016/J.SCITOTENV.2022.160118

Hubadillah SK, Othman MHD, Tai ZS, Jamalludin MR, Yusuf NK, Ahmad A, Rahman MA, Jaafar J, Kadir SHSA, Harun Z (2020) Novel hydroxyapatite-based bio-ceramic hollow fiber membrane derived from waste cow bone for textile wastewater treatment. Chem Eng J 379. https://doi.org/10.1016/j.cej.2019.122396

Ikram M (2022) Transition toward green economy: technological Innovation's role in the fashion industry. Curr Opin Green Sustain Chem 37:100657. https://doi.org/10.1016/j.cogsc.2022.100657

Imran M, Crowley DE, Khalid A, Hussain S, Mumtaz MW, Arshad M (2015) Microbial biotechnology for decolorization of textile wastewaters. Rev Environ Sci Biotechnol 14:73–92. https://doi.org/10.1007/S11157-014-9344-4

Iqbal M (2008) Textile dyes, 1st edn. Rehbar Publishers Karachi

Islam T, Islam T, Islam T, Repon R (2022) Synthetic dyes for textile colouration: process, factors and environmental impact. Textile Leather Rev 5:327–373. https://doi.org/10.31881/TLR.2022.27

Islam M, Mostafa M (2019) Textile dyeing effluents and environment concerns-a review. J Environ Sci Nat Res 11:131–144. https://doi.org/10.3329/jesnr.v11i1-2.43380

Izdebska J, Sabu T (2016) Printing on polymers. Elsevier. https://doi.org/10.1016/C2014-0-02411-2

Judkins Jr JF, Hornsby JS (1978) Color removal from textile dye waste using magnesium carbonate. Water Pollut Control Fed 2446–2456

Jain R, Thakur A, Garg N, Devi P (2021) Impact of industrial effluents on groundwater. In: Groundwater geochemistry: pollution and remediation methods, pp 193–211. https://doi.org/10.1002/9781119709732.ch10

Jeong S, Lim J, Il Hong S, Kwon SC, Shim JY, Yoo Y, Cho H, Lim S, Kim J (2023) A framework for environmental production of textile dyeing process using novel exhaustion-rate meter and multi-layer perceptron-based prediction model. Process Saf Environ Prot 175:99–110. https://doi.org/10.1016/j.psep.2023.05.009

Jnr EB-A, Acquaye R (2022) Mitigating risks in inappropriate handling of chemicals by small-scale textile producers in Sekondi-Takoradi, Ghana. J Textile Sci Technol 08:1–14. https://doi.org/10.4236/jtst.2022.81001

Kamenická B (2024) Chemical degradation of azo dyes using different reducing agents: a review. J Water Process Eng 61:105350. https://doi.org/10.1016/j.jwpe.2024.105350

Kang S, Qin L, Zhao Y, Wang W, Zhang T, Yang L, Rao F, Song S (2020) Enhanced removal of methyl orange on exfoliated montmorillonite/chitosan gel in presence of methylene blue. Chemosphere 238. https://doi.org/10.1016/j.chemosphere.2019.124693

Kant R (2012) Textile dyeing industry an environmental hazard. Nat Sci (Irvine) 04:22–26. https://doi.org/10.4236/ns.2012.41004

Kargar H, Fallah-Mehrjardi M, Behjatmanesh-Ardakani R, Bahadori M, Moghadam M, Ashfaq M, Munawar KS, Tahir MN (2022) Spectroscopic investigation, molecular structure, catalytic activity with computational studies of a novel Pd(II) complex incorporating unsymmetrical tetradentate Schiff base ligand. Inorg Chem Commun 142:109697. https://doi.org/10.1016/j.inoche.2022.109697

Karthik T, Gopalakrishnan D (2014) Environmental analysis of textile value chain: an overview, pp 153–188. https://doi.org/10.1007/978-981-287-110-7_6

Karunanidhi D, Subramani T, Roy PD, Li H (2021) Impact of groundwater contamination on human health. Environ Geochem Health 43:643–647. https://doi.org/10.1007/S10653-021-00824-2/METRICS

Kaur J, Bhatti SS, Bhat SA, Nagpal AK, Kaur V, Katnoria JK (2021) Evaluating potential ecological risks of heavy metals of textile effluents and soil samples in vicinity of textile industries. Soil Syst 5:63. https://doi.org/10.3390/SOILSYSTEMS5040063/S1

Khan I, Hou F, Zakari A, Tawiah VK (2021) The dynamic links among energy transitions, energy consumption, and sustainable economic growth: a novel framework for IEA countries. Energy 222:119935. https://doi.org/10.1016/j.energy.2021.119935

Khan WU, Ahmed S, Dhoble Y, Madhav S (2023) A critical review of hazardous waste generation from textile industries and associated ecological impacts. J Indian Chem Soc 100:100829. https://doi.org/10.1016/J.JICS.2022.100829

Khattab TA, Abdelrahman MS, Rehan M (2020) Textile dyeing industry: environmental impacts and remediation. Environ Sci Pollut Res 27:3803–3818. https://doi.org/10.1007/s11356-019-07137-z

Kishor K, Singh MK (2024) Functionalization of wool fabric by rice stubble extract dyeing and bio-mordanting. Bioresour Technol Rep 26:101829. https://doi.org/10.1016/j.biteb.2024.101829

Kishor R, Purchase D, Saratale GD, Saratale RG, Ferreira LFR, Bilal M, Chandra R, Bharagava RN (2021) Ecotoxicological and health concerns of persistent coloring pollutants of textile industry wastewater and treatment approaches for environmental safety. J Environ Chem Eng 9:105012. https://doi.org/10.1016/j.jece.2020.105012

Koç E, Çinçik E (2010) Analysis of energy consumption in woven fabric production. In: Fibres and textiles in Eastern Europe, vol 79, pp 14–20

Khude P (2017) A review on energy management in textile industry. Innov Energy Res 06. https://doi.org/10.4172/2576-1463.1000169

Kumar A, Konar A (2011) Dyeing of textiles with natural dyes. In: Natural dyes, InTech. https://doi.org/10.5772/21341

Kumar A, Sharma G, Naushad M, Al-Muhtaseb AH, García-Peñas A, Mola GT, Si C, Stadler FJ (2020) Bio-inspired and biomaterials-based hybrid photocatalysts for environmental detoxification: a review. Chem Eng J 382. https://doi.org/10.1016/j.cej.2019.122937

Kumelachew DM, Wagaye BT, Adamu BF (2023) Digital textile printing innovations and the future. In: Digital textile printing, Elsevier, pp 241–259. https://doi.org/10.1016/B978-0-443-15414-0.00006-6

Lacasse K, Baumann W (2004) Textile chemicals. Springer, Berlin, Heidelberg. https://doi.org/10.1007/978-3-642-18898-5

Langhals H (2004) Color chemistry. synthesis, properties and applications of organic dyes and pigments. 3rd revised edn. In: Zollinger H (ed) Angewandte chemie international edition, vol 43, pp 5291–5292. https://doi.org/10.1002/anie.200385122.

Lellis B, Fávaro-Polonio CZ, Pamphile JA, Polonio JC (2019) Effects of textile dyes on health and the environment and bioremediation potential of living organisms. Biotechnol Res Innov 3:275–290. https://doi.org/10.1016/j.biori.2019.09.001

Li K, Liu Q, Fang F, Wu X, Xin J, Sun S, Wei Y, Ruan R, Chen P, Wang Y, Addy M (2020) Influence of nanofiltration concentrate recirculation on performance and economic feasibility of a pilot-scale membrane bioreactor-nanofiltration hybrid process for textile wastewater treatment with high water recovery. J Clean Prod 261:121067. https://doi.org/10.1016/j.jclepro.2020.121067

Lin B, Bai R (2020) Dynamic energy performance evaluation of Chinese textile industry. Energy 199:117388. https://doi.org/10.1016/j.energy.2020.117388

Lin B, Chen Y, Zhang G (2018) Impact of technological progress on China's textile industry and future energy saving potential forecast. Energy 161:859–869. https://doi.org/10.1016/j.energy.2018.07.178

Lu X, Liu L, Liu R, Chen J (2010) Textile wastewater reuse as an alternative water source for dyeing and finishing processes: a case study. Desalination 258:229–232. https://doi.org/10.1016/j.desal.2010.04.002

Madhav S, Ahamad A, Singh P, Mishra PK (2018b) A review of textile industry: wet processing, environmental impacts, and effluent treatment methods. Environ Qual Manag 27:31–41. https://doi.org/10.1002/TQEM.21538

Madhav S, Ahamad A, Singh P, Mishra PK.-E.Q., undefined 2018, A review of textile industry: Wet processing, environmental impacts, and effluent treatment methods. Wiley Online Library. https://doi.org/10.1002/tqem.21538. (Madhav S, Ahamad A, Singh P, Mishra PK (2018) Environmental quality management, vol 27. Wiley Online Library, pp 31–41)

Mahapatra NN (2016) Textile dyes. WPI Publishing. https://doi.org/10.1201/b21336

Malik AN, Kuznetsov A, Ali A, Ashfaq M, Tahir MN, Siddique A (2022) Imine-based Zwitterion: synthesis, single-crystal characterization, and computational investigation. J Mol Struct 1253:132237. https://doi.org/10.1016/j.molstruc.2021.132237

Manzoor S, Shah MH, Shaheen N, Khalique A, Jaffar M (2006) Multivariate analysis of trace metals in textile effluents in relation to soil and groundwater. J Hazard Mater 137:31–37. https://doi.org/10.1016/j.jhazmat.2006.01.077

Markandeya D, Mohan SP (2022) Shukla, Hazardous consequences of textile mill effluents on soil and their remediation approaches. Clean Eng Technol 7:100434. https://doi.org/10.1016/J.CLET.2022.100434

Marszałek J, Żyłła R (2021) Recovery of water from textile dyeing using membrane filtration processes. Processes 9:1833. https://doi.org/10.3390/pr9101833

Melati I, Rahayu G, Effendi H, Henny C, Yanto DHY (2023) Biodecolorization of anthraquinone and azo dyes by dark septate endophytic fungi. Bioresour Technol Rep 22:101427. https://doi.org/10.1016/j.biteb.2023.101427

Mia R, Mofasser AZM, Bhat MA, Howlader MI, Sayed NI, Bakar MA (2024) Utilizing neem wood waste as a natural dye in the development of bio-based functionalized colored jute fabric. Sustain Chem Pharm 38:101489. https://doi.org/10.1016/j.scp.2024.101489

Moyo S, Makhanya BP, Zwane PE (2022) Use of bacterial isolates in the treatment of textile dye wastewater: a review. Heliyon 8:e09632. https://doi.org/10.1016/j.heliyon.2022.e09632

Mudhoo A, Paliya S, Goswami P, Singh M, Lofrano G, Carotenuto M, Carraturo F, Libralato G, Guida M, Usman M, Kumar S (2020) Fabrication, functionalization and performance of doped photocatalysts for dye degradation and mineralization: a review. Environ Chem Lett 18:1825–1903. https://doi.org/10.1007/s10311-020-01045-2

do Nascimento NR, Borges FF, de Luana FM, da Sliva Junior AG (2017) Sistema de baixo custo baseado em resíduos têxteis para irrigação subterrânea no semiárido e mitigação das consequências da "fast-fashion." Anais Do Congresso Brasileiro de Gestão Ambiental e Sustentabilidade

Nair GP (2011) Methods and machinery for the dyeing process. Woodhead Publishing Limited. https://doi.org/10.1533/9780857093974.1.245

Natarajan S, Bajaj HC, Tayade RJ (2018) Recent advances based on the synergetic effect of adsorption for removal of dyes from waste water using photocatalytic process. J Environ Sci 65:201–222. https://doi.org/10.1016/j.jes.2017.03.011

Newman MC (2009) Fundamentals of ecotoxicology, 3rd edn, pp 1–551. https://doi.org/10.1201/9781439883129

Nicolai S, Tralau T, Luch A, Pirow R (2021) A scientific review of colorful textiles. J Consum Prot Food Saf 16:5–17. https://doi.org/10.1007/s00003-020-01301-1

Nieminen-Kalliala E (2003) Environmental indicators of textile products for ISO (type III) environmental product declaration. Autex Res J 3:206–218. https://doi.org/10.1515/AUT-2003-030407/HTML

O'Neill C, Hawkes FR, Hawkes DL, Lourenço ND, Pinheiro HM, Delée W (1999) Colour in textile effluents-sources, measurement, discharge consents and simulation: a review. J Chem Technol Biotechnol 74:1009–1018. https://doi.org/10.1002/(SICI)1097-4660(199911)74:11%3c1009::AID-JCTB153%3e3.0.CO;2-N

OECD (2023) eChemPortal-The global portal to information on chemical substances, Organisation for Economic Cooperation and Development-OECD. https://www.echemportal.org/echemportal/substance. https://www.echemportal.org/echemportal/substance-search

Ogugbue CJ, Sawidis T (2011) Bioremediation and detoxification of synthetic wastewater containing triarylmethane dyes by aeromonas hydrophila isolated from industrial effluent. Biotechnol Res Int 2011:1–11. https://doi.org/10.4061/2011/967925

de Oliveira CRS, da Silva Júnior AH, Mulinari J, Immich APS (2021) Textile re-engineering: eco-responsible solutions for a more sustainable industry. Sustain Prod Consum 28:1232–1248. https://doi.org/10.1016/j.spc.2021.08.001

de Oliveira CRS, de Aguiar CRL, Missner MEP, Aragão FV, da Silva Júnior AH, Mapossa AB (2023) A comprehensive guide to textile process laboratories: risks, hazards, preservation care, and safety protocol. Laboratories 1:1–33. https://doi.org/10.3390/laboratories1010001

de Oliveira CRS, de Oliveira PV, Pellenz L, de Aguiar CRL, da Silva Júnior AH (2023) Supercritical fluid technology as a sustainable alternative method for textile dyeing: an approach on waste, energy, and CO2 emission reduction. J Environ Sci. https://doi.org/10.1016/j.jes.2023.06.007

de Oliveira CRS, da Silva Júnior AH, Mulinari J, Ferreira AJS, da Silva A (2023) Fibrous microplastics released from textiles: occurrence, fate, and remediation strategies. J Contam Hydrol 104169. https://doi.org/10.1016/j.jconhyd.2023.104169

de Oliveira CRS, da Silva Júnior AH, Mulinari J, Ferreira AJS, da Silva A, Fibrous microplastics released from textiles: occurrence, fate, and remediation strategies. J Contam Hydrol 256:104169. https://doi.org/10.1016/J.JCONHYD.2023.104169

Öner E, Sahinbaskan BY (2011) A new process of combined pretreatment and dyeing: REST. J Clean Prod 19:1668–1675. https://doi.org/10.1016/j.jclepro.2011.05.008

Orts F, del Río AI, Molina J, Bonastre J, Cases F (2018) Electrochemical treatment of real textile wastewater: trichromy Procion HEXL®. J Electroanal Chem 808:387–394. https://doi.org/10.1016/j.jelechem.2017.06.051

Ozturk E, Karaboyaci M, Yetis U, Yigit NO, Kitis M (2015) Evaluation of integrated pollution prevention control in a textile fiber production and dyeing mill. J Clean Prod 88:116–124. https://doi.org/10.1016/j.jclepro.2014.04.064

Palamutcu S (2010) Electric energy consumption in the cotton textile processing stages. Energy 35:2945–2952. https://doi.org/10.1016/j.energy.2010.03.029

Parisi ML, Fatarella E, Spinelli D, Pogni R, Basosi R (2015) Environmental impact assessment of an eco-efficient production for coloured textiles. J Clean Prod 108:514–524. https://doi.org/10.1016/j.jclepro.2015.06.032

Patra AK, Pariti SRK (2022) Restricted substances for textiles. Text Prog 54:1–101. https://doi.org/10.1080/00405167.2022.2101302

Peña-Pichardo P, Martínez-Barrera G, Martínez-López M, Ureña-Núñez F, dos Reis JML (2018) Recovery of cotton fibers from waste Blue-Jeans and its use in polyester concrete. Constr Build Mater 177:409–416. https://doi.org/10.1016/j.conbuildmat.2018.05.137

Piribauer B, Bartl A (2019) Textile recycling processes, state of the art and current developments: a mini review. Waste Manag Res 37:112–119. https://doi.org/10.1177/0734242X18819277

Pizzicato B, Pacifico S, Cayuela D, Mijas G, Riba-Moliner M (2023) Advancements in sustainable natural dyes for textile applications: a review. Molecules 28:5954. https://doi.org/10.3390/molecules28165954

Qiu Y, Zhou S, Zhang C, Zhou Y, Qin W (2022) Soil microplastic characteristics and the effects on soil properties and biota: a systematic review and meta-analysis. Environ Pollut 313:120183. https://doi.org/10.1016/J.ENVPOL.2022.120183

Rade KA, Pharande VA, Saini DR (2017) Effect of change in process parameters on energy consumption during textile dyeing process. Int J Theoret Appl Mech 12:579–588

Rahman SS, Siddiqua S, Cherian C (2022) Sustainable applications of textile waste fiber in the construction and geotechnical industries: a retrospect. Clean Eng Technol 6. https://doi.org/10.1016/j.clet.2022.100420

Ranganathan K, Karunagaran K, Sharma DC (2007) Recycling of wastewaters of textile dyeing industries using advanced treatment technology and cost analysis—case studies. Resour Conserv Recycl 50:306–318. https://doi.org/10.1016/j.resconrec.2006.06.004

Rapo E, Tonk S (2021) Factors affecting synthetic dye adsorption; desorption studies: a review of results from the last five years. Mol 26:5419. https://doi.org/10.3390/molecules26175419

Rauf MA, Meetani MA, Hisaindee S (2011) An overview on the photocatalytic degradation of azo dyes in the presence of TiO2 doped with selective transition metals. Desalination 276:13–27. https://doi.org/10.1016/j.desal.2011.03.071

Rehman K, Shahzad T, Sahar A, Hussain S, Mahmood F, Siddique MH, Siddique MA, Rashid MI (2018) Effect of reactive black 5 azo dye on soil processes related to C and N cycling, PeerJ 2018. https://doi.org/10.7717/peerj.4802

Robinson PR, Hsu CS (2017) Petroleum and its products. Springer International Publishing. https://doi.org/10.1007/978-3-319-52287-6_2

Robinson T, McMullan G, Marchant R, Nigam P (2001) Remediation of dyes in textile effluent: a critical review on current treatment technologies with a proposed alternative. Bioresour Technol 77:247–255. https://doi.org/10.1016/S0960-8524(00)00080-8

Rodríguez Couto S (2009) Dye removal by immobilised fungi. Biotechnol Adv 27:227–235. https://doi.org/10.1016/j.biotechadv.2008.12.001

Rosa JM, Tambourgi EB, Vanalle RM, Carbajal Gamarra FM, Curvelo Santana JC, Araújo MC (2020) Application of continuous H2O2/UV advanced oxidative process as an option to reduce the consumption of inputs, costs and environmental impacts of textile effluents. J Clean Prod 246. https://doi.org/10.1016/j.jclepro.2019.119012

Rose NM, Yves T, Junie Lauraine T, Dieudonné A, Charles F (2023) Evaluation of toxicological risks among screen printing workers: the case of the city of Yaoundé (Cameroon). Pharm Pharmacol Int J 11:29–34. https://doi.org/10.15406/ppij.2023.11.00398

Roy Choudhury AK (2013) Green chemistry and the textile industry, Text Prog 45:3–143. https://doi.org/10.1080/00405167.2013.807601

Roy Choudhury, AK (2014) Environmental impacts of the textile industry and its assessment through life cycle assessment. In: Muthu, S. (eds) Roadmap to sustainable textiles and clothing. Textile

Science and Clothing Technology. Springer, Singapore. https://doi.org/10.1007/978-981-287-110-7_1

Salem V (2010) Tingimento têxtil, 1st edn. Blucher, São Paulo-SP

Samanta AK, Agarwal P (2009) Application of natural dyes on textiles-the traditional way in Rajasthan, India. J Fibre Text Res 34:384–399

Samantaa AK, Agarwal P (2009) Application of natural dyes on textiles. Indian J Fibre Text Res 34:384–399

Santiago D, Cunha J, Cabral I (2023) Chromatic and medicinal properties of six natural textile dyes: a review of eucalyptus, weld, madder, annatto, indigo and woad. Heliyon 9:e22013. https://doi.org/10.1016/j.heliyon.2023.e22013

Saravanan P, Chandramohan G (2011) Dyeing of silk with ecofriendly natural dye obtained from barks of ficus Religiosa. L. Univ J Environ Res Technol 1:1–6

Saremi R, Rai S, Sharma S (2022) Dyeing of fibers and impact on the environment, green sustainable process for chemical and environmental engineering and science: green composites: preparation, properties and allied applications, pp 517–543. https://doi.org/10.1016/B978-0-323-99643-3.00021-8

Sax I (1986) Hazardous chemicals information annual, 1st edn. Van Nostrand Reinhold Information Services, New York

Saxena G, Kishor R, Zainith S, Bharagava RN (2021) Environmental contamination, toxicity profile and bioremediation technologies for treatment and detoxification of textile effluent. Elsevier. https://doi.org/10.1016/B978-0-12-820524-2.00017-1

Sharma J, Sharma S, Soni V (2021a) Classification and impact of synthetic textile dyes on aquatic flora: a review. Reg Stud Mar Sci 45:101802. https://doi.org/10.1016/j.rsma.2021.101802

Sharma K, Sharma P, Dhiman SK, Chadha P, Saini HS (2022) Biochemical, genotoxic, histological and ultrastructural effects on liver and gills of fresh water fish Channa punctatus exposed to textile industry intermediate 2 ABS. Chemosphere 287:132103. https://doi.org/10.1016/J.CHEMOSPHERE.2021.132103

Sharma J, Sharma S, Soni V (2021) Classification and impact of synthetic textile dyes on Aquatic Flora: a review. Reg Stud Mar Sci 45. https://doi.org/10.1016/j.rsma.2021.101802

Sher F, Hanif K, Iqbal SZ, Imran M (2020) Implications of advanced wastewater treatment: electrocoagulation and electroflocculation of effluent discharged from a wastewater treatment plant. J Water Process Eng 33. https://doi.org/10.1016/j.jwpe.2019.101101

Sigma-Aldricht (2006) Safety data sheet: hydroxymethanesulfinic acid monosodium salt dihydrate

Sivaram NM, Gopal PM, Barik D (2018a) Toxic waste from textile industries. Woodhead Publishing. https://doi.org/10.1016/B978-0-08-102528-4.00004-3

Sivaram NM, Gopal PM, Barik D (2018b) Toxic waste from textile industries. In: Energy from toxic organic waste for heat and power generation, Woodhead Publishing, pp 43–54. https://doi.org/10.1016/B978-0-08-102528-4.00004-3

Speight JG, Exall DI (2014) Refining used lubricating oils. In: Refining used lubricating oils, pp 1–443. https://doi.org/10.1201/B16745

Strand V, Lotfi M, Flynn A, Walker H (2024) A systematic literature review of modern slavery in supply chain management: state of the art, framework development and research opportunities. J Clean Prod 435:140301. https://doi.org/10.1016/j.jclepro.2023.140301

Sweet DV, Fang J, Shiang D (1986) Registry of toxic effects of chemical substances (RTECS), Cincinnati, OH (USA)

Textiles AS-S (undefined 2009) Achieving sustainable textiles: a designer's perspective, Elsevier (n.d.)

Uddin F (2014) Energy management and energy crisis in textile finishing. Am J Energy Res 2:53–59. https://doi.org/10.12691/ajer-2-3-2

Uddin F (2021) Environmental hazard in textile dyeing wastewater from local textile industry. Cellulose 28:10715–10739. https://doi.org/10.1007/s10570-021-04228-4

Uddin MA, Begum MS, Ashraf M, Azad AK, Adhikary AC, Hossain MS (2023) Water and chemical consumption in the textile processing industry of Bangladesh. PLOS Sustain Transf 2:e0000072. https://doi.org/10.1371/journal.pstr.0000072
UNECE (2018) UNECE-Annual report
Varadarajan G, Venkatachalam P (2016) Sustainable textile dyeing processes. Environ Chem Lett 14:113–122. https://doi.org/10.1007/s10311-015-0533-3
Wan Y, Wu C, Xue Q, Hui X (2019) Effects of plastic contamination on water evaporation and desiccation cracking in soil. Sci Total Environ 654:576–582. https://doi.org/10.1016/J.SCITOTENV.2018.11.123
Wanyama PAG, Kiremire BT, Ogwok P, Murumu JS (2011) Indigenous plants in Uganda as potential sources of textile dyes. Afr J Plant Sci 5:28–39
Weetman C (2019) Economia circular: conceitos e estratégias para fazer negócios de forma mais inteligente e lucrativa., 1st edn. Autêntica Business, São Paulo
Wijannarong S, Aroonsrimorakot S, Thavipoke P, Kumsopa C, Sangjan S (2013) Removal of reactive dyes from textile dyeing industrial effluent by ozonation process. APCBEE Proc 5:279–282. https://doi.org/10.1016/j.apcbee.2013.05.048
Xie X, Hong Y, Zeng X, Dai X, Wagner M (2021) A systematic literature review for the recycling and reuse of wasted clothing. Sustainability 13:13732. https://doi.org/10.3390/su132413732
Yadav S, Tiwari KS, Gupta C, Tiwari MK, Khan A, Sonkar SP (2023) A brief review on natural dyes, pigments: Recent advances and future perspectives. Results Chem 5:100733. https://doi.org/10.1016/j.rechem.2022.100733
Yaseen DA, Scholz M (2019) Textile dye wastewater characteristics and constituents of synthetic effluents: a critical review. Springer, Berlin, Heidelberg. https://doi.org/10.1007/s13762-018-2130-z
Yaseen DA, Scholz M (2019b) Textile dye wastewater characteristics and constituents of synthetic effluents: a critical review. Int J Environ Sci Technol 16:1193–1226. https://doi.org/10.1007/s13762-018-2130-z
Yusuf RO, Sonibare JA (2004) Characterization of textile industries' effluents in Kaduna, Nigeria and pollution implications. Global Nest J 6:212–221. https://doi.org/10.30955/gnj.000284
Zdohcp Z (2016) Textile industry wastewater discharge quality standards: literature review
Zhang YQ, Lykaki M, Markiewicz M, Alrajoula MT, Kraas C, Stolte S (2022) Environmental contamination by microplastics originating from textiles: emission, transport, fate and toxicity. J Hazard Mater 430:128453. https://doi.org/10.1016/J.JHAZMAT.2022.128453
Zhou Q, Van Le Q, Meng L, Yang H, Gu H, Yang Y, Chen X, Lam SS, Sonne C, Peng W (2022) Environmental perspectives of textile waste, environmental pollution and recycling. Environ Technol Rev 11:62–71. https://doi.org/10.1080/21622515.2021.2017000
Zonatti WF (2016) Geração de resíduos sólidos da indústria brasileira têxtil e de confecção: materiais e processos para reuso e reciclagem, Universidade de São Paulo. https://doi.org/10.11606/T.100.2016.tde-26042016-192347

Chapter 2
Sustainable Innovations in Textile Coloration Machinery and Process Enhancements

Md. Sakib Al Montasir, Md. Belal Uddin Rabbi, Alal Ahmed Sani, and Md. Rashedul Islam

Abstract The recent shifts in the textile coloration industry toward sustainable machinery development and process improvement are a direct result of substantial environmental obstacles, economic disadvantages, and processing complexities. Prior pertinent research concentrated on particular practices and procedures for textile coloring machinery. However, a comprehensive analysis of all processes and machinery associated with textile coloration—including pretreatment, coloration, finishing, and more—and their sustainable transition toward future possibilities has been absent. This chapter investigates the progress in sustainable coloration practices by assessing machinery renovations and process enhancements. It compares and contrasts the conventional methods, recognizes the potential environmental consequences of the textile coloration sector, and evaluates the viability of integrating optimal sustainable principles into contemporary machinery design and process optimization. This chapter has also documented the advancements in printing and dying machinery and process improvements to achieve sustainability throughout the pretreatment, coloration, and finishing stages. Through a meticulous analysis of the interplay between current methodologies and forthcoming developments, the chapter has proposed a more sustainable trajectory in textile coloration. It underscores the capacity for substantial environmental impact reductions without compromising the quality of color performance, thereby establishing a pathway toward a more conscientious and environmentally aware textile sector.

Keywords Textile coloration · Sustainability · Textile machinery · Dyeing · Pretreatment · Finishing

Md. S. Al Montasir (✉) · Md. B. U. Rabbi · A. A. Sani · Md. R. Islam
Department of Wet Process Engineering, Faculty of Textile Chemical Engineering, Bangladesh University of Textiles, Tejgaon, Bangladesh
e-mail: sakibalmontasir29@gmail.com

Md. R. Islam
e-mail: rashedul.islam@wpe.butex.edu.bd

S. Maiti et al. (eds.), *Sustainable Coloration Techniques in Textiles*,
https://doi.org/10.1007/978-981-96-4975-4_2

2.1 Introduction

Textile wet processing refers to a range of chemical and mechanical processes used in the textile industry to treat materials like fabrics, yarns, and fibers. These treatments involve the use of various chemicals and auxiliaries. The process includes a pretreatment, which prepares the fabric for subsequent treatments by removing impurities like sizing agents, natural oils, and waxes; a coloration process where color is applied to the fabric either through dyeing or printing using dyes or pigments and finishing treatments, which are applied to enhance the fabric's performance and appearance (Broadbent 2001; Wanget al. 2020; Wardman n.d.). Coloration in textiles provides an aesthetically pleasing and marketable appearance and enhances its performance properties (Kalaoglu and Paul 2015; Lee 2017). The textile coloration industry, despite its economic significance, has been facing significant sustainability challenges due to limitations such as excessive resource consumption, energy inefficiency, and usage of harmful chemicals and emissions of toxic substances (Khattab et al. 2020). The research focus has currently been shifted to coloration machinery and processes, which are currently being renovated to prioritize sustainable dyeing technologies. These include the exploration of eco-friendly dyes, closed-loop water systems, and advanced machines that use less water and energy. Additionally, efforts have been made to optimize water and energy usage, adopt cleaner production methods, and minimize waste generation through process improvements. Transparency in the supply chain is being promoted to better understand dyes and chemicals' environmental and social impact. Responsible sourcing and sustainable alternatives are being encouraged, and collaboration among industry stakeholders, academia, and government is being fostered to accelerate the research and development of sustainable solutions (Hossain et al. 2024; Khairul Akter et al. 2022; Todeschini et al. 2020). The industry is constantly innovating toward eco-friendly and sustainable processes using recyclable resources, reducing water consumption, and minimizing harmful chemicals (Choudhury 2017). This chapter deals with the present condition and recent sustainable advancements in textile machinery and process enhancements along with these innovations' technical efficacy, environmental benefits, and economic viability, evaluating their potential for widespread adoption and transformative impact on the textile industry. Finally, it is going to identify key challenges and future research directions toward achieving genuinely sustainable textile coloration practices. Figure 2.1 describes the general process diagram of the textile dyeing industry.

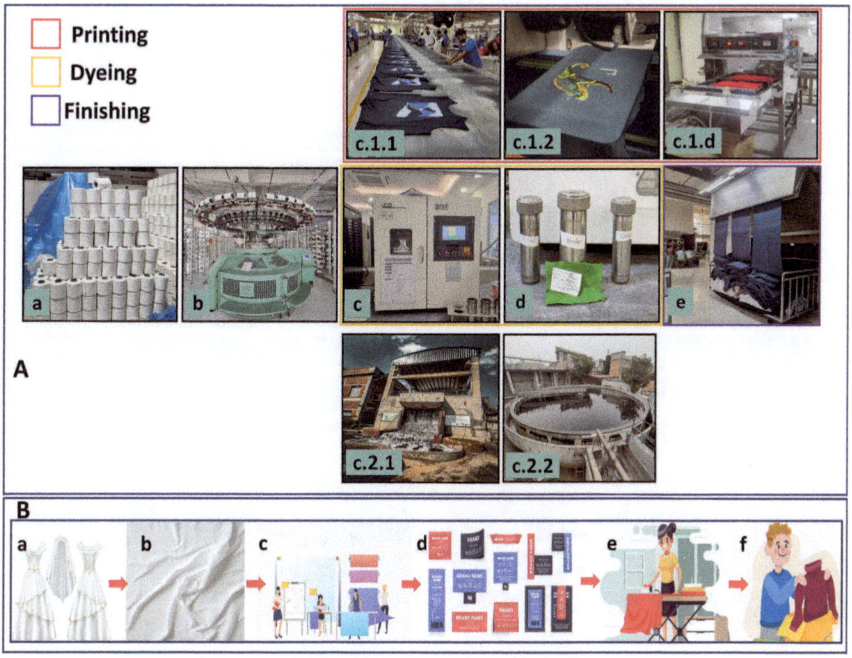

Fig. 2.1 General process flow of a textile garment manufacturing industry **A** (**a**) yarn, (**b**) knitting, (**c.1.1**; **c.1.2**) screen printing, (**c.1.d**) drying and curing, (**c**) dyeing, (**d**) lab dyed fabric, (**e**) finishing, (**c.2.1**) effluent water, (**c.2.2**) treatment process. **B** (**a**) Design development, (**b**) fabric spreading, (**c**) pattern making, cutting, sewing, (**d**) label attaching, (e) ironing, (**f**) packing and delivery

2.2 Common Textile Coloration Process

2.2.1 Textile Coloration Process

Textile coloration, which encompasses dyeing and printing, is the fundamental technique for producing a wide range of vibrant, colorful clothing. Dyeing and printing are two distinct methods of adding color to textiles. Dyeing involves the infusion of color throughout the fabric by allowing dyes to penetrate the fiber structure, resulting in consistent coloration. On the other hand, printing selectively applies colored designs to specific areas of the fabric. Due to the different design requirements of fabric, it is essential to consider factors such as fabric type, desired colorfastness, and environmental effects, which affect water usage, chemicals, and energy consumption. The textile industry progressively prioritizes sustainable practices such as optimizing processes and adopting utility-saving technologies. Additionally, a growing emphasis is being placed on using dyes and pigments with reduced toxicity, such as biodegradable alternatives. Furthermore, the industry is adapting to advancements in digital printing technology, which enable higher resolution, faster printing speeds, and a

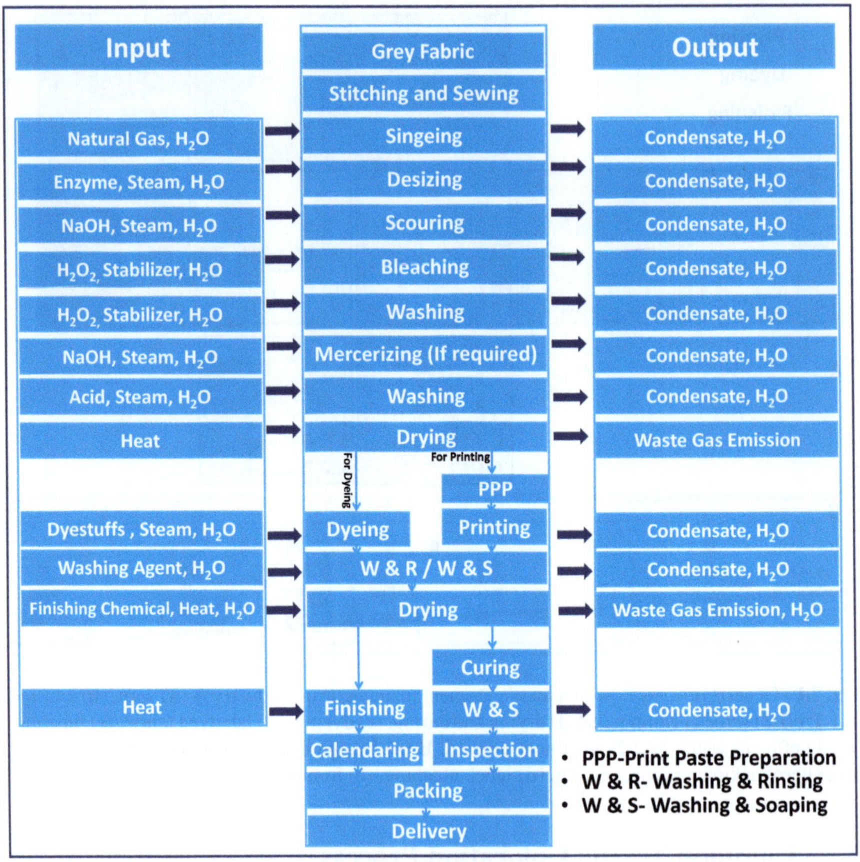

Fig. 2.2 General process flow diagram in a textile industry

reduced environmental footprint (Azeem et al. 2019; Halim et al. 2021). Figure 2.2 represents a general process diagram of the textile industry.

2.2.1.1 Fabric Pre-treatment

Fabric pre-treatment is a necessary process in textile coloration, as it significantly contributes to the dyed and printed design's overall quality, longevity, and vividness. Treating fabric correctly can lead to various advantages, such as enhanced penetration and adhesion of dyes and pigments and absorption of color by eliminating impurities and foreign substances, improved resistance to fading, increased uniformity of the fabric, better washing performance, improved ability to print on the fabric, and reduced shrinkage of the fabric (Fan et al. n.d.; Kan 2007). Fabric pretreatment encompasses multiple procedures, with each stage including utilizing different technologies. Pretreatment in textile coloration plays a significant role by providing cleanliness, color evenness, color depth and vibrancy, color fastness, and material performance (fiber, yarn, fabric). By removing contaminants, pre-treatment ensures the dye has a clean, responsive surface and enables uniform dye-fabric interaction. This results in a dyeing of the textile that is more even and consistent throughout. Colors are less likely to fade or bleed when the dye adheres well to clean fibers, allowing for better dye uptake and resulting in more affluent, deeper, and more vibrant colors than skipping pre-treatment. To enhance ink adherence, scouring and desizing machines eliminate contaminants, sizing agents, and natural waxes from the cloth. Different machinery brands are working in this regard e.g. Monforts Textilmaschinen provides industrial-grade scouring ranges incorporating advanced features such as efficient chemical application and water recycling, Maschinenfabrik offers uninterrupted scouring lines for large-scale production. Bleaching machines are utilized to lighten and bleach brighter prints, Ahiba Tepla supplies continuous bleaching ranges designed for large-scale manufacturing, while Benninger offers batch bleaching machines ideal for different fabrics. Neutralization and washing machines are used to eliminate any remaining chemicals and regulate the pH level of the fabric to create ideal printing conditions e.g. Mettler Toledo Albi provides industrial washing lines incorporating precise chemical dosing and water conservation capabilities, Brückner Textile Technologies offers effective washing systems for different types of fabrics, whereas Ramón Camps specializes in batch washing machines designed for smaller print studios. Drying machines extract surplus moisture from the fabric before printing, for example, Mönch supplies industrial dryers with multiple heating options and tension control features, Maba Textilmaschinen offers efficient dryers suitable for varied fabric weights and volumes, Monforts Textilmaschinen offers stenters that integrate drying with fabric width adjustment.

2.2.1.2 Dyeing

Textile dyeing imparts color to textile materials, such as fibers, yarns, or fabrics, converting raw materials into a visually appealing and higher-value end product. Textile dyeing has multiple functions since dyes enhance the color and vitality of fabrics, making them visually attractive. Dyeing machines used for coloring textile materials can be classified according to their shape: U, V, J, and L. They also vary in capacity, ranging from 5 to 50 kg for laboratory use and 500–2000 kg for bulk production. These machines are designed to dye various materials such as fiber, yarn, and fabric. The dyeing process could be classified as continuous methods (such as pad dry and pad stream), semi-continuous methods (such as pad batch and pad jig), and discontinuous methods (such as jet, jigger, and Winch, etc.). Dyes can augment the functionality of fabric, such as providing UV protection and concealing minor flaws in the original color of the fibers. Various categories of dyes are classed according to their chemical composition and how they are applied. For example, natural dyes are obtained from plants, animals, or minerals, while artificial dyes provide a broader spectrum of colors and have superior resistance to fading (Gobalakrishnan et al. 2020). The specific dye, types of fiber, dyeing circumstances, and

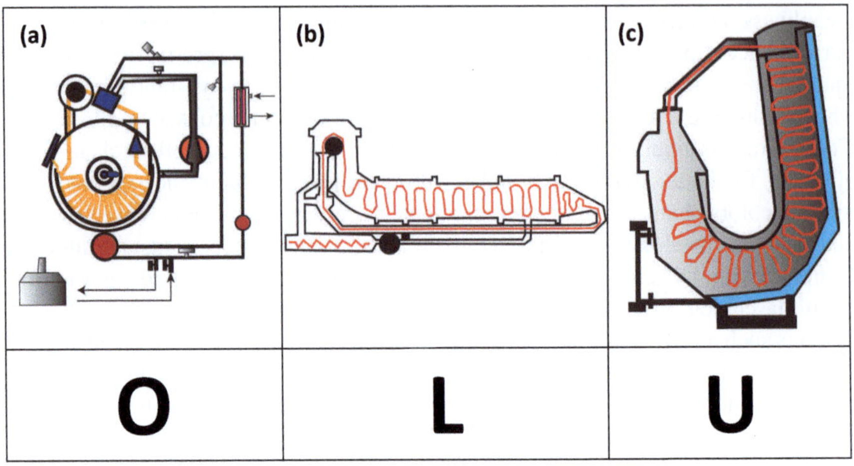

Fig. 2.3 Different shapes of the dyeing machine. Reproduced (adapted) with permission. Copyright 2025, TextileSchool (School 2010)

desired results cumulatively determine the selection of specific dyeing methods. Commonly employed dyes include reactive, direct, vat, basic, and acid dyes. Additionally, using biodegradable dyes is being explored as a more environmentally friendly option (Varadarajan and Venkatachalam 2016). Figure 2.3 illustrates the shape of dyeing machines: O, L, and U.

Some well-known manufacturers of dyeing machines include Sclavos (Greece), AK (Thailand), MCS (Italy), and Canlar (Italy), which are shown in Fig. 2.4. Sclavos produces industrial dyeing machines with features such as double soft floating dyeing system, integrated control of liquor ratio, fast process time, and batch-to-batch repeatability. AK offers batch dyeing machines suitable for various fabrics and dyes. MCS manufactures dyeing ranges that can be integrated with other wet processing machines. Canlar specializes in continuous dyeing lines for high-volume products.

2.2.1.3 Printing

Printing is the process of applying dyes or pigments in specific patterns or designs onto textiles. Printing opens up a wide range of creative opportunities in fashion and other applications by allowing the creation of elaborate designs, logos, patterns, and images on fabrics. Printing processes provide more exact and vibrant colors than dyeing, enabling the replication of intricate artwork with an accurate color arrangement. Printing can be utilized on a wide range of fabrics, including natural and synthetic materials, creating diverse designs for various purposes. Commonly employed printing techniques include block printing, engraved roller printing, screen printing (flat screen and rotary type screen), digital printing, transfer printing,

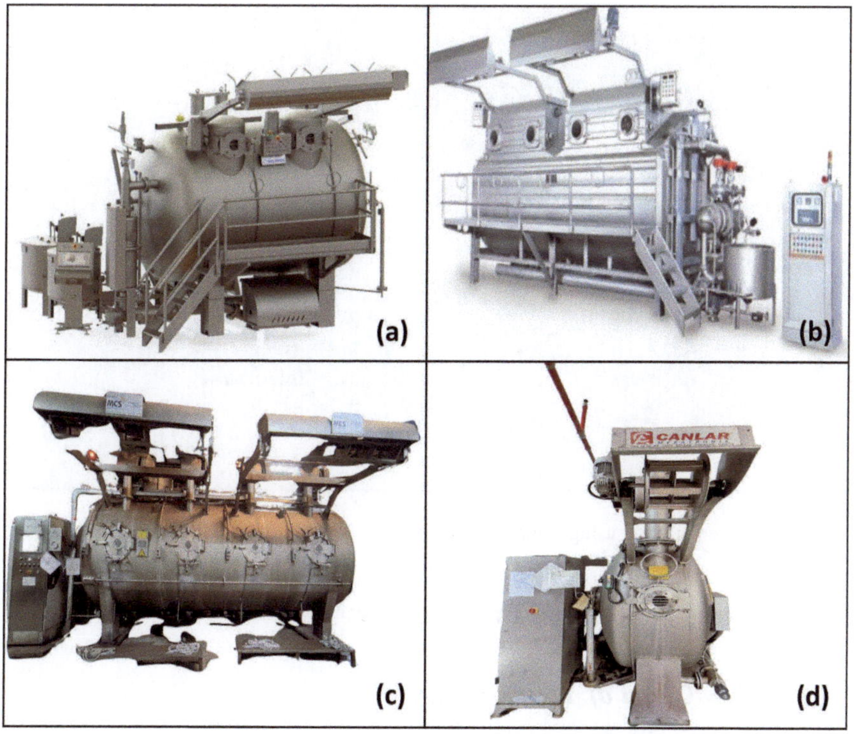

Fig. 2.4 Dyeing machine of **a** Sclavos (Greece), **b** AK (Thailand). Reproduced (adapted) with permission. Copyright 2025, Asia Kingdom Machinery (AK Dyeing Machine n.d.), **c** MCS (Italy), and **d** Canlar (Italy)

roller printing, flock printing, burn-out printing, and digital printing methods. In recent practices, substituting solvent-based pigments with water-based alternatives has effectively decreased the release of harmful volatile organic compounds during printing. Print pastes made from renewable resources such as soybean oil or natural sources also provide a sustainable option to synthetic inks (Gobalakrishnan et al. 2020). Some well-known manufacturers of printing machines include A.B.I. Labelling Machine, Reggiani Macchine, Stork, and Mimaki, specializes in rotary screen-printing machines known for their fast speed, precise printing, and versatility. Reggiani Macchine offers rotary screen-printing machines with advanced features for high-quality printing on various textiles. Stork utilizes rotary screen-printing machines specifically designed for large-scale production. Mimaki offers digital printing machines suitable for printing on small and large textile batches. E.F.I. (Jeti) provides digital printing machines with advanced features for high-resolution printing and efficient textile pigment usage. Figure 2.5 shows screen printing using an Octopus screen printing machine in the textile industry.

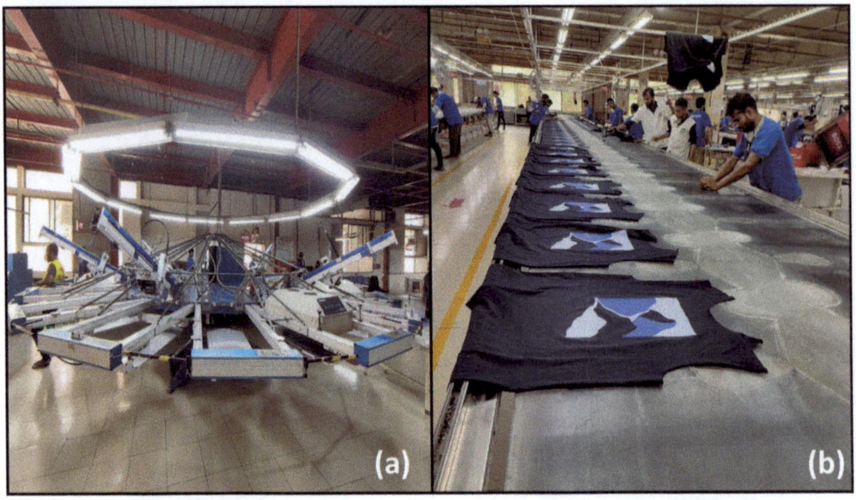

Fig. 2.5 **a** Screen printing using octopus screen printing machine, **b** screen printing

2.3 Existing Coloration Process Flow

2.3.1 Flow Chart of Yarn Dyeing

The sewing thread dyeing process includes receiving grey yarns from the store, which are then subjected to soft winding, batch preparation, and pressing. Afterwards, the sewing thread goes through the dyeing process. After the dyeing process, hydro-extraction is carried out to remove any remaining water from the yarn. Conning involves the transformation of yarn into a cone shape. Finally, the packing and delivery process of the sewing thread dyeing process is completed. The process of package dyeing for yarn is similar to that of dyeing sewing thread. However, the dyeing process is carried out in a packaged form. After the dyeing process, the hydro-extracting and drying processes are completed. The yarn is now finally prepared for delivery. Hank dyeing involves the initial step of receiving grey fabric from the store. Following that, the process of yarn singeing and reeling is carried out. Next, yarn mercerization is performed to eliminate any impurities present in the yarn. The hank dyeing procedure involves dyeing the yarn in its hank form. Once the dyeing process is complete, the drying process follows. The yarn transforms from a hank form to a cone shape throughout the hank dyeing process. Next, the yarn is transferred to the knitting section, where the knitting process occurs. Once the processes of knitting, singeing, and mercerization have been completed. At last, the yarn is prepared for delivery. Figure 2.6 shows the three dyeing processes, and Fig. 2.7 shows the sewing thread, package, and hank dyeing machine.

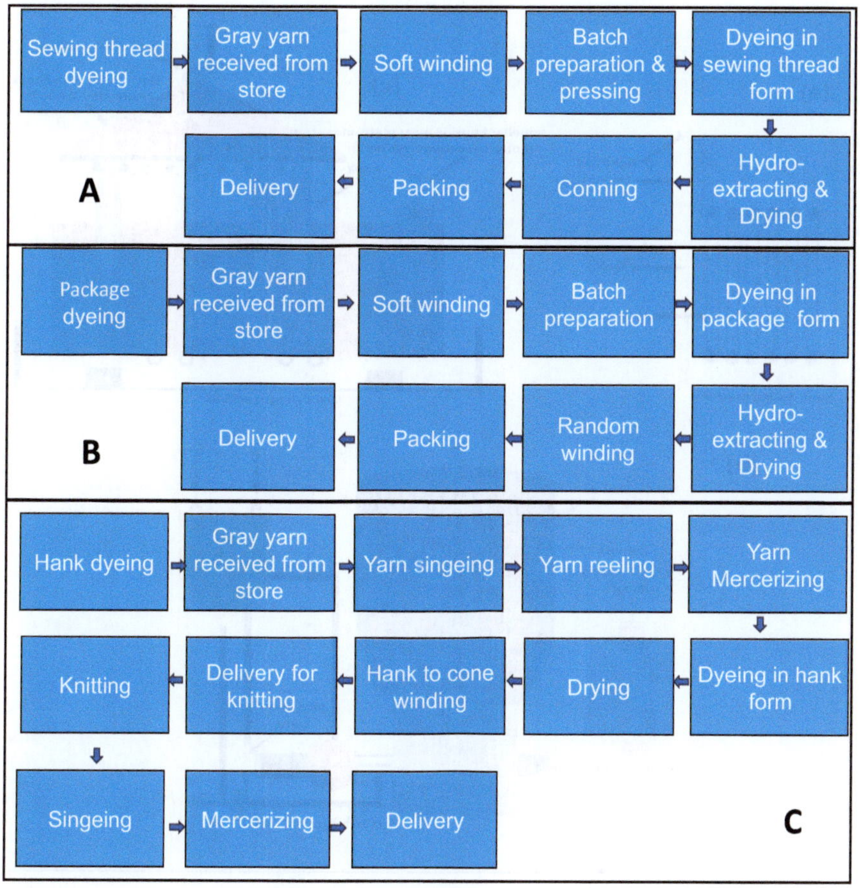

Fig. 2.6 Flowchart of yarn dyeing **a** sewing thread dyeing, **b** package dyeing, **c** hank dyeing

2.3.2 *Flow Chart of Fabric Dyeing*

The fabric dyeing process begins by collecting grey fabrics from the knitting area. The fabric is pretreated after turning and back sewing in the batching section. Then, fabric undergoes scouring and bleaching processes, followed by a neutralization process to regulate the pH level. Subsequently, the fabric undergoes peroxide treatment to eliminate organisms, followed by dyeing and neutralization processes. Once the dyed materials have been washed and treated with soap, they are removed from the dying machine. Figure 2.8 depicts the fabric dyeing process in a dyeing machine.

Table 2.1 highlights the commonly used dyeing parameters in the textile industry for light, medium, and dark shades, and Table 2.2 highlights the commonly used production parameters in factories.

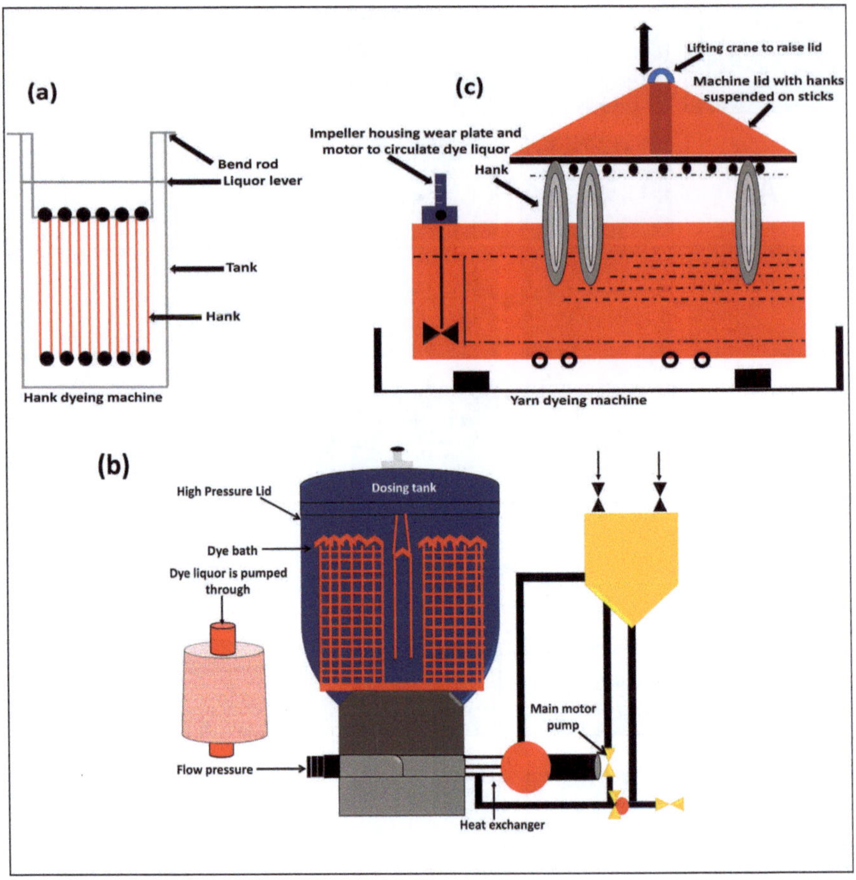

Fig. 2.7 Yarn dyeing machine **a** sewing thread dyeing machine, **b** package dyeing machine, **c** hank dyeing machine

2.4 Commonly Used Dyeing Machine Types in the Industry

Dyeing machines in textiles are classified as continuous, semi-continuous, and discontinuous variants, which can be further divided into subgroups. However, in this section, we are going to discuss some of the most used dyeing machine types in knit and woven industries with the comparative analysis of exhaust, semi-continuous, and continuous dyeing machines.

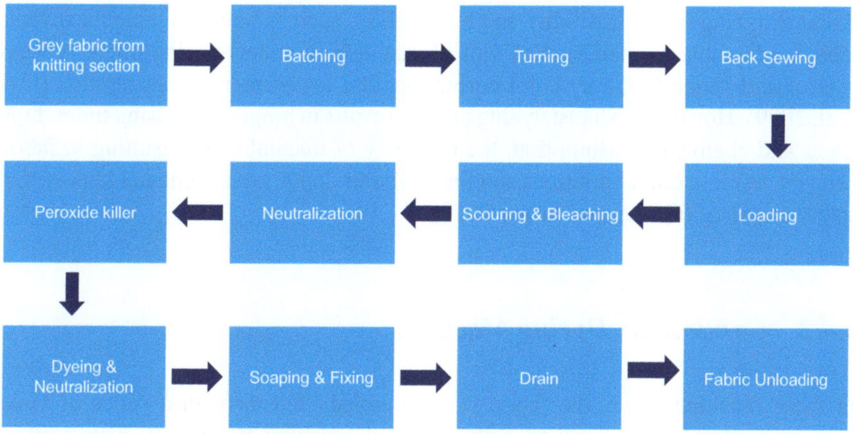

Fig. 2.8 Flow chart of fabric dyeing

Table 2.1 Commonly used dyeing parameters in factories (for 100% cotton)

Parameter	Light Shade	Medium Shade	Dark Shade
Water temperature (°C)	50	50	50
Dye (%)	<0.5	0.5–1.5	>1.5
M:L	1:6	1:6	1:6
Water consumption (L/kg)	30–40	70–80	100–120
Mixing time (min)	10	15	20
Dyeing time (h)	8	10	10

Table 2.2 Commonly used production parameters in factories

Particulars	p^H	Time (min)	Temperature (°C)
Scouring and bleaching	11–12	40	80
Dye bath	Light shade: 10.5 Medium shade: 10.8 Dark shade: 11.2	Depends on shade check	60
Soaping bath	6–6.5	10	80
Neutralization	6.5–7	5	80
Peroxide killer	6–7	10	60
Enzyme bath	Acid: 4.5 Neutral: 6	60	60

2.4.1 Exhaust Dyeing Machine

In the exhaust dyeing machines, dyes are gradually added and exhausted onto the fabric in the dye bath, offering good control over color depth (Cay et al. 2009).

Exhaust dyeing offers flexibility and adaptability, such as being suitable for diverse fabrics, adaptable to small batches with affordable equipment offering simple process control and versatility, easy color correction, and lower energy consumption (Cay et al. 2009). However, exhaust dyeing usually results in longer processing times, high water and chemical consumption, the necessity of manual tasks resulting in fabric damage, random color variations and limited color depth, environmental impact, and safety concerns (Günay and Jasper 2010).

2.4.2 *Continuous Dyeing Machine*

In these machine types, the fabric is continuously fed through a series of tanks containing dye liquor, washing solutions, and finishing treatments. This method is efficient for high-volume production but requires significant investment and may not be suitable for all fabric types (Nair 2011). Continuous dyeing machines have increased efficiency with faster processing, reduced labor costs along with higher production capacity, improved color uniformity with better control over the process, reduced risk of human error, reduced water and chemical usage along with lower energy consumption (Ammayappan et al. 2016). However, continuous dyeing machines require high initial investment and maintenance costs, reduced adaptability and flexibility to frequent changes in colors or styles compared to batch dyeing methods, limited product variety, and complexity of operation in the case of downtime on any technical problems with higher environmental impact (Khattab et al. 2020). Figure 2.9 illustrates different types of continuous dyeing machines.

2.4.3 *Semicontinuous Dyeing Machine*

At first, the fabric is immersed into the dyeing solution in a padding machine, and then the fabric is subjected to batch-wise treatment. However, high temperatures are used when the pad roll machine is used to dye the fabric in a warming chamber. In a semicontinuous dyeing machine, padding is applied to the dye continuously, fixation and washing, two necessary dyeing steps, remain discontinuous. Pad jig, pad roll, and pad batch are examples of semicontinuous dyeing machines. Adaptability, versatility, simplicity, flexibility, enhanced handle, and cleaner surface appearance are the advantages of the pad batch dyeing process, whereas the advantages of pad-jig dyeing machines are even dyeing, good penetration of the dye, and good absorbing capacity (Muthu 2018). Figure 2.10 represents the semi-continuous dyeing machine.

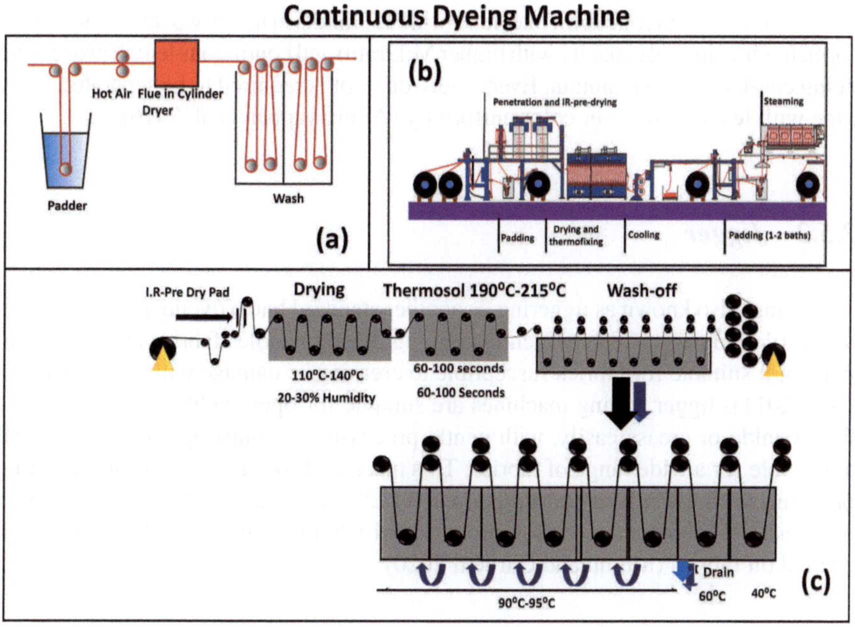

Fig. 2.9 Continuous dyeing machine. Reproduced (adapted) with permission. Copyright 2025, IntechOpen (Omerogullari Basyigit 2021)

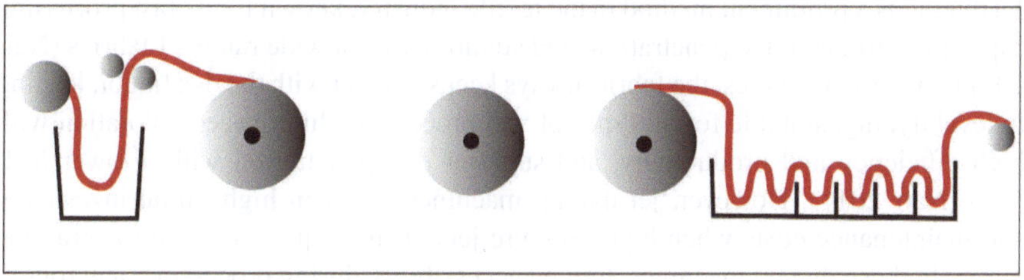

Fig. 2.10 Semi-continuous dyeing machine

2.5 Commonly Used Dyeing Machines in Textile Coloration

2.5.1 Winch

Winch dyeing is a typical batch dyeing process used in the textile industry to dye fabrics in rope form. In this machine type, several ropes or loops of fabrics are immersed in the liquor of the dye bath (Nair 2011). Winch machines are suitable for all types of wet processing and fabrics. These machines require comparatively less pressure, being cost-effective with gentle processing (Senthil Kumar and Yaashikaa 2018). However, this machine type can also have some drawbacks as fabrics processed in rope form may lead to crease marks in heavy woven and synthetic form, having limited capacity with higher M:L ratio and continuous length processing being challenging to maintain. Every individual lot is required to be separated every time with less control over color uniformity (Ammayappan et al. 2016).

2.5.2 *Jigger*

Jig dyeing, also known as jiggering, is another standard batch dyeing process used in the textile industry. Unlike winch dyeing, jig dyeing handles fabrics in open width, making it suitable for fabrics susceptible to creasing or damage with other methods (Nair 2011). Jigger dyeing machines are suitable for open-width fabrics or fabrics that wrinkle or crease easily, with gentle processing minimizing fabric stress, and applicable for a wide range of fabrics. This machine type efficiently avoids creating problems with lower heat consumption and M:L ratio (Nair 2011). However, jigger machines are labor-intensive, slow processes with limited capacity and higher tension exerted on fabrics (Muthu and Gardetti 2020).

2.5.3 *Jet*

Jet dyeing is a prominent method in the textile industry, known for its fast processing capacities, efficient dye penetration, and suitability for a wide range of fabrics (Nair 2011). In these machines, the fabric always keeps contact with the dye liquor, leading to level dyeing, suitable for all types of wet processing, high-speed coloration with high efficiency and productivity, and superior dye penetration with a lower M:L ratio (Nair 2011). However, jet dyeing machines result in high initial investment and maintenance costs when high-pressure jets of dye liquor are used to spray the fabric. In these dyeing machines, limited accessibility during processing, any roughness and sharpness can cause severe damage; rethreading is complicated in case of thread breakage (Hasanbeigi and Price 2015). Figure 2.11 shows commonly used dyeing machines in the textile industry, such as Jiggers, Winches, and Jet dyeing machines.

2.5.4 *Cold Pad Batch (CPB) Dyeing Machine*

Cold Pad Batch (CPB) dyeing machines offer a sustainable option for dyeing textiles, particularly cellulosic fibers like cotton, linen, and rayon, involving general fabric preparation, padding, batching, washing, and calendaring workflow. Compared to traditional dyeing methods involving heating large volumes of water, CPB uses significantly less water where the dye liquor is applied directly to the fabric, and

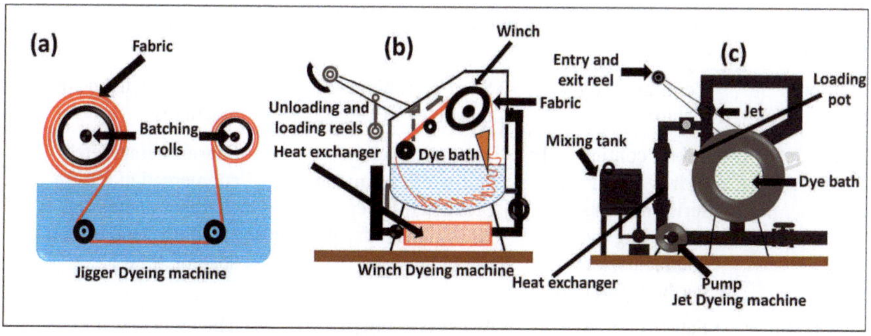

Fig. 2.11 **a** Jigger, **b** Winch, **c** Jet dyeing machine. Reproduced (adapted) with permission. Copyright 2025, IntechOpen (Omerogullari Basyigit 2021)

the batching stage often occurs at room temperature minimizing water evaporation. Since the process operates at lower temperatures, CPB dyeing requires less energy than methods involving boiling or steaming the fabric, and it is gentler on fabrics compared to high-heat dyeing methods, making CPB suitable for delicate materials. Additionally, the process does not require salt, which is commonly used in some dyeing methods, and requires additional wastewater treatment for removal. Figure 2.12 shows the cold pad batch dyeing machine.

Table 2.3 illustrates the comparative analysis of continuous, semi-continuous, and discontinuous dyeing machines.

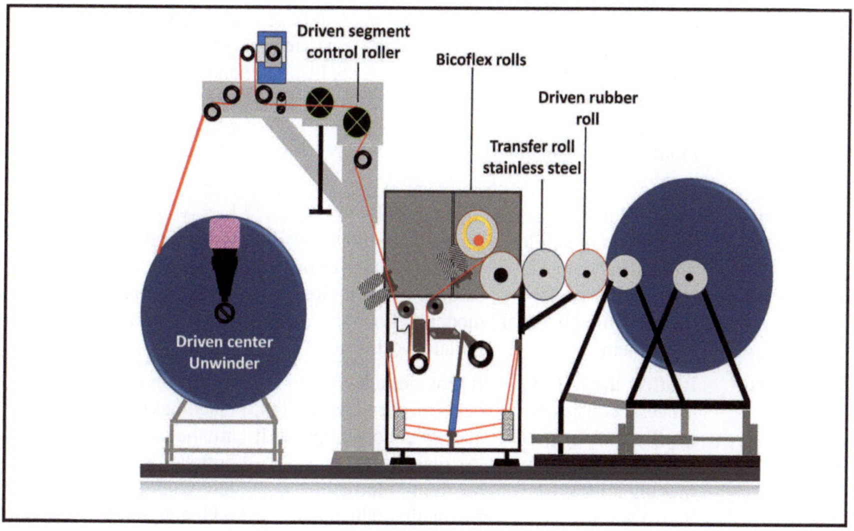

Fig. 2.12 Cold pad batch dyeing machine. Reproduced (adapted) with permission. Copyright 2025, Jagenberg Group ("Dyeing—Kusters Calico" n.d.)

Table 2.3 Comparative analysis of continuous, semi-continuous, and discontinuous dyeing machines

Factors	Continuous dyeing machines	Semi-continuous dyeing machines	Discontinuous dyeing machines
Process	Fabric continuously moves through a series of tanks containing dyes, chemicals, and washing solutions	Fabric is dyed in full-width format and processed in separate washing and drying stages	Fabric is completely submerged in a large chamber containing dye and chemicals for a specific time
Sustainability	Fabric does not require constant re-wetting, and it consumes less water. It also uses less energy and has optimized heating and cooling cycles. Additional dosing systems can minimize dye and chemical usage, resulting in resource optimization	Water consumption can be moderate, as fabric might require washing after the pad dyeing stage. Energy usage varies depending on the number and efficiency of separate machines in the post-dyeing stages. Compared to continuous machines, there is a higher potential for chemical waste	The highest water consumption is needed to wet the fabric for each dyeing cycle. Compared to other processes, higher energy is required to heat water for each batch. It is difficult to precisely control dye and chemical usage for each batch, which has the highest potential for chemical waste
Efficiency	The fastest dyeing process for high-volume production compared to the rest of the processes. Consistent quality through control over process parameters. Requires skilled operators to monitor and maintain systems	It is faster than batch dyeing, offering total immersion in a batch with moderator operator requirements for process application and handling fabric through separate machines	The slowest dyeing speeds are suitable for low-volume production or custom dyeing. They require careful monitoring and adjustment for each batch of variable quality and less operator intervention than continuous machines
Production suitability	High volume production of similar fabrics and colors, requiring large floor space with high initial investment	Medium-volume production with some variation in fabrics or colors which require moderate floor space and initial investment	Low-volume production with a high color variation or custom dyeing requires the least space and the lowest initial investment
Advantages	It offers the fastest process, consistent quality, and lowest water consumption, making it ideal for high-volume production of similar fabrics and colors	It provides a balance between speed, flexibility, and water consumption. It is suitable for medium-volume production with some variation in fabrics or colors	Discontinuous dyeing machines (batch dyeing) are the most versatile option, ideal for low-volume production with a high color variation or custom dyeing needs

(continued)

Table 2.3 (continued)

Factors	Continuous dyeing machines	Semi-continuous dyeing machines	Discontinuous dyeing machines
Disadvantages	Complex operation requires specialized training for operators with limited flexibility, as adjusting for frequent color changes is difficult	It offers moderate flexibility to adjust for color changes compared to continuous dyeing. However, proper pad application techniques are required due to the potential for uneven dyeing	They have the slowest speeds, the highest water consumption, and the potential for inconsistent quality. Due to the risk of fabric damage, they require careful fabric handling during loading and unloading
Examples	Pad dye, pad steam	Pad batch, pad jig, CPB	Jigger, Winch, Jet

2.6 Printing

2.6.1 Screen Preparation

Screen preparation is a vital aspect of textile screen printing, as it directly affects the quality of the final printed pattern. Choosing the right screen preparation machinery for textile printing depends on numerous processes, e.g. selecting the proper mesh, degreasing the screen, screen coating, drying the emulsion, burning the design onto the screen, washing off the screen, drying the screen, and screen inspection. Several textile printing screen preparation machinery brands offer equipment to expedite the process via manual and automation (Ali et al. 2019). Manual screen stretching machines are ideal for smaller print companies or low-volume manufacturing, such as Vastex FSR is a versatile manual screen stretching machine that can accommodate a wide range of screen sizes featuring a sturdy steel frame and a micro-registration system for precise screen alignment; Triangle S-100 Manual offers a simple and effective design with adjustable levelling and tensioning mechanisms. Automatic screen stretching devices boost efficiency and consistency in high-volume manufacturing scenarios. Agme offers a series of automatic screen stretching machines with varied capacities, giving durability, precision, and ease of operation; M&R Falcon delivers consistent tension control and efficient processing for various screen sizes. Emulsion coating machines ensure uniform and efficient application of light-sensitive emulsion onto the screens. Wilcom WS-610 is ideal for both water-based and solvent-based emulsions, including adjustable coating thickness and speed controls. Kiwo offers different emulsion coating machines, including manual and automatic choices. Exposure units are a vital aspect of screen preparation for giving regulated light exposure to harden the emulsion and generate the desired screen design. Nuarc S1000 Exposure Unit is famous for its durability and constant UV exposure. The M&R Maverick LED Exposure Unit is utilized for energy-efficient and controlled exposure.

2.6.2 Print Paste Preparation

The preparation of the print paste directly impacts the printability, color fastness, and final appearance of the drawings. The main components of print paste include thickeners, which give the paste the desired susceptibility for proper application through screens or printing heads. Binders are also added to help the pigment or dye stick to the fabric fibers, improving durability and resistance to washing. Colorants, such as pigments or dyes, provide the desired color to the printed design. Additionally, auxiliaries or other chemicals are added to the paste to achieve specific functions, such as leveling, wetting, and anti-foaming agents (Hussien et al. 2020; Kan et al. 2011). The preparation procedure for printing paste often relies on the printing technique, fabric composition, and the intended print outcome. Various machinery options can be utilized to manufacture print paste in textile printing to achieve consistent quality. These options include dissolving machines, mixing tanks, agitators, and dispensing systems. Dissolving and dispersion machines manufactured by various companies such as IKA, Netzsch, Cowles Dissolvers, Graco, Exair Corporation, and Asytec are utilized to equally distribute the components of print paste, resulting in a smooth and homogeneous product. Mixing tanks and agitators are utilized to effectively blend all components of print paste to achieve a homogeneous texture. Many brands, such as Ross, Lightnin, and Gea Group, produce this equipment.

2.6.3 Washing

Washing is crucial in textile printing as it eliminates unfixed pigments and dyes at two critical stages: after printing and curing (Kašiković et al. 2018). After printing, washing eliminates surplus chemicals and unfixed dyes from the printed cloth to enhance colorfastness, minimize stains, and make the fabric feel softer. Following the curing or fixing procedure, washing eliminates water-soluble residues, boosting printability, lowering fabric brittleness, and minimizing chemical odor. Several well-known brands, including Ahiba, produce washing machines specifically designed for textile printing, Tepla provides industrial washing machines with delicate fabric handling, optimal water use, and accurate process management, Monforts Textil-maschinen offers a variety of washing machines that can handle delicate fabrics and have features such as multi-stage washing and integrated chemical dosing systems, Maschinenfabrik provides continuous washing lines specifically designed for high-volume production in textile printing. These washing lines are very efficient and capable of handling big fabric batches. On the other hand, Metler Toledo Albi specializes in industrial washing lines.

2.6.4 Drying

The drying process in textile printing significantly impacts the final product's quality, functionality, and durability. It improves odor control, enhances colorfastness, preserves fabric integrity, fixes prints, provides a softer and more comfortable fabric handle, and enables efficient storage, resulting in sharper and more defined print quality (Tyler 2005). Drying is crucial before and after curing/fixing to provide a durable and enduring printed outcome. The purpose of drying after washing is to inhibit the growth of microorganisms, prevent the bleeding of colors, and avoid the formation of rust stains. Before curing/fixing, drying is performed to achieve efficient curing, minimize energy usage, and enhance print quality. Drying after curing/fixing is performed to enhance the texture, decrease the drying duration for subsequent procedures, and optimize storage efficiency. Several well-known brands provide drying machines specifically designed for textile printing e.g. Mönch provides industrial dryers with diverse heating options and tension control features designed explicitly for efficient drying and delicate handling of printed textiles. On the other hand, Maba Textilmaschinen offers efficient dryers tailored for various fabric weights and volumes commonly used in textile printing. These dryers have adjustable temperature control and optimized airflow to ensure uniform drying. Brückner Textile Technologies offers a selection of dryers for different fabric types and production volumes in textile printing. These dryers include efficient heat recovery systems and integrated air filtration. Xenia Machinery provides batch dryers suitable for smaller print shops or for drying specialty fabrics with specific drying needs. These dryers are affordable and easy to operate.

2.6.5 Curing

Curing is subjecting the printed cloth to heat, light, or a mixture of both to activate the binders included in the print paste. Curing is crucial for achieving colorfastness and wash resistance in fabric and strengthening its durability and utility (Bahria and Erbil 2016). The printed fabric can acquire various capabilities, such as water resistance, wrinkle resistance, or flame retardancy, depending on the curing procedure and additives employed. Specific curing procedures are typically necessary to obtain the appropriate visual and tactile features of specific print effects, such as metallic finishes or raised textures. The selection of a curing machine is contingent upon the particular pigment or dye employed, the required rate of curing, the production capacity, the energy efficiency, and the temperature regulation capabilities. Several well-known brands provide drying equipment specifically designed for textile curing, including heat curing and UV curing. Heat curing machines utilize heat to activate binders in the print paste to fix pigments and dyes. Monforts Textilmaschinen delivers industrial heat curing ranges with advanced features such as accurate temperature control and efficient airflow to ensure consistent and uniform curing. On the

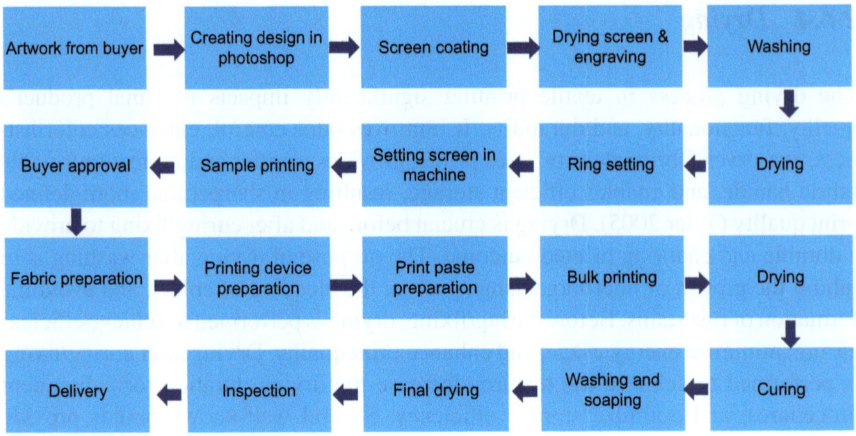

Fig. 2.13 Flow chart of fabric printing

other hand, Maschinenfabrik specializes in continuous heat curing lines. UV curing machines utilize ultraviolet radiation to fix pigments and dyes. Heraeus Noble Light provides industrial UV curing systems featuring diverse lamp configurations and high-intensity output to facilitate rapid curing. The American Ultraviolet Company offers UV curing systems with various lamp types and conveyor belt options to ensure efficient processing. MGI specializes in UV curing units designed for textile applications, often integrated into printing machines. Some machines produced by firms such as Brückner Textile Technologies, EMB, and Stenter Texstar utilize the synergistic impact of heat and UV curing.

Figure 2.13 shows a flow chart of the fabric printing process used in the textile industry, and Table 2.4 shows different prints available in the textile industry.

2.7 Problems with Existing Textile Coloration Machinery and Processes

Although the textile coloring sector is developing rapidly with higher productivity, efficiency, and effectiveness, textile coloration machinery-related drawbacks are significantly hindering the sector. Addressing these challenges is crucial for the textile coloring industry to become more sustainable, efficient, and responsible. Implementing cutting-edge machinery technologies will advance the industry toward a more efficient and responsive future. These difficulties fall into three primary categories: environmental and sustainability issues, efficiency and productivity-related issues, and quality and consistency-related issues.

Table 2.5 represents existing problems in textile coloration and machinery.

All these sustainability challenges associated with existing textile coloration machines contribute to various negative consequences such as depletion of water

Table 2.4 Various prints available in the industry

Case No.	Name of the print	Type of print
1	Rubber print	Water base
2	Pigment print	Water base
3	High-density (HD) print	Plastisol
4	Photo print	Water base
5	Puff print	Water base
6	Reflective print	–
7	Discharge print	Water base
8	Foil print	Water base
9	Flock print	Water and fiber base
10	Glitter print	Water base
11	Sublimation print	Ink base
12	Burn out print	Acidic base
13	Digital print	Ink base
14	Crack print	Water base
15	Silicon print	Plastisol
16	Gel print	–

resources, increased environmental negative impact and global warming, and health risks for workers and surrounding communities (De Oliveira et al. 2021; Hossain et al. 2018; Khattab et al. 2020).

2.8 Sustainable Advancements in Textile Coloration Machinery and Processes

2.8.1 Environmental Concerns on Textile Coloration

While the textile coloring market is constantly changing, several issues with machinery-related processes still exist. These issues must be resolved if the textile coloring sector is to improve its efficiency and sustainability. Dyes, chemicals, and auxiliaries used in the dyeing process can be harmful pollutants if not adequately treated before releasing the wastewater, posing a risk to aquatic ecosystems and human health as many traditional dyes and chemicals used in the coloration process are hazardous, leading to air and water pollution, as well as potential health risks for people handling them, polluting water and air during dyeing, and harming surrounding communities (Kishor et al. 2021; Khan and Malik 2014). In the current coloration processes, harsh chemical handling and processing conditions can damage delicate fabrics impacting product quality and yield (Choudhury 2017). Heating and cooling large volumes of water for dyeing requires significant energy that contributes

Table 2.5 Existing problems on textile coloration machinery

Existing problems	Description	Refs.
Older model	• Older model machines require comparatively more significant volumes of water for various stages of the processes, putting strains on water resources in water-stressed regions • Lack of effective emission control systems allowing harmful volatile organic compounds to be released into the atmosphere, contributing to air pollution and smoke formation • Older machinery often requires frequent maintenance and repairs, leading to increased operational costs	Hasanbeigi and Price (2015), Ozturk et al. (2016), Hynes et al. (2020)
Inefficient heat recovery	• Lack of efficient heat recovery systems leading to wasted energy and increased overall energy consumption, also inefficient water recycling systems in textile coloration machines lead to significant water waste and increased discharge of contaminated wastewater • Heating and cooling large volumes of water during the textile coloration process in the existing coloration machinery require significant energy, often generated from fossil fuels, contributing to greenhouse gas emissions and climate change	Ozturk et al. (2020), Hasanbeigi and Price (2015), Muthu and Gardetti (2020), Terinte et al. (2014)
Lack of adaptability	• Machines are designed for specific colors or fabrics, limiting their versatility and adaptability to changing market demands	Khatri et al. (2015)

to greenhouse gas emissions. Also, improper disposal of spent dyes and chemicals poses significant environmental and health risks (Hasanbeigi and Price 2012; Mani et al. 2019; Hussain and Wahab 2018). Also, reliance on synthetic dye resources raises concerns about sustainability and long-term availability as they are often derived from non-renewable sources (Roy Choudhury 2013). Sourcing practices for dyes and chemicals may involve unethical labor practices or environmental damage in other countries (Guarnieri and Trojan 2019). Additionally, the complex supply chains of dyes and chemicals often lack transparency, making it difficult to assess their environmental and social impact (Cesar Da Silva et al. 2021). The problem is worsening in the fast fashion industry as textile industries emphasize quick turnaround times and low prices, resulting in unsustainable and wasteful dyeing practices (Zanjirani Farahani et al. 2022). The overall situation is not getting any significant turnaround as consumers often lack awareness about the environmental and social impacts of the textile industry, hindering demand for sustainable alternatives (Islam et al. 2021).

2.8.2 *Sustainable Approach in Textile Coloration*

By adopting these innovations, the industry may minimize the ecological footprint of the textile coloration sector and establish sustainable technological processes, guaranteeing responsible resource management and a more environmentally friendly setting. It is crucial to acknowledge the ongoing research and developments in this field continuously improving the sustainability of textile coloration processes. These advancements focus on reducing water usage, implementing energy-efficient technologies, using sustainable dyes and chemicals, and employing advanced process control through automation. Recent innovations in textile coloration, such as closed-loop systems, low liquor ratio, and foam dying, have improved the process flow and led to sustainable long-term effects. Systems that include modern textile dyeing machines effectively reduce water consumption by recycling and reusing water, hence limiting freshwater use and forming wastewater (Muthu and Gardetti 2020).

2.8.2.1 Utilizing Low Liquor Ration Dyeing

Low liquor ratio textile dyeing processes employ low quantities of water compared to conventional procedures (Ammayappan et al. 2016). Furthermore, using jet dyeing machines equipped with specialized nozzles and optimal spray patterns enhances the effectiveness of dye penetration while minimizing water usage (Jiménez-González and Guerrero-Beltrán 2021). Innovative foam dyeing techniques utilize a stable foam instead of water to transport the dye onto the cloth. This method dramatically reduces water usage and provides good color retention and penetration, particularly for sensitive fabrics (Mohsin and Sardar 2019).

2.8.2.2 Incorporating an Advanced Heat Recovery System

Advanced machines incorporate heat recovery systems that effectively catch and repurpose waste heat produced during dyeing. This waste heat is then utilized to heat water in subsequent stages, significantly reducing overall energy usage (Kim et al. 2022). Integrating more energy-efficient motors and implementing adequate insulation in textile coloration machines can significantly reduce the energy lost and the overall energy requirements (Hasanbeigi and Price 2015).

2.8.2.3 Using Renewable Energy Source

By incorporating renewable energy sources such as solar panels into the dyeing process, there can be a substantial reduction in reliance on fossil fuels and a decrease in greenhouse gas emissions (Chen et al. 2020).

2.8.2.4 Digitalization and Automation Implementation

Implementing digitalization and automation in the textile dyeing processes can revolutionize the decade-old system. Implementing real-time advanced monitoring of critical parameters such as temperature, pH, and dye concentration allows for precise control and optimization of the textile coloration process. This approach results in reduced resource waste and ensures consistent color quality. Digital coloration systems employ computer-controlled dosing and application of dyes, which minimizes errors, optimizes dye/pigment usage, and reduces water and chemical waste (Khattab et al. 2020). Furthermore, automation, digital control systems, and improved processes in sophisticated coloration equipment are establishing new standards for efficiency and consistency in the future (Khalil et al. 2023).

2.8.2.5 Using Bio-based Dyes and Pigment

On the other hand, bio-based dyes/pigments have the potential to enhance circularity and sustainability greatly. These dyes/pigments are obtained from natural sources such as plants, insects, and microorganisms, making them more environmentally friendly than synthetic dyes. Bio-based dyes are often biodegradable and have a reduced environmental impact (Da Silva et al. 2021). Furthermore, sustainable textile coloration technologies use environmentally friendly dyes and pigments (Senthil Kumar and Yaashikaa 2018). On-demand digital printing minimizes water and chemical consumption, allowing for more extensive color customization (Polston et al. 2015).

2.8.2.6 Using Enzymatic Dyeing Technique

Biotechnology and enzymatic dyeing are investigating eco-friendly alternatives to conventional dyes and chemicals that can serve as promising substitutes (Chatha et al. 2017). The traditional textile coloration industries need more accountability, transparency, and maintenance of standard operating procedures. Therefore, promoting transparency and accountability and establishing standardized SOPs in the supply chain is crucial. This will lead to the environmental and social consequences of dyes and chemicals, enabling responsible sourcing and the adoption of sustainable alternatives. By promoting cooperation and ingenuity among industry, academia, and government, we may expedite the study and development of sustainable solutions, ultimately hastening the anticipated revolution in machines. Conducting a life cycle assessment of dyeing machines, which includes evaluating their environmental impact from manufacturing and operation to disposal, can help identify areas that can be improved and highlight sustainable options currently available. Implementing circular economy principles in the textile business entails investigating possibilities for recycling utilized textiles and dyes, advocating for resource sustainability, and minimizing waste output.

2.9 Latest Sustainable Development in Textile Coloration Pieces of Machinery

2.9.1 *Dyeing Machines*

2.9.1.1 Low Liquor Ratio Dyeing Machine

ACE Machinery and Co. (China) has recently launched a range of dyeing machines, such as the BIG JUMBO, which features a specialized soft flow design to enable very efficient and huge-volume dyeing. The conveyor system effectively prevents tangles and kinks when moving fabric because of its ultra-low liquor ratio. The JET machine conserves water and energy, reducing the cost of wastewater treatment and eliminating fabric entanglement through its back basket design with a 3-step structure. It also maximizes energy savings by utilizing direct cooling and a DCW washing device. JET Reeless offers a simple method of inserting fabric using two nozzles, ensuring that the fabric remains tensionless and frictionless between the reel and the nozzle. This is achieved by eliminating the driving reel found in traditional jet dyeing machines. Jumbo Reel maximizes energy efficiency through direct cooling and a DCW washing device. It also prevents reel friction and slippage through inverter control and saves resources using a low-liquor ratio for dyeing. HIGH SOFT prevents fabric twisting and significantly reduces dyeing creases and pilling. This is achieved by lowering fabric tension using a fabric-aligning device installed on the rear side. Additionally, it enhances the uniformity of dyeing and improves fabric quality through a soft touch. Xiamen Rapid Precision Machinery Co. Ltd. (China) has introduced the "Eco-Dyer", an electronic heat conduction machine with a low liquor ratio. This machine produces a smooth, wrinkle-free finish from powder or liquid input. This machine does not require a sensor pot to detect temperature, avoiding issues related to temperature detection. It also has a deficient energy consumption and requires no additional water piping. Additionally, it has a twin air-cooling system to achieve faster cooling, has created a dual-function device called the "Ez Doser" that combines stock preparation and recipe dispensing. This device utilizes advanced robotic technology to achieve great accuracy in volumetric measuring without tubes or solenoid valves. It also features an independent injection design for optimal precision. Monforts Textilmaschinen GmbH & Co. KG (Germany) provides sustainable dyeing equipment, such as the TexCoat® ECOMpact, designed to minimize water and energy usage. The TexCoat® ECOMpact system achieves a low liquor ratio of 1:4, resulting in substantially reduced water usage compared to conventional dyeing methods. It incorporates a closed-loop water management system that efficiently recycles and reuses process water, lowering freshwater consumption. The TexCoat® ECOMpact utilizes optimized dyeing cycles and insulating techniques to minimize energy usage during the dyeing process. Heat recovery systems can catch waste heat and use it to preheat water, which improves energy efficiency. The machine's advanced control systems enable optimal utilization of dyes and chemicals, reducing waste and potential environmental pollution. This makes it suitable for eco-friendly

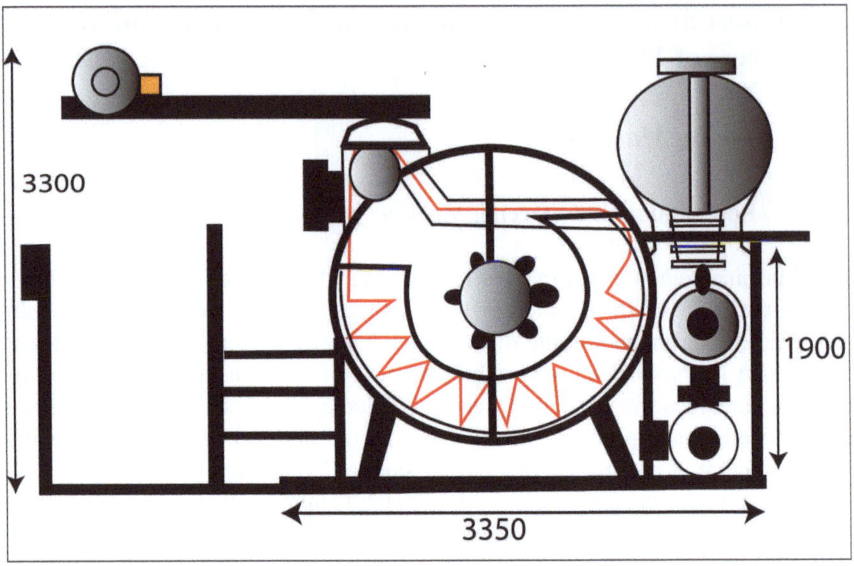

Fig. 2.14 Low liquor ratio dyeing machine. Reproduced (adapted) with permission. Copyright 2025, Ministry of Environment and Energy, Denmark, Danish Environmental Protection Agency (Wenzel et al. n.d.)

colors with less environmental impact than traditional dyes. The Ecodye® system (England) is renowned for its low liquor ratio technology, which uses considerably less water than conventional dyeing methods, reducing freshwater use and wastewater production. The system could integrate water recycling and reuse functionalities to decrease the need for freshwater. The Ecodye® system utilizes accurate dosing and application methods for dyes and chemicals, which effectively minimizes waste and guarantees optimal dye absorption by the fabric. This, in turn, reduces the quantity of chemicals required. The Ecodye® system employs optimized dyeing cycles and insulating techniques to minimize energy usage throughout the dyeing process. In addition, incorporating heat recovery systems can capture and repurpose wasted heat for preheating water or other industrial operations, improving energy efficiency. Figure 2.14 illustrates the low liquor ratio dyeing machines.

2.9.1.2 Fully Automated Dyeing Machine

Roaches and Co. (England) have created the jigger dyeing machine, an advanced machine with fully automated direction maintenance and control over dyeing parameters such as temperature and pressure. It also has adjustable fabric tension and large-diameter draw rollers with low heat capacity to reduce the cooling effect of the dyeing process. Additionally, the machine is equipped with castors for easy mobility.

2.9.1.3 Digital Waterless Dyeing Machine

Coloreel (Switzerland) is at the forefront of digital waterless dyeing technology with its ColorDose® system for on-demand, water-free coloration. The fundamental idea behind ColorDose® is using supercritical carbon dioxide (CO_2) as a dyeing medium, eliminating the requirement for vast amounts of water typically used in the dyeing process. This approach dramatically reduces water usage and the formation of effluent. The ColorDose® method utilizes accurate dosing and application of dyes. The supercritical CO_2 functions as a medium for the dyes, enabling them to penetrate the fabric fibers more deeply. This may result in a reduced amount of dye needed compared to conventional procedures. ColorDose® operates at lower temperatures than standard dyeing methods, typically heating significant amounts of water. This has the potential to save energy. Due to the elimination of water usage, the process results in the formation of little wastewater, simplifying the requirements for wastewater treatment. The ColorDose® technology is more adaptable to a broader spectrum of materials than specific water-based dyeing procedures.

2.9.1.4 Optimum Process-Controlled Dyeing Machine

Noya Mekatronik (Turkiye) has developed the "Noya Multidye" machine, incorporating several innovative features. These include a new-generation water tunnel with patent processing, a second nozzle for quick washing, and a second pump for shower dyeing. The machine is designed to maintain a full water level inside, even against gravity, and allows vertical and horizontal flow. Additionally, the machine utilizes two-dished edges to eliminate unnecessary areas, resulting in improved output with a reduced liquor ratio. Enmos (Turkiye) has created a dyeing machine that is fully traceable and digitalized. These machines are integrated with the current enterprise central monitoring systems, such as Simplex 520, Simplex 540, and Simplex 620. The company has also created the "Profix FW/CAR/DD" system, which automates powdered dyestuffs' management, storage, weighing, and dispersing/dissolving. Additionally, they have developed the "Profix BD" machine, which is highly accurate and traceable for bulk dyeing. The company also offers the "Profix DS & R/S" system for dispensing and dissolving salt-soda and the "Profix C" system for managing liquid dyestuffs.

2.9.1.5 Multi-color Warp Dyeing Machine

Jupiter (India) introduced the "Mult-tank Indigo Rope Dyeing" method to dye yarn of varying thicknesses using indigo in a continuous warp dyeing machine. This process results in a rich indigo color that may be used for topping or dyeing with different dyes. The recent developments in this regard are transitioning from conventional denim to a more lightweight denim fabric. The utilization of multi-tank-based continuous multi-color warp dyeing has facilitated enhanced process flexibility and improved optimization of dyes and chemical consumption in an environmentally friendly manner.

2.9.1.6 Modern Yarn Dyeing Machine

MacChem Corporation (Bangladesh) has introduced a new line of beam dyeing machines. These machines control the flow rate of dye liquid using a main pump regulated by a frequency conversion speed system. This allows for efficient energy consumption and precise control of dyeing. Manoj Engineering (Bangladesh) is developing a high-pressure–temperature yarn/package dyeing machine. This machine has a liquor ratio ranging from 1:3 to 1:5.2 and is equipped with single and dual-direction pumps. It is suitable for dyeing various yarns and materials, such as cross-wound bobbins, warp beams, hank yarns, loose stock, and combed stock. This machine is versatile and can bleach and dye surgical and loose fibers such as hemp, abaca, banana, jute, and more. It has a duplex sensor and a highly reliable temperature control system that prevents leaks. The system also includes a differential level transmitter, ensuring the liquor is filled 75–100 mm above the package.

2.9.2 Printing Machines

DGI (Korea) has created the "Fabrijet FT-3240X", an advanced 3.2 m Hybrid textile printer with a dual roll printing mechanism. DGI provides printers that utilize eco-solvent inks, which are generally less environmentally damaging than solvent-based inks. However, these may have some limits in terms of wash fastness. These state-of-the-art printers minimize water and energy usage while minimizing waste output. Brother (USA) has recently launched a new direct-to-garment printer called the GTXpro, which utilizes a piezo head for volume printing. The machine incorporates automated cleaning procedures, reducing water consumption and maintenance duration compared to manual cleaning techniques. The GTXpro may have features such as sealed printheads or mist reduction technologies, effectively reducing ink mist and potential overspray. This results in less ink wastage and a more hygienic printing environment. Spgprints (Netherlands) has introduced a new printing system called "Kapok" based on the RD8 technology. This system offers excellent dependability and high-value operations. It also features a uniform-optimized glue position, consistent operation, exceptional reproducibility, and an adaptable dual-driver option. Additionally, the print heads of this system are equipped with lifetime-lubricated bearings. In addition, they introduced Randomscreen Eco and Pentascreen 125 Mesh, which allow for more excellent surface printing and offer a more cost-effective alternative to penetrative printing with a reduced volume of print paste. Y-JET has created advanced oval screen-printing machines with a full servo rotation and printing system. These machines are designed to be inconspicuous on the surface and offer grayscale printing capabilities. They also include a built-in print head capping system, eliminating the need for a printing frame. Additionally, they provide precise control over temperature and pressure. HUMAN DIGITAL has introduced "E-Press", a cutting-edge printing technology that offers high-resolution printing with a minimum 3.5 pL variable printing capability. This technology also includes a user-friendly mesh design program with a stick belt. This series of printing machines is both cost-effective and highly prolific. K.B. Engineers have enhanced a completely automated flatbed screen printing machine with a digital double servo motor blanket-location system. This system allows for flexible stroke, speed, direction, and a changeable screen frame. EFI (Italy) has introduced environmentally-friendly printing solutions such as the EFI Reggiani Colors line, which uses digital textile printing techniques that minimize water and energy consumption. The EFI Reggiani COLORS printer uses water-based pigment inks rather than solvent-based inks, thereby eliminating the release of hazardous solvents and minimizing the possible health hazards connected with solvents. EFI Reggiani's technique features an efficient one-step process that combines pigment ink with binder and pre-treatment chemistry. This can reduce water consumption compared to conventional multi-stage printing methods requiring distinct pre-treatment and cleaning procedures. These printers can reduce energy usage by operating at lower temperatures than standard textile printing technologies that require heating significant volumes of water.

Stäubli's (Switzerland) automation solutions can enhance accuracy and productivity, decreasing ink and printing waste. For instance, their systems may optimize squeegee strokes and printing pressure, thereby eliminating ink waste. Stäubli's systems enhance printing processes through improved material handling efficiency, reduced setup times, increased production speeds, consistent print quality, optimized machine operation, automatic standby modes to minimize energy usage during idle periods, enhanced worker safety, and data-driven printing. M&R, Vastex International, and Akers are attempting to include electrically heated, fully automatic flatbed screen printing machines. Electric heating systems can reduce greenhouse gas emissions and air pollution typically associated with oil-based heating systems. Electric heating systems generally have a lower fire danger than oil-heated systems and require less maintenance regarding oil changes and leak checks. Contemporary electric heating technology provides accurate temperature regulation and can be more energy-efficient than oil heating, particularly when combined with adequate insulation. Electrically heated, fully automatic flatbed screen printing machines can enhance their sustainability benefits by utilizing water-based inks, eco-friendly pre-treatment and after-treatment chemicals, long-lasting screens, and recycled materials. Mimaki Engineering Co. (Japan) specializes in developing digital printing machines, such as the TX3000P-3000, designed for high-speed, water-based textile printing. For its unidirectional printing capabilities, this machine can print a whole pattern on fabric in a single pass of the print head. Mimaki has also created digital printing machines, such as the Mimaki TX500P-3000. This model focuses on water-based textile printing, offering high speed and low ink use. The Geo Knight double heat press machine, Powermatic pm3 dual heat press, and Hotronix dual platen heat presses can apply heat-transfer designs onto different substrates such as fabrics, mugs, and phone cases. These machines utilize heat and pressure to activate a unique adhesive layer on the transfer material, creating a permanent bond between the design and the substrate. Stenter, Bruckner, and Fongs' National are currently working on the development of continuous transfer printing equipment that can handle both chopped pieces and roll-fed fabrics. These machines utilize a combination of printing techniques, employing a digital printing machine for detailed designs on individual pieces and a specialized rotary printing machine for extensive, uniform color regions on fabrics fed in rolls. Various manufacturers have customizable continuous transfer printing equipment that may be adjusted to accommodate both cut pieces and roll-fed materials. However, these machines may be less easily accessible and perhaps more costly.

2.10 Challenges and Recommendations

Although promising advancements in sustainable textile coloration are acquired, several obstacles in technology, marketing, and societal acceptance, are needed to be addressed to achieve broad implementation. Sustainable technologies in textile colors may necessitate a more significant initial investment and additional refinement to get equivalent levels of colorfastness, durability, and tactile qualities as conventional approaches. Furthermore, implementing new sustainable processes on a large scale for high-volume production can take time, as further modifications may be required to integrate them seamlessly into current production lines. In addition, many sustainable printing and dyeing techniques may provide a unique range of colors compared to conventional alternatives. The need for consumer awareness hinders the progress of sustainable improvements in textile coloration machinery. Consequently, it can be challenging to teach customers about the importance of sustainable textiles and their potentially higher costs. Adopting sustainable practices may necessitate sourcing materials from environmentally friendly suppliers and establishing partnerships with them. It also involves implementing standardized sustainable practices and enforcing clear regulations on eco-friendly certifications to ensure transparency and gain consumer trust. Socio-economic aspects are crucial in implementing sustainability in the textile coloration business. Training the workforce on new sustainable technologies and processes is particularly demanding for successful implementation.

Furthermore, the fast-fashion industry's prioritization of affordability and quick manufacturing cycles can conflict with the emphasis on excellence and environmental accountability in sustainable coloration procedures. To address these challenges, textile machinery manufacturers, chemical suppliers, garment brands, governments, and consumers must work together. This collaboration should involve government support and incentives, consumer education to increase awareness, emphasis on research and development, and cooperation among industry stakeholders. By doing so, we can promote investment in sustainable technologies and infrastructure and ultimately move toward a future where we can achieve vibrant and innovative designs while prioritizing environmental responsibility and social well-being.

2.11 Future Trends

The textile coloration machinery industry is experiencing a surge in sustainable innovations, driven by advancements in intelligent and automated machinery systems, the integration of hybrid technologies, the adoption of biotechnology and sustainable chemistry for process optimization, and a shift in consumer preferences and industry practices toward sustainability. Soon, digital printing is expected to establish further dominance by utilizing advanced inkjet heads that will provide exceptional precision and minimize ink wastage. Single-pass printing technology will eliminate the requirement for multiple printing passes, increasing efficiency. Additionally, on-demand printing will reduce the need for excessive production and minimize textile waste. The integration of AI-powered color matching and recipe optimization in textile dyeing is attracting interest for its ability to reduce water and chemical usage. Internet of Things (IoT) connected machines to monitor and control processes in real-time, reducing mistakes and maximizing resource use, is becoming increasingly popular. Hybrid technologies that combine digital printing with other sustainable methods, such as laser printing, are being used to apply colors precisely and locally while minimizing water usage. Cold plasma treatments also modify fabric surfaces, improve dye adhesion, and reduce water requirements in dyeing processes. The textile coloration industry also incorporates biotechnology and sustainable chemistry to develop more efficient enzymes targeted explicitly for certain colorants and fabrics. Furthermore, bio-based colorants derived from natural sources like algae or bacteria are being utilized to offer a broader range of vibrant and eco-friendly colors. Enzymes are also biocatalysts in dyeing, improving color fixation with minimal environmental impact. Utilizing advanced wastewater treatment technologies to reduce environmental pollution, increase water reuse, and incorporate life cycle assessment practices into the textile coloration process to identify areas for sustainability improvement can be crucial in developing optimized coloration processes that promote water, energy, and resource recovery and conservation. It is essential to prioritize the shift in consumer expectations and industry practices. This is because customers will increasingly demand openness and traceability in the supply chain, explicitly emphasizing ethical sourcing and sustainable methods. Prioritizing, implementing, and enforcing more stringent laws and certifications is crucial to establishing clear standards for sustainable textile coloration practices. This will facilitate the transition toward a circular economy model emphasizing recycling and upcycling textiles to minimize waste. The industry may achieve high-quality colored textiles while reducing environmental effects and promoting social responsibility by adopting innovative technology, bio-based solutions, and closed-loop systems. This requires a shift in consumer behavior and a solid commitment to sustainability throughout the sector.

2.12 Conclusion

By all counts, along with proven findings, textile coloration machinery developments and process enhancements in the textile coloration industry are going to experience radical changes in the coming days toward sustainability due to significant environmental challenges, economic drawbacks, and processing complications. This chapter has examined the possibilities for sustainable advancements in textile coloration by comparing traditional practices to both machinery renovations and process improvements, taking into account the potential environmental effects of the textile coloration industry, and determining whether it is feasible to integrate the most sustainable practices into the design of contemporary machinery and process optimization. This chapter has proposed a path toward a more sustainable future for textile coloration by critically examining the connection between current practices and upcoming advancements, focusing on the potential for significant reductions in environmental impact while maintaining high-quality color performance, paving the way for a more responsible and environmentally conscious textile industry. However, there are further relevant sustainability scopes to explore, e.g., structural outline, brand, and model-specific technology, efficiency, and productivity analysis, etc., which can pave the way to a further in-depth understanding of normalizing efficient and effective sustainable machinery development and process enhancement in the field of textile coloration and beyond.

2.12.1 Declaration

The authors declare that they have no known competing financial interests or personal relationships that could have appeared to influence the work reported in this chapter. Generative AI and AI-assisted technologies have been used in the writing process to improve the readability and language of the work only. After using this tool, the authors carefully reviewed and revised the content to ensure accuracy and quality. The authors take full responsibility for the final version of this publication.

References

Ali M, Lin L, Faisal S, Sahito IA, Ali SI (2019) Optimization of screen printing process for functional printing. Pigment Resin Technol 48:456–463. https://doi.org/10.1108/PRT-05-2019-0043

Ammayappan L, Jose S, Arputha Raj A (2016) Sustainable production processes in textile dyeing. In: Muthu SS, Gardetti MA (eds) Green fashion, environmental footprints and eco-design of products and processes. Springer Singapore, Singapore, pp 185–216. https://doi.org/10.1007/978-981-10-0111-6_8

Azeem M, Iqbal N, Mir RA, Adeel S, Batool F, Khan AA, Gul S (2019) Harnessing natural colorants from algal species for fabric dyeing: a sustainable eco-friendly approach for textile processing. J Appl Phycol 31:3941–3948. https://doi.org/10.1007/s10811-019-01848-z

Bahria H, Erbil Y (2016) UV technology for use in textile dyeing and printing: Photocured applications. Dyes Pigment 134:442–447. https://doi.org/10.1016/j.dyepig.2016.07.042
Broadbent AD (2001) Basic principles of textile coloration. Society of Dyers and Colourists, Bradford, West Yorkshire
Cay A, Tarakçıoğlu I, Hepbasli A (2009) Assessment of finishing processes by exhaustion principle for textile fabrics: an exergetic approach. Appl Therm Eng 29:2554–2561. https://doi.org/10.1016/j.applthermaleng.2008.12.032
Cesar Da Silva P, De Oliveira C, Neto G, Ferreira Correia JM, Pujol Tucci HN (2021) Evaluation of economic, environmental and operational performance of the adoption of cleaner production: survey in large textile industries. J Clean Prod 278:123855. https://doi.org/10.1016/j.jclepro.2020.123855
Chatha SAS, Asgher M, Iqbal HMN (2017) Enzyme-based solutions for textile processing and dye contaminant biodegradation—a review. Environ Sci Pollut Res 24:14005–14018. https://doi.org/10.1007/s11356-017-8998-1
Chen G, Li Y, Bick M, Chen J (2020) Smart textiles for electricity generation. Chem Rev 120:3668–3720. https://doi.org/10.1021/acs.chemrev.9b00821
Choudhury AKR (2017) Sustainable chemical technologies for textile production. In: Sustainable fibres and textiles. Elsevier, pp 267–322. https://doi.org/10.1016/B978-0-08-102041-8.00010-X
Da Silva CJG, De Medeiros ADM, De Amorim JDP, Do Nascimento HA, Converti A, Costa AFS, Sarubbo LA (2021) Bacterial cellulose biotextiles for the future of sustainable fashion: a review. Environ Chem Lett 19:2967–2980. https://doi.org/10.1007/s10311-021-01214-x
De Oliveira CRS, Da Silva Júnior AH, Mulinari J, Immich APS (2021) Textile re-engineering: eco-responsible solutions for a more sustainable industry. Sustain. Prod. Consum. 28:1232–1248. https://doi.org/10.1016/j.spc.2021.08.001
Fan Q, Kim YK, Perruzzi MK, Lewis AF (n.d.) Fabric pretreatment and digital textile print quality
Gobalakrishnan M, Saravanan D, Das S (2020) Sustainable finishing process using natural ingredients. In: Muthu SS, Gardetti MA (eds) Sustainability in the textile and apparel industries, Sustainable textiles: production, processing, manufacturing and chemistry. Springer International Publishing, Cham, pp 129–146. https://doi.org/10.1007/978-3-030-38545-3_5
Guarnieri P, Trojan F (2019) Decision making on supplier selection based on social, ethical, and environmental criteria: a study in the textile industry. Resour Conserv Recycl 141:347–361. https://doi.org/10.1016/j.resconrec.2018.10.023
Günay M, Jasper WJ (2010) Limitations in predicting dyebath exhaustion using optical spectroscopy. Color Technol 126:140–146. https://doi.org/10.1111/j.1478-4408.2010.00238.x
Halim AFMF, Islam MT, Hoque MMU (2021) Chemistry of sustainable coloration of textile materials. In: Green chemistry for sustainable textiles. Elsevier, pp 57–67. https://doi.org/10.1016/B978-0-323-85204-3.00003-8
Hasanbeigi A, Price L (2015) A technical review of emerging technologies for energy and water efficiency and pollution reduction in the textile industry. J Clean Prod 95:30–44. https://doi.org/10.1016/j.jclepro.2015.02.079
Hasanbeigi A, Price L (2012) A review of energy use and energy efficiency technologies for the textile industry. Renew Sustain Energy Rev 16:3648–3665. https://doi.org/10.1016/j.rser.2012.03.029
Hossain L, Sarker SK, Khan MS (2018) Evaluation of present and future wastewater impacts of textile dyeing industries in Bangladesh. Environ Dev 26:23–33. https://doi.org/10.1016/j.envdev.2018.03.005
Hossain MT, Shahid MA, Limon MGM, Hossain I, Mahmud N (2024) Techniques, applications, and challenges in textiles for a sustainable future. J Open Innov Technol Mark Complex 10:100230. https://doi.org/10.1016/j.joitmc.2024.100230
Hussain T, Wahab A (2018) A critical review of the current water conservation practices in textile wet processing. J Clean Prod 198:806–819. https://doi.org/10.1016/j.jclepro.2018.07.051
Hussien A, Abdelrahman M, Wahab S, Mashaly H, Ali S, Maamoun D, Khattab T (2020) Review in textile printing technology. Egypt J Chem. https://doi.org/10.21608/ejchem.2020.23726.2418

Hynes NRJ, Kumar JS, Kamyab H, Sujana JAJ, Al-Khashman OA, Kuslu Y, Ene A, Suresh Kumar B (2020) Modern enabling techniques and adsorbents based dye removal with sustainability concerns in textile industrial sector—a comprehensive review. J Clean Prod 272:122636. https://doi.org/10.1016/j.jclepro.2020.122636

Islam MM, Perry P, Gill S (2021) Mapping environmentally sustainable practices in textiles, apparel and fashion industries: a systematic literature review. J Fash Mark Manag Int J 25:331–353. https://doi.org/10.1108/JFMM-07-2020-0130

Jiménez-González O, Guerrero-Beltrán JÁ (2021) Extraction, microencapsulation, color properties, and experimental design of natural pigments obtained by spray drying. Food Eng Rev 13:769–811. https://doi.org/10.1007/s12393-021-09288-7

Kalaoglu F, Paul R (2015) Finishing of jeans and quality control. In: Denim. Elsevier, pp 425–459. https://doi.org/10.1016/B978-0-85709-843-6.00014-7

Kan CW (2007) The use of plasma pre-treatment for enhancing the performance of textile ink-jet printing. J Adhes Sci Technol 21:911–921. https://doi.org/10.1163/156856107781393884

Kan CW, Yuen CWM, Tsoi WY (2011) Using atmospheric pressure plasma for enhancing the deposition of printing paste on cotton fabric for digital ink-jet printing. Cellulose 18:827–839. https://doi.org/10.1007/s10570-011-9522-2

Kašiković N, Vladić G, Milić N, Novaković D, Milošević R, Dedijer S (2018) Colour fastness to washing of multi-layered digital prints on textile materials. J Natl Sci Found Sri Lanka 46:381. https://doi.org/10.4038/jnsfsr.v46i3.8489

Khairul Akter MMd, Haq UN, Islam MdM, Uddin MA (2022) Textile-apparel manufacturing and material waste management in the circular economy: a conceptual model to achieve sustainable development goal (SDG) 12 for Bangladesh. Clean Environ Syst 4:100070. https://doi.org/10.1016/j.cesys.2022.100070

Khalil E, Sarkar J, Rahman MdM, Shamsuzzaman Md, Das D (2023) Advanced technology in textile dyeing. In: Rahman MdM, Mashud M, Rahman MdM (eds) Advanced technology in textiles: fibre to apparel, Textile science and clothing technology. Springer Nature, Singapore, pp 97–138. https://doi.org/10.1007/978-981-99-2142-3_4

Khan S, Malik A (2014) Environmental and health effects of textile industry wastewater. In: Malik A, Grohmann E, Akhtar R (eds) Environmental deterioration and human health. Springer Netherlands, Dordrecht, pp 55–71. https://doi.org/10.1007/978-94-007-7890-0_4

Khatri A, Peerzada MH, Mohsin M, White M (2015) A review on developments in dyeing cotton fabrics with reactive dyes for reducing effluent pollution. J Clean Prod 87:50–57. https://doi.org/10.1016/j.jclepro.2014.09.017

Khattab TA, Abdelrahman MS, Rehan M (2020) Textile dyeing industry: environmental impacts and remediation. Environ Sci Pollut Res 27:3803–3818. https://doi.org/10.1007/s11356-019-07137-z

Kim Y, Lim J, Shim JY, Lee H, Cho H, Kim J (2022) Optimizing wastewater heat recovery systems in textile dyeing processes using pinch analysis. Appl Therm Eng 214:118880. https://doi.org/10.1016/j.applthermaleng.2022.118880

Kishor R, Purchase D, Saratale GD, Saratale RG, Ferreira LFR, Bilal M, Chandra R, Bharagava RN (2021) Ecotoxicological and health concerns of persistent coloring pollutants of textile industry wastewater and treatment approaches for environmental safety. J Environ Chem Eng 9:105012. https://doi.org/10.1016/j.jece.2020.105012

Lee KE (2017) Environmental sustainability in the textile industry. In: Muthu SS (ed) Sustainability in the textile industry, Textile science and clothing technology. Springer Singapore, Singapore, pp 17–55. https://doi.org/10.1007/978-981-10-2639-3_3

Mani S, Chowdhary P, Bharagava RN (2019) Textile wastewater dyes: toxicity profile and treatment approaches. In: Bharagava RN, Chowdhary P (eds) Emerging and eco-friendly approaches for waste management. Springer Singapore, Singapore, pp 219–244. https://doi.org/10.1007/978-981-10-8669-4_11

Mohsin M, Sardar S (2019) Multi-criteria decision analysis for textile pad-dyeing and foam-dyeing based on cost, performance, productivity and sustainability. Cellulose 26:4143–4157. https://doi.org/10.1007/s10570-019-02337-9

Muthu SS (ed) (2018) Sustainable innovations in textile chemical processes, Textile science and clothing technology. Springer Singapore, Singapore. https://doi.org/10.1007/978-981-10-8491-1

Muthu SS, Gardetti MA (eds) (2020) Sustainability in the textile and apparel industries: production process sustainability, Sustainable textiles: production, processing, manufacturing and chemistry. Springer International Publishing, Cham. https://doi.org/10.1007/978-3-030-38545-3

Nair GP (2011) Methods and machinery for the dyeing process. In: Handbook of textile and industrial dyeing. Elsevier, pp 245–300. https://doi.org/10.1533/9780857093974.1.245

Ozturk E, Cinperi NC, Kitis M (2020) Improving energy efficiency using the most appropriate techniques in an integrated woolen textile facility. J Clean Prod 254:120145. https://doi.org/10.1016/j.jclepro.2020.120145

Ozturk E, Koseoglu H, Karaboyaci M, Yigit NO, Yetis U, Kitis M (2016) Sustainable textile production: cleaner production assessment/eco-efficiency analysis study in a textile mill. J Clean Prod 138:248–263. https://doi.org/10.1016/j.jclepro.2016.02.071

Polston K, Parrillo-Chapman L, Moore M (2015) Print-on-demand inkjet digital textile printing technology: an initial understanding of user types and skill levels. Int J Fash Des Technol Educ 8:87–96. https://doi.org/10.1080/17543266.2014.992050

Roy Choudhury AK (2013) Green chemistry and the textile industry. Text Prog 45:3–143. https://doi.org/10.1080/00405167.2013.807601

Senthil Kumar P, Yaashikaa PR (2018) Sustainable dyeing techniques. In: Muthu SS (ed) Sustainable innovations in textile chemical processes, Textile science and clothing technology. Springer Singapore, Singapore, pp 1–29. https://doi.org/10.1007/978-981-10-8491-1_1

Terinte N, Manda BMK, Taylor J, Schuster KC, Patel MK (2014) Environmental assessment of coloured fabrics and opportunities for value creation: spin-dyeing versus conventional dyeing of modal fabrics. J Clean Prod 72:127–138. https://doi.org/10.1016/j.jclepro.2014.02.002

Todeschini BV, Cortimiglia MN, De Medeiros JF (2020) Collaboration practices in the fashion industry: environmentally sustainable innovations in the value chain. Environ Sci Policy 106:1–11. https://doi.org/10.1016/j.envsci.2020.01.003

Tyler DJ (2005) Textile digital printing technologies. Text Prog 37:1–65. https://doi.org/10.1533/tepr.2005.0004

Varadarajan G, Venkatachalam P (2016) Sustainable textile dyeing processes. Environ Chem Lett 14:113–122. https://doi.org/10.1007/s10311-015-0533-3

Wang H, Memon H (eds) (2020) Cotton science and processing technology: gene, ginning, garment and green recycling, Textile science and clothing technology. Springer Singapore, Singapore. https://doi.org/10.1007/978-981-15-9169-3

Wardman RH (n.d.) An introduction to textile coloration

Zanjirani Farahani R, Asgari N, Van Wassenhove LN (2022) Fast fashion, charities, and the circular economy: challenges for operations management. Prod Oper Manag 31:1089–1114. https://doi.org/10.1111/poms.13596

Chapter 3
Waterless Dyeing of Textiles—A Sustainable Alternative to Conventional Wet Processing

Ragini Mukherjee, Ritwick Halder, and Sankar Roy Maulik

Abstract The conventional textile dyeing process uses vast amounts of water and hazardous chemicals, negatively impacting the environment. Waterless dyeing technologies, such as supercritical CO_2 dyeing, air dyeing, and dope dyeing, are gaining popularity as sustainable alternatives. These methods drastically reduce water usage and eliminate the need for harmful chemicals, offering both environmental and economic benefits by lowering operational costs. While waterless dyeing presents a promising future for the textile industry, challenges like high initial equipment costs remain a barrier to widespread adoption. Nonetheless, waterless dyeing is set to shape the industry's future as demand for sustainability grows.

Keywords Air dyeing · Dope · Dyeing · Environment · Supercritical · Textile · Waterless

3.1 Introduction

Textiles play a significant role in human life. They are considered one of the largest growing sectors that manufacture various products like apparel, home furnishing and technically finished textiles. It involves a sequence of operations, i.e. production of the fibres, spinning yarns from the fibres, construction of the fabric, carrying out dyeing, printing, and finishing. These processes, especially the dyeing of textiles require an enormous volumes of water, dyes, chemicals, and energy, and also emitting large volumes of wastewater in the form of by-products, commonly known as effluent. This effluent used to be highly toxic and contains several chemicals that are non-biodegradable, like residual dyes, salts, finishing chemicals, etc. These generated wastes can pollute the environment at an extreme level if not appropriately treated (Abate and Tadesse 2021). Immense adverse effects can occur not only to the water and soil but also to the marine life. Thus, attention has been given to stewardship

R. Mukherjee · R. Halder · S. Roy Maulik (✉)
Department of Silpa-Sadana, Visva-Bharati (A Central University), Sriniketan, India
e-mail: s_r_moulik@yahoo.co.in; sankar.raymaulik@visva-bharati.ac.in

S. Maiti et al. (eds.), *Sustainable Coloration Techniques in Textiles*,
https://doi.org/10.1007/978-981-96-4975-4_3

towards the environment for a long time in the manufacturing industries, particularly the textile industry, which utilises almost several hundred million gallons of water daily to develop commercial products (Luna et al. 2014).

Furthermore, countries like India produce almost 13,00,000 tonnes of dyes, where the textile industry needs nearly 80% of the consumption. Environmentally conscious people promote green chemistry in the textile industry, focussing on reducing harmful substance consumption and emissions. Traditionally, the initiation of green chemistry in the textile industry has concentrated on measures to prevent pollution that further help reduce the environmental influence of wastewater discharged from wet processing (Zaidy et al. 2019).

3.2 Drawbacks of the Conventional Dyeing Process

The textile industry is responsible for environmental pollution; the colouration process produces almost 20% of the total water pollution. Textile dyeing is an evolving process that incorporates many interrelated aspects, such as the kind of fibres used, the structure of the textile substrate, the chemical composition of the goods, the type of dyes used, the method, and the kind of equipment utilised. Each of these components immediately affects the processes. The standard dyeing methods now used by the textile industry have a detrimental effect on the environment, particularly on the quantity and quality of water pollution, toxic chemicals, emissions into the atmosphere, and excessive energy consumption (Lara et al. 2022).

Dyes may be defined as substances that have an affinity for textiles and are soluble in water at the time of application. Dyes are extensively utilised in various industries, viz., textile, pharmaceutical, food, cosmetics, plastic, photographic, and paper (Gunay 2013). In addition to having a colour, textile effluents exhibit high Biological Oxygen Demand (BOD), Chemical Oxygen Demand (COD), total dissolved solids, total suspended solids, and low dissolved oxygen (DO) values. The colour also reduces the visibility of the discharged water, which lowers the amount of light available to aquatic plants. Salts cover almost 50% of the effluent, which is responsible for the contribution of a considerable amount of solids in a dissolved form, increasing the BOD value and the demand for oxygen required to oxidise all the pollutants in water (Lara et al. 2022).

Water is the world's most valuable natural resource comprising about 2.5% fresh water and over 97% saltwater. Of the 2.5% freshwater, over two-thirds (68.7%) are frozen as snow and ice, and the remaining one-third is held underground as groundwater. It appears that surface water in lakes, bogs, rivers, and streams makes up only 0.3% of the world's total freshwater. Thus, the textile industry must focus on creating alternate dyeing methods to preserve this finite supply of freshwater (Mahmud and Kaiser 2020). The effluent from the textile sector produces numerous pollutants, such as surfactants, salts, heavy metals, enzymes, oxidising and reducing agents, etc. Discharging highly contaminated effluents into the surrounding groundwater and surface waters causes quality degradation over time. The ecosystem suffers when harmful, hazardous, and persistent chemicals are used in wet processing and untreated effluents are released (Madhav et al. 2018). Conventional dyeing requires a huge volume of water, and the treatment cost of effluent is also high. Approximately 100–150 L of water is required for dyeing one kilogramme of textile, which may vary depending on the fibre type, colour, and types of machinery used. It is also observed that a significant amount (~10–15%) of the dye is lost in the wastewater after dyeing and is released into the environment (Abate and Tadesse 2021).

3.3 Development of Waterless Dyeing Techniques

The enormous amount of freshwater consumption as well as wastewater disposal has prompted the importance of the development of remedial solutions to save the resources of the earth. One of them is the dyeing technique, which involves zero water consumption or minimal water required to dye the textiles. The dyeing technique with a non-aqueous medium has drawn much attention as it offers more promising alternatives. This dyeing technique includes supercritical carbon dioxide for dyeing hydrophobic fibres like polyester (Zaidy et al. 2019).

In an attempt to minimise water consumption and wastewater production in textile wet processing, low wet-pick-up equipment was developed, which used 80–90% less water to complete the process (Bach et al. 2002). The common waterless dyeing processes where alternatives of water are utilised as the dyeing medium are discussed below.

3.3.1 Air Dyeing Technology

This process was patented in 1978 by Wilhelm Christ, the man who developed the fabric colouration method using air as the dyeing medium (Abate and Tadesse 2021), and since then, various research has been carried out. The air dyeing technology is considered one of the sustainable methods of the textile colouration process that reduces the consumption of water by up to 95%, energy by up to 86%, and emission of greenhouse gases by up to 84% when compared to the traditional dyeing methods in the textile industries (Mahmud and Kaiser 2020). This dyeing technology only requires negligible water to dissolve the dye particles.

3.3.1.1 Mechanism of Air Dyeing

The aqueous medium of dye solution, containing dyestuff and auxiliaries, is converted to free gaseous atoms, which are consequently mixed with airflow of high pressure that is emitted from the blower of the air dyeing machine and sprayed over the fabric through several nozzles.

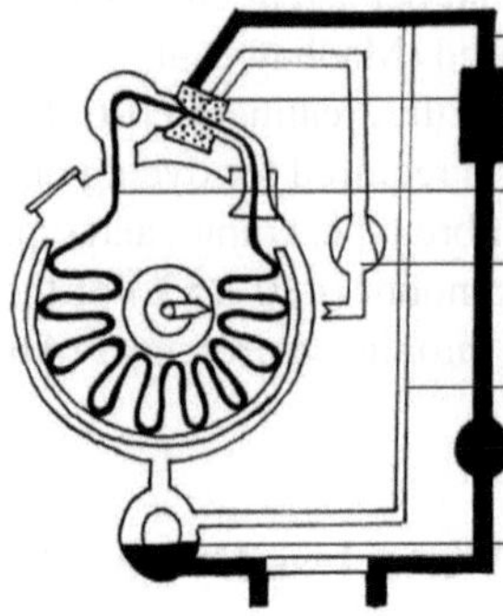

Fig. 3.1 Air dyeing mechanism

The fabric in the machine is circulated and moved forward by this highly pressurised mixture of air and atomised dye particles, as shown in Fig. 3.1. Thus, airflow dyeing can be described similarly to the jet dyeing technique, where air is utilised as an alternative to water. The atomised form of dye possesses a strong infusion into the fibres and wide contact regions with the fabric when it is sprayed; this activity enhances the diffusion rate of the dye. With the airflow dyeing technique, it is possible to reach lower liquor ratios (1:2 to 1:5) and a fabric speed of up to 1,000 m/min without causing damage to the fabric. After high-temperature dyeing, the wastewater is finally drained without coming into contact with the dyed fabric. Since there is no tension on the fabric during processing, this method can be used to process elastane fibres (Yusuf and Shahid 2022).

3.3.2 *Supercritical Fluid Dyeing Using CO_2*

Supercritical CO_2 dyeing is an innovative and sustainable dyeing method that uses supercritical carbon dioxide as the dyeing medium instead of water. This technology, characterised by its high diffusivity and low viscosity, allows for efficient and even distribution of dye on both synthetic and natural fibres (Goñi et al. 2021). The process is environmentally friendly as it significantly reduces water usage and eliminates harmful wastewater. Despite the high initial investment for the necessary equipment, the potential for lower operating costs and CO_2 recycling can make it a cost-effective solution. The resulting dyed textiles often exhibit vibrant and fade-resistant colours, making this technology a promising alternative to traditional dyeing methods in the textile industry (Agrawal 2015).

3.3.2.1 Efficiency of CO_2 as Supercritical Fluid

Solids melt when heated, causing the motion of the molecules to increase and the formation of a liquid and vapour phase. When a vapour is compressed below its critical temperature, the vapour-liquid equilibrium line is crossed, and the vapour condenses. No matter how much pressure is added, condensation is impossible above the critical temperature due to the considerable thermal energy of the vapour molecules. At the critical temperature, the vapour-liquid equilibrium line stops (Bach et al. 2002). A fluid is called supercritical when it is above its critical temperature and the accompanying critical pressure. At this point, it cannot be classified as either a vapour or a liquid (Wu and Han 2019).

Carbon dioxide is the most commonly used supercritical fluid due to its very mild critical point, which can be easily achieved. To make the solid dye particle soluble, the temperature required for CO_2 is much higher, which is 600 kg/m^3 and 120°C, respectively. This condition can only be achieved at a pressure of around 300 bar, so that both the temperature and pressure are far above the critical point (31ºC, 74 bar) (Varma et al. 2020).

Several attributes of CO_2, like non-inflammatory, non-toxicity, and cost-effectiveness, make it a perfect choice as a supercritical fluid for textile colouration. Additionally, no other process is required for the production of CO_2 as it can be obtained as a bi-product of major industrial practices, viz., combustion, fermentation, and ammonia synthesis. Moreover, its low surface tension and viscosity facilitate better and faster diffusion of the dye molecules into the fibres. The unused dye particles and pure CO_2 gas can be reused further, making it eco-friendly (Knez et al. 2023).

Supercritical CO_2 dyeing is most effective in dyeing polyester with non-polar disperse classes of dyes since it does not involve dispersing agents. The dyeing of polyester using water results in the agglomeration of the dye particles on the fabric, causing uneven dyeing, and hence dispersing agents are used. In the case of supercritical CO_2 dyeing, both the dye and the solvent have the same polarity, eliminating the use of dispersing agents for achieving uniform dyeing (Sicardi et al. 2000).

3.3.2.2 Dyeing Mechanism with Supercritical CO_2

The fabric to be dyed is loaded in the dyeing machine, primarily an autoclave. During the dyeing process, the carbon dioxide is released into the heat exchanger chamber, where the dyestuff dissolves with the supercritical fluid. On dissolving the dye, the supercritical fluid moves towards the dyeing chamber. The colouration occurs in the dyeing vessel, where the dye particles get diffused in the swollen polyester polymer matrix. Upon getting the desired shade, the vessel is depressurised, which causes excess CO_2 and dye particles to come out of the shrinking fibres. The CO_2 is returned to its gaseous state in the pressure-reducing chamber, and the unused dye and CO_2 are again restored and completing the dyeing cycle. The dye and gaseous CO_2 obtained can be reused for further dyeing (Saus et al. 1993).

3.3.2.3 Commercialisation of Dyeing with Supercritical CO_2

The mechanism of dyeing textiles using supercritical fluid was first documented in Germany in 1994. The energy and operating costs of the apparatus were much higher than traditional dyeing processes (Goñi et al. 2021). The circulation pumps and the heat transfer mechanism also posed several challenges in manufacturing the dyeing machinery. However, DyeCoo Textile Systems was founded in 2008 by its parent company, Feyecon, in collaboration with Stork and the Delft University of Technology. It was the first to manufacture a supercritical CO_2 dyeing machine. Tong Siang Co. Ltd., a division of the Yeh Group, in collaboration with DyeCoo, was the first polyester textile industry to set up a commercial supercritical CO_2 dyeing machine, branding the process as 'DryDye' (Tabata et al. 2001).

3.3.2.4 Advancements in Supercritical CO_2 Dyeing

The supercritical CO_2 dyeing was primarily designed to dye polyester fibre with disperse dyes. There were several limitations to dye natural and regenerated cellulosic fibre with other dyes using this method. The presence of hydroxyl groups in cellulosic fibres and amino groups in protein fibres make the natural fibres polar, which causes difficulties in dyeing using non-polar supercritical CO_2. Though other dyes behave similarly in supercritical CO_2 as in water, reactive dyes being polar causes the aggregation of dye particles on the fabric (Kraan 2005).

In recent years, researchers have tried to dye natural fibres with reactive dyes using supercritical CO_2 and reverse micelles or dispersing agents. The reverse micellar systems form stable micro-emulsions of water dispersed in the continuous supercritical phase (Tang et al. 2023). However, this involves the usage of fluorocarbons, which are, in turn, not environmentally friendly. Reactive disperse dyes are also used to dye natural fibres with supercritical CO_2 as the solvent. The dyes are formed by introducing specific reactive groups to CO_2 soluble disperse dyes, which allows improved dye transfer on polar fibres. Commercial textile processes using supercritical CO_2 are expected to have various benefits compared to the traditional methods of water-based colouration processes. Supercritical CO_2 is expected to be used in a process that can be successfully commercialised (Schmidt et al. 2003).

3.3.3 Dope Dyeing

Significant advancements in the extraction and manufacturing of polymeric fibres have flooded the world textile markets with polyester (PET), nylon, acrylic (polyacrylonitrile), and many more in the past decades, owing to its versatile attributes and cost-effectiveness (Nova-Institute 2018). In contradiction to these added functionalities, some processes of manufacturing these fibres pose severe threats to the environment and add up to the rate of carbon emissions. However, recent growth in sustainable and eco-friendly processes has inclined manufacturers towards environment-friendly production processes. The colouration of these fibres involves the introduction of the textiles into high-temperature dye baths. The treatment of the effluents from these dye baths is a concerning issue and is a significant factor in water pollution. On the contrary, bio-based polymers like polylactic acid (PLA) tend to degrade at higher temperatures (Balakrishnan et al. 2022).

Dope or solution dyeing has recently been widely practised as an alternative to traditional exhaust dyeing to overcome the problems associated with the conventional dyeing of synthetic fibres and meet sustainability parameters. The dope dyeing involves the introduction of pigment granules in the polymer solution resulting in the formation of coloured filaments, thus eliminating the dyeing process and reducing the probability of water contamination by industrial effluents (Hussain and Wahab 2018).

Dope Dyeing Mechanism

The mass colouration of polyester fibres by dope dyeing has been widely prevalent since the 1980s. Primarily, solution dyeing in earlier stages was well appreciated for dyeing dark shades, but with the advent of technology, it is possible to dye any desired colour with solution dyeing. The solution dyeing process is done using several principles depending on its effectiveness (Srivastava et al. 1987). Colourants, stable under high temperatures of about 260–290°C and high pressures of less than 1 torr, are mixed with the polymer melt to produce the coloured filaments. The cost of dyeing any shade using this technique is almost 5% of dyeing the fibres in the same shade with traditional disperse dyeing following the exhaust method (Rakshit 2000).

The dyeing of polyester chips with suitable disperse dyes is an alternative to conventional exhaust dyeing. The coloured polyester chips are dried and then melted and spun. This method is appropriate only for small batches and cannot produce a wide range of shades (Gurudatt et al. 2005). Moreover, the complete drying of the polyester chips is very difficult, resulting in the hydrolysis of the fibre and loss of strength.

The masterbatch addition method is the most widely used method to obtain solution-dyed fibre. In this method, the coloured granules are generated using polymers like polyethylene or low-melting polyester. These coloured chips are masterbatch pellets and contain about 10 to 50% of colourants. Polyethylene terephthalate can also be used to manufacture these coloured granules. These coloured granules are mixed with the white polymer chips in desired proportions and spun to get the coloured fibres (Srivastava et al. 2023). This method is much more effective than any other dope dyeing technique.

Masterbatch Addition Polymer Dyeing

All the ingredients are initially mixed homogeneously in the high-speed mixer at specific proportions and temperatures. This mixture includes pigments, polymer-soluble dyes, and the carrier polymer. Wetting agents, optical whitening agents, and other special additives are mixed into this process. The solution obtained from mixing is cooled at room temperature and then fed into the hopper of the twin-screw extruder (Barbas et al. 2014). After the extrusion, the coloured molten polymer is obtained from the die head of the extruder and placed into a water bath in the form of polymer strands. These polymer strands, when cooled, are cut into small chips using a pelletiser. This process is commonly referred to as pelletising. These pellets are stored and used with white polymer chips during melt spinning. Depending on the required shade, the amount needed for masterbatch is measured and dried at about 120°C. The dried pellets are then fed to the extruder mouth by the gravimetric K-Tron feeder. The pellets melt in the extruder and are mixed with the virgin polymer. After mixing, the coloured polymer is passed to the spinning system for fibre formation (Yücetürk and Seydibeyoglu 2024).

There are mainly two processes of masterbatch mixing with the virgin polymer, i.e. the joint feed process and the injection process. In the joint feed process, both the virgin polymer and the coloured chips are mixed in the extruder and then passed to spinning systems, whereas in the injection process, a separate extruder is used to melt the masterbatch granules. The molten granules are then injected into the molten virgin polymer coming out from the mainstream extruder using a dynamic stirrer installed at the mouth of the extruder (Sisilia 2024). The factors affecting the quality of masterbatch pellets include the dispersion of the colourants in the carrier polymer. The screw profile of the twin extruder and other process parameters has to be optimised to obtain better quality masterbatch pellets. During extrusion, the colourants and the additives can be added at any point between the feeding zone and the extruder outlet. The quality of the masterbatch is highly dependent on the appropriate position of mixing the elements in the polymer melt. The choice of carrier polymer used for mixing the colourants also plays an important role in determining the quality of the masterbatch pellets (Barbas et al. 2014). Polybutylene terephthalate (PBT) is considered to be the most high-quality carrier polymer due to its low sensitivity to hydrolytic degradation, high thermal stability and low melting point. The drying of the pellets before feeding them in the extruder is a crucial process since residual moisture can degrade the polymer (Srivastava et al. 2023).

Advantages of Dope Dyeing Over Conventional Dyeing

Conventional polyester dyeing causes most of the freshwater pollution and has been a growing concern in recent times. Dope dyeing serves as the most suitable alternative to the HTHP exhaust dyeing process as it eliminates the use of water. It also serves as an economical method of dyeing as it does not involve any post-dyeing processes. Moreover, dope-dyed fibres give excellent rubbing fastness and facilitate high production rates resulting in significant savings in utilities, energy, and manpower services. Texturisation of the yarns is also possible during dope dyeing without affecting the quality and shade (Srivastava et al. 1987).

3.3.4 Nanotechnology in Waterless Dyeing

The emergence of nanotechnology has occurred as the 'key' technology, the perspective of which has been proven to be the novel technique of developing a range of new value-added materials that involve textiles and polymers. This technology utilises materials that range from 1 to 100 nm in length. This allows for the engineering, synthesis, and alteration of the material characteristics to develop better materials in the next generation. The utilisation of nanotechnology contributes numerous functional qualities to textiles, which include strength, structure of the surface, smooth feel, durability, water repellency, fire retardancy, and antimicrobial qualities that can be created at the molecular level (Sawhney et al. 2008).

3.3.4.1 Dyeing with Silver Nanoparticles

The dyeing of cotton fibres can be performed by assembling anisotropic silver nanoparticles to examine the properties obtained after the dyeing process. The Localised Surface Plasmon Resonance (LSPR) characteristics of silver nanoparticles are responsible for developing several colours on the fabric. The morphology of the nanoparticles remains unchanged during the process of dyeing, with superior colourfastness to washing of cotton dyed with silver nanoparticles (Tang et al. 2012).

3.3.4.2 Dyeing with Silicon Nano-Micelles

The consumption of large volumes of water also takes place significantly in the conventional process of dyeing polyester fibres. The wastewater that is discarded after dyeing results in severe problems in the environment, resulting in the necessity for the development of sustainable dyeing techniques. The silicone waterless dyeing system is an emerging method to save energy and water, thus solving the problems associated with conventional dyeing. This method does not require a dispersing agent and does not need a reduction-clearing process (Chen et al. 2023).

3.3.4.3 Nano-Micelles

Micelles are spherical structures of diameters ranging from 5 – 100 nm containing hydrophilic as well as hydrophobic regions that act as carriers for various drugs to

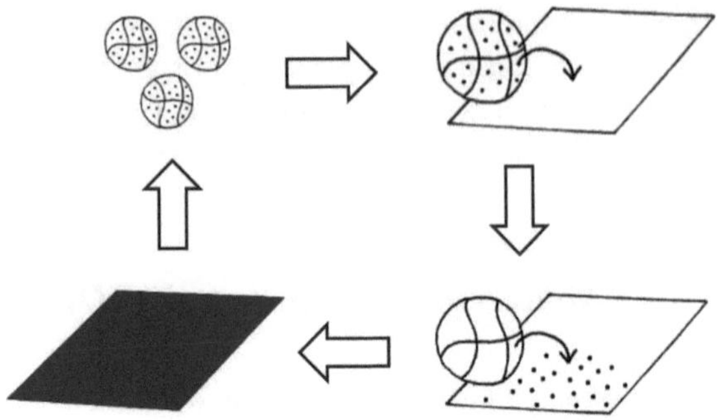

Fig. 3.2 Dyeing mechanism with silicone nano-micelles

the targeted regions. Nano-micelles exhibit specific unusual or unexpected properties depending on their size, solubility, customised surface, or environmental exposure, making them multifunctional and essential in various other domains, including biomedical applications (Bose et al. 2020).

Silicone nano-micelle dyeing can be carried out by preparing nano-emulsion through solvent evaporation composed of organic silicon oil that remains in its polymeric phase. Nano-micelles showed significant characteristics when used for dyeing polyester fabric. The dye remains highly dispersed in the prepared nano-emulsion and is gradually released from the nano-micelle in the dyeing process that gets uniformly distributed in the fibre. The yield of colour has been proven to be higher than that sample in which the conventional technique of the colouration process is carried out on polyester fabrics (Gao et al. 2018). Figure 3.2 illustrates the mechanism of dyeing polyester with silicone nano-micelles.

3.3.4.4 The Reverse Micelles

Researchers examined the reverse micelles using biodegradable non-ionic surfactants acting as building blocks (Tang et al. 2023). These reverse micelles can be utilised for the colouration of cotton fabric with reactive dye in a waterless octane medium. The process was optimised based on the molar ratio of surfactant to water as well as surfactant to co-surfactant, amount of dye and sodium carbonate and alkali to substrate, and finally the time and temperature of dyeing. The waterless dyed fabric samples showed better outcomes to colour strength, improved levelness, fastness, and physical properties without significant change in surface damage to the dyed fibre than those dyed using the conventional method. The Transmission Electron Microscope (TEM) showed the structure as spherical shaped with a core–shell arrangement, confirming its existence as a carrier of dye particles and the efficiency of the whole system (Wang et al. 2019a).

3.3.4.5 Nanocellulose-Based Dyeing

Nanocellulose-based cotton fibre dyeing is a promising approach to sustainable textile production with minimised water pollution. This method involves using nano-fibrillated cellulose (NFC) hydrogels, which are aqueous dispersions of the NFC pigment. The use of these hydrogels as a sustainable carrier of dye molecules results in a six-time reduction in the consumption of water (Kim et al. 2017). The conventional dyeing of cellulose with reactive dyes involves about 80 to 85% of the dye particles having a covalent bond to the primary –OH groups of the cellulose. The remaining dye and other additives are discharged into the waterbodies, which metabolise to form mutagenic products, leading to water pollution. Furthermore, the treatment of this wastewater is very costly (Ghosh and Netravali 2012).

NFC is a non-toxic dye carrier that can be obtained abundantly from various cellulosic substances and is effective for transferring reactive dye particles to cellulosic fibres, forming covalent bonds (Missoum et al. 2013). The NFC hydrogel is made by mechanically homogenising cellulose obtained from powdering chopped wood pulp. This process helps the defibrillation of cellulose into nanofibrils, resulting in the formation of a hydrogel. This NFC hydrogel is then dyed with reactive dyes under the traditional dyeing condition. The dyeing of the nano-fibrillated cellulose occurs when the hydroxyl groups present on the surface of the nano fibres react with the functional groups, i.e. the chlorotriazine and ethyl sulfonyl of the reactive dyes (Kim et al. 2017).

The presence of sodium sulphate and sodium carbonate at higher temperatures facilitates better dye absorption and fixation by the nanofibrils. When dyed, these nanofibrils act as coloured dye pigments, which are, in turn, applied over the surface of the cellulosic textile with the aid of a conventional deposition technique at the required temperature (Ghosh and Netravali 2012).

In the conventional reactive dyeing methods, the hydrolysed dyes are formed with water, whereas in NFC dyeing, these soluble dyes react on the surface of NFC, and with polycarboxylic acid applied to the fabric. Moreover, polycarboxylic acid post-treatment is effective in imparting a wrinkle-free effect on the fabric. Maleic acid and citric acid establish the cross-linkage and the entire cross-linking occurs in the presence of a poly-esterification catalyst (Liyanapathiranage et al. 2020).

3.3.5 Dyeing with Silicon Solvent Medium

Disperse dye can be used to colour polyester by using Decamethylcyclopentasiloxane (D5) as the solvent, maintaining optimum temperature and the dosage of dye acceleration. The dyed fabric showed comparable fastness properties and dye uptake to that of traditional water-based dyeing with a uniform dispersion on the textile substrate. The selection of the proper variety of disperse dye plays an important role as it affects various parameters like the uptake of dye particles, affinity, etc. The levelling and colour fastness properties achieved are remarkable in the case of the colouration process in silicon medium. Hence, it can be considered an environment-friendly solution that is capable of providing the required output of dyeing fabric (Cheng et al. 2021).

3.3.6 Plasma Dyeing Technology

The exploration of waterless processing of textiles to encourage green chemistry resulted in the development of plasma treatment that eliminates water consumption. This advancement in textile technology is also considered an efficient method of minimising waste. It is presented as an efficient alternative to traditional dyeing techniques and is also ecologically beneficial. Since there is no need for the use of solvents in this technique, it differs from conventional dyeing with water. It utilises much less time as well as the cost to pursue its effectiveness as a textile processing technique. The effectiveness of plasma technology depends upon certain aspects, such as the composition of substrates and the operational parameters of the treatment. It acts on the surface of the substrate without altering its characteristics (Yusuf and Shahid 2022). Figure 3.3 illustrates the interaction behaviour of plasma on the textile surface.

Plasma is considered to be the fourth state of matter that develops when a gas is ionised (Bogaerts et al. 2002). A gas usually has the property of an electrical insulator. However, disintegration occurs when the exposure of gas takes place at a high voltage that consequently conducts electricity. This is because the gas containing atoms or molecules that are neutral split into ions of positive charge and electrons of a negative charge, resulting in an ionised gas commonly referred to as a discharge or plasma. An ionised gas is popularly known as plasma, which consists of an equal amount of positive and negative charges, along with the presence of neutral active species and UV rays. Textile surfaces can be processed through this plasma and improvements in the development of plasma have led to the advancement of chemical processing in textiles (Graham 2007). Processing of textiles by plasma offers a promising future as a replacement for traditional, labour-intensive, sluggish, and environmentally harmful processes. Plasma application is ecologically friendly and cost-effective as it uses

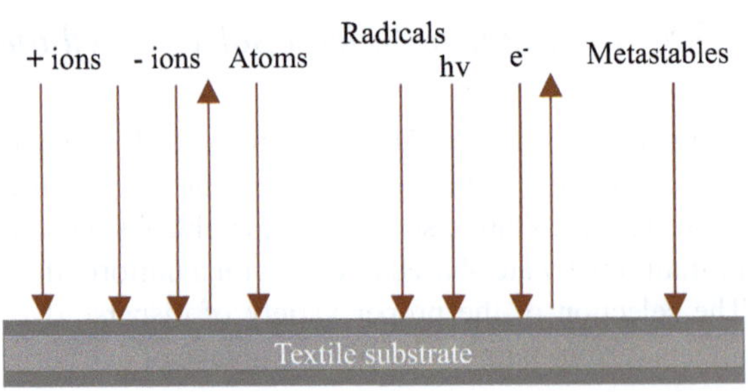

Fig. 3.3 Interaction of the species of plasma with the surface of a textile substrate

less energy and requires fewer processing steps. Additionally, textile finishes can be obtained with plasma treatment without affecting the essential textile characteristics. Various natural as well as synthetic fibres are being successfully treated with the help of plasma to impart different finishes. An increase in the efficiency of finishing and dyeing processes by a sustainable alternative for wet finishing is being effectively provided by plasma treatment (Ahmed 2012).

3.3.6.1 Categorisation of Plasma

Plasmas are classified into two categories viz., thermal and non-thermal according to the temperature of their working systems. These categories have certain merits and demerits. In thermal plasma, the heat is produced by an electric current through a conductor; as a result, thermal ionisation enables high power delivery at high operating pressures. Electrode complications, significantly high gas temperature, and the need for serious quenching lead to inadequate energy efficacy, limiting the application of hot plasma or thermal plasma (Kan 2014). On the other hand, cold or non-thermal plasma encourages higher energy efficiency to pursue the chemical reactions on the substrate. Since the operation can occur at low temperatures, no specific stimuli are required for chemical processing. The applications of cold plasmas have been investigated in various fields like controlling the rate of pollution, exclusion of volatile organic compounds (VOCs), controlling the emission rate of car exhaust as well as pre-treatment processes of polymer surfaces to enhance wettability, dye uptake, printability etc. Furthermore, plasma application has also gained popularity for generating ozone gas to purify water. The characteristic of plasma interacting with the polymers of textile substrates by modifying them without any changes to their bulk properties is significant (Nema and Jhala 2015).

3.3.6.2 Dyeing of Textile with Plasma

There have been consequent investigations and commercialisation regarding the development of thin films on the substrate by the activities of plasma species. The electronics industry utilised the technique that led to the generation of innovative products possessing notable surface properties (Tomasino et al. 1995). The process of dyeing with plasma medium can be pursued on textile fibres by various methods of plasma technology, namely, plasma polymerisation and vapour deposition.

Plasma Polymerisation

The novel technique of modifying the surface of the textile substrate by deposition of a thin layer or film of polymer is known as plasma polymerisation (Wertheimer and Schreiber 1981). This technique is advantageous due to several characteristics of the deposited films. The formation and deposition of a thin film of thickness up to 1 μm is a simple process. The nature of this developed film is remarkable in both physical and chemical aspects. Since they are comprised of structures with high cross-linking of pinholes, they can be employed as immensely active barriers. The establishment of layers can be done on several substrates like ceramics, glass, metals, etc. (Kan 2014). The polymerisation process with plasma is generally referred to as the simplest technique of obtaining efficient bonding of the deposited film and polymer of the substrate. However, particular difficulty prevails in this process of plasma polymerisation due to the complexity of the structure of the thin film. This complicated structure results in various parameters depending on which the film is generated, namely the structure, flow rate, frequency, pressure of the monomer, temperature of the substrate, level of power generated, etc. (Nema and Jhala 2015).

The polymerisation process takes place in generally two ways, one of which involves the activation of plasma initiating polymerisation of the monomers at the surface level of either liquid or solid substance. The structure of the monomer should meet specific conditions for this type of polymerisation process. Another technique comprises disruption of bonds that occurs by the reactive species, viz., electrons and others, as they contain sufficient energy to do so. The type of structure does not play any significant role in plasma state polymerisation as that of plasma-induced polymerisation. Organic compounds can be successfully utilised despite not having a structure that could be polymerised. Thus, the polymerisation rates are independent of the structure of every monomer (Wertheimer and Schreiber 1981).

Vapour Deposition

Vapour deposition has shown great potential in various domains of scientific studies, such as electronics, medical diagnosis, and material engineering, and also in textiles. Several novel properties can be achieved through this plasma-enhanced vapour deposition technique on the textile substrate and different forms of nanostructure of various designs can be produced on woven, nonwoven, knitted, and even fibres (Chan et al. 1996).

The vapour deposition technique comprises the method of condensation that the materials undergo after being in a vapour state. Besides the condensation process, the conversion of the vapour state of the materials is also carried out by chemical reactions to produce solid materials. This process changes the substrates in various aspects, such as thermal, mechanical, optical, electrical, corrosion resistance, and wear properties (Miśkiewicz et al. 2020). Vapour deposition processes carried out in a vacuum chamber occur in two ways. Vapour deposition in physical form involves a coating material of a thin layer of 10 microns that is condensed on the substrate in the vacuum chamber, whereas in the case of chemical vapour deposition, a layer of 1000 microns is deposited on the substrate by condensation. In this way, metals and non-metals can be deposited depending on the techniques utilised. The deposition process is also selected depending on the thickness of the coating film, reducing any undesirable interactions with voids and quickly adjusting the film's composition (Chan et al. 1996). Excellent performance and quality solid materials are produced via vapour deposition through a chemical process. The technique has several uses in a wide range of industries. Using this technique, volatile precursors on the substrate surface can react or break down to form a suitable coating. Several by-products produced by this process are removed through the chamber using a gas flow technique (Shahidi et al. 2013).

3.3.7 Dyeing Technology

The foam dyeing technique evolved and gained popularity as an energy-efficient and low-liquor process (Zhu et al. 2022). The emergence of foam dyeing technology has enabled the treatment of 1100 m of pre-treated fabric with only 400 L of dye solution, with no need to wash the dyed fabrics, resulting in the saving of a minimum of 90% of energy and water. This method is used in textile industries for pre-treatment, dyeing, and finishing, offering a cost-effective alternative to traditional methods (Wang et al. 2019b).

3.3.7.1 Characteristics of Foam

Foam is a gas bubble-like substance that can be used in the foam dyeing system to replace traditional padding processes. This sustainable technique maintains a controlled dye solution in the fabric, preventing excess materials from absorbing. However, it has drawbacks such as high thermal drying costs, dye uniformity, and chemical dispersal across the fabric. This can reduce the fabric's durability to wrinkles and shrinkage after washing (Elbadawi and Pearson 2009). Foam wets a textile substrate slowly, spreading over the fabric and breaking down. Its formation and collapse during dyeing cause bubbles to break and transfer dye to the fabric. After penetration, fixation, and washing are performed using conventional methods, ensuring effective dyeing and fixation (Capponi et al. 2008).

Foam dyeing requires constant maintenance of homogeneous medium, bubble size, and fineness, evenly distributed with appropriate application methods. The collapse of foam is an immediate requirement for dye fixation and can be done in batches or continuously (Elbadawi and Pearson 2009). Figure 3.4 depicts the process of foam dyeing through padding.

The process of foam dyeing began in 1930 and was patented in the US. It replaced water in batch dyeing with foam in 1972. The introduction of foam dyeing took place in the textile industry in 1979, with the dyeing of carpets being the first trial conducted successfully in 1981 (Capponi et al. 2008).

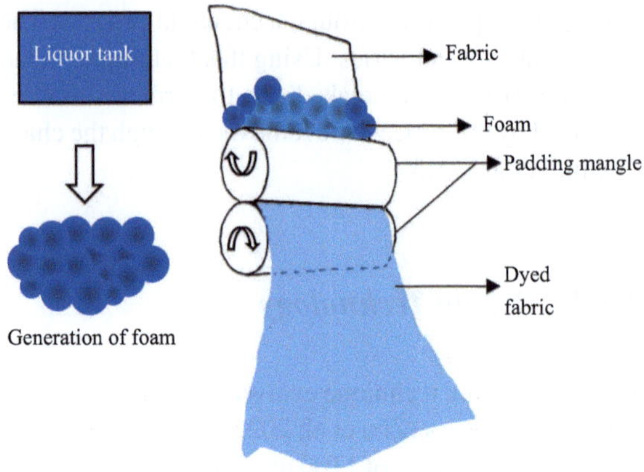

Fig. 3.4 Foam dyeing technology through padding process

3.3.8 Application of Natural Dye through a Waterless Medium

The importance of utilising natural dyes as an eco-friendly alternative is growing gradually due to the increase in pollution by the effluent generated after dyeing with synthetic dyes. Though synthetic dyes gained popularity due to their wide variations of colours, excellent fastness, and cost-effectiveness, their consequences for the ecology and health of humans result in awareness of a healthy and environment-friendly solution. Moreover, the value-added quality, i.e. characteristic of providing various functionalities like protection from harmful ultraviolet rays, resisting bacterial attack, repelling insects, antioxidants, etc., have significantly caused the development of natural dyes. This encourages researchers and experts to investigate advanced processes to implement dyes and pigments obtained from natural sources without or with minimal water utilisation to decrease the detrimental effect on the environment (Cai et al. 2023).

3.3.9 Application of Microbial Fermented Nano-Micelles

Prodigiosin, the secondary metabolite of a microbe called *Serratia macescens*, was produced by conducting microbial fermentation. The pigment of prodigiosin has been utilised to develop dyed cotton fabrics and generate antimicrobial properties without affecting the morphology of the dyed fabric (Gong et al. 2017).

3.4 Impact of Waterless Dyeing

Waterless dyeing technology, which involves zero or very little water use during the dyeing process, has gained momentum in the textile industry due to its environmental benefits. Conventional dyeing processes are water-intensive and generate polluted wastewater. In contrast, waterless dyeing offers several advantages. It eliminates chemicals and ensures no effluent (Gassert et al. 2015). Dyed products gain a commercial advantage, appealing to environmentally conscious consumers. The textile industry is increasingly aware of the environmental impact of traditional dyeing processes. Water scarcity, pollution, and energy consumption drive the need for sustainable alternatives, and researchers, manufacturers, and startups actively develop and promote waterless dyeing technologies. Several brands like Nike, H&M, IKEA, etc. (Vogue 2024). have initiated pilot projects to test waterless dyeing methods. These trials assess feasibility, quality, and scalability. Certifications like Bluesign and OEKO-TEX recognise these sustainable practices of waterless dyeing. Brands prioritise certified suppliers and encourage the adoption of this technology. Challenges remain, including initial implementation costs and industry-wide adoption, but waterless dyeing continues to grow as a sustainable alternative (Gassert et al. 2015).

3.5 Market Adoption

Nike, recognising the potential for sustainability, partnered with DyeCoo in 2012 to implement waterless dyeing in one of their Taiwanese factories. Three other Taiwanese factories (supplying brands like Adidas) are also investing in the technology, and two machines produce 920,000 kg of fabric annually, resulting in an impressive 8,256,000 m^3 reduction in water withdrawals compared to conventional dyeing methods (Graham 2007).

3.6 Conclusion

Implementing waterless dyeing technologies requires an upfront investment. Manufacturers must adopt these methods, which can be costly for the initial expenses of setting up the machinery. Despite the initial investment, waterless dyeing ultimately brings down production costs. Reduced energy and chemical usage lead to approximately 30 to 50% cost savings. Encouraging all textile producers to embrace waterless dyeing methods remains a challenge. The dyeing of fibres other than manufactured fibres is quite challenging for waterless dyeing techniques. Major global brands play a crucial role in promoting waterless dyeing. By emphasising the environmental credentials of fabrics produced using this technology, they encourage other factories and brands to adopt it. The difficulty of CO_2 in swelling the hydrophilic natural fibres is one reason that often reduces the degree of diffusion of the dye molecules, which further results in poor colouration. Further research is necessary to ensure that waterless dyeing processes maintain product quality and environmental sustainability over time. However, several technological advancements in waterless dyeing have made it as efficient as possible. The dyeing of natural fibres faces many challenges when subjected to the waterless technique of dyeing. Various manufacturers are developing advanced machinery, making waterless dyeing applicable to various dyes and fibres on a pilot scale. Dyes extracted from plant extracts use enormous quantities of water, which can effectively be reduced or eliminated by dyeing polyester fabrics using supercritical CO_2 fluid dyeing.

The revolution in textile dyeing through waterless techniques represents a pivotal shift towards a more sustainable and efficient industry. Embracing innovation and collaboration will drive the continued transformation of textile dyeing processes.

Declaration The authors declare that they have no known competing financial interests or personal relationships that could have appeared to influence the work reported in this chapter. Generative AI and AI-assisted technologies have been used in the writing process to improve the readability and language of the work only. After using this tool, the authors carefully reviewed and revised the content to ensure accuracy and quality. The authors take full responsibility for the final version of this publication.

References

Abate MT, Tadesse MG (2021) Innovative and emerging technologies for textile dyeing and finishing. Scrivener Publishing LLC, Ethiopia

Agrawal BJ (2015) Supercritical carbon-dioxide assisted dyeing of textiles: an environmental benign waterless dyeing process. Int J Innov Res Creat Technol 1:201–202

Ahmed HAM (2012) Plasma treatment in textile substrates—a review. J Eng Fibers Fabr 7:1–7

Bach E, Cleve E, Schollmeyer E et al (2002) Past, present and future of supercritical fluid dyeing technology—an overview. Rev Prog Color 32:88–102

Balakrishnan NK, Siebert S, Richter C, Groten R, Seide G et al (2022) Effect of colorants and process parameters on the properties of dope-dyed polylactic acid multifilament yarns. Polymers 14:5021

Barbas JM, Machado AV, Covas JA (2014) Processing conditions effect on dispersion evolution in a twin-screw extruder: polypropylene-clay nanocomposites. Chem Eng Technol 37(2):257–266

Bogaerts A, Neyts E, Gijbels R, Mullen JVD et al (2002) Gas discharge plasmas and their applications. Spectrochim Acta, Part B 57:609

Bose A, Burman DR, Sikdar B, Patra P et al (2020) Nanomicelles: types, properties and applications in drug. IET Nanobiotechnol 15:19

Cai Y, Xiao L, Ehsan MN, Jiang T, Pervez MN, Lin L, Stakelbeck HP, Tschudin P, Naddeo V et al (2023) Green penetration dyeing of wool yarn with natural dye mixtures in D5 medium. J Market Res 22:6524–6541

Capponi M, Flister A, Hasler R, Oschatz C, Robert G, Robinson T, Stakelbeck HP, Tschudin P, Vierlina JP et al (2008) Foam technology in textile processing. Rev Prog Coloration 12:48–57

Chan CM, Ko TM, Hiraoka H et al (1996) Polymer surface modification by plasmas and photons. Surface Sci Rep 24:1–54
Chen J, Pei L, Shi W, Yi J, Wang J et al (2023) Effect of dispersant on disperse dyeing in silicone waterless dyeing system. Polymers 15:1–12
Cheng W, Pei L, Saleem MA et al (2021) Sustainable pilot scale disperse dyeing in a waterless silicone medium dyeing system for no discharge of wastewater. J Clean Prod 321:1–11
Elbadawi AM, Pearson JS (2009) Foam technology in textile finishing. Text Prog 33:1–31
Gao A, Hu L, Zhang H, Fu D, Hou A, Xie K et al (2018) Silicone nanomicelle dyeing using the nanoemulsion containing highly dispersed dyes for polyester fabrics. J Clean Prod 200:48–53
Gassert F, Reig P, Shiao T, Luck M, Landis M et al (2015) Waterless dying technology in textile processing. Taiwan
Ghosh TD, Netravali AN (2012) "Green" crosslinking of native starches with malonic acid and their properties. Carbohydr Polym 90:1620–1628
Gong J, Ren Y, Fu R, Li Z, Zhang J et al (2017) PH-Mediated antibacterial dyeing of cotton with prodigiosins nanomicelles produced by microbial fermentation. Polymers 9:1–3
Goñi LM, Gañán NA, Martini RE et al (2021) Supercritical CO2-assisted dyeing and functionalization of polymeric materials: a review of recent advances (2015–2020). J CO2 Util 54:101760
Graham WG (2007) The physics and chemistry of plasmas for processing textiles and other materials. In Shishoo R (ed) Plasma technologies for textiles, Woodhead publishing, New York
Gunay M (2013) Eco-friendly textile dyeing and finishing. INTECH open science
Gurudatt K, De P, Rakshit AK, Bardhan MK et al (2005) Dope-dyed polyester fibers from recycled PET wastes for use in molded automotive carpets. J Ind Text 34:167–179
Hussain T, Wahab A et al (2018) A critical review of the current water conservation practices in textile wet processing. J Clean Prod 198:806–819
Kan C (2014) A novel green treatment for textiles: plasma treatment as a sustainable technology. In: Cann MC (ed) Sustainability: contributions through science and technology, CRC Press, Boca Raton
Kim Y, McCoy LT, Lee E, Lee H, Saremi R, Feit C, Hardin IR, Sharma S, Mani S, Minko S et al (2017) Environmentally sound textile dyeing technology with nanofibrillated cellulose. Green Chem 19:4031–4035
Knez Ž, Lütge C et al (2023) What are supercritical fluids? In: Product, process and plant design using subcritical and supercritical fluids for industrial application. Springer, Cham
Kraan, MVD (2005) Process and equipment development for textile dyeing in supercritical carbon dioxide. Dissertation, Delft University of Technology
Lara L, Cabral I, Cunha J et al (2022) Ecological approaches to textile dyeing: a review. Sustainability 14:1–17
de Luna LAV, da Silva THG, Nogueira RFP, Kummrow F, Umbuzeiro GA et al (2014) Aquatic toxicity of dyes before and after photo-fenton treatment. J Hazard Mater 276:332
Liyanapathiranage A, Peña MJ, Sharma S, Minko S et al (2020) Nanocellulose-based sustainable dyeing of cotton textiles with minimized water pollution. ACS Omega 5:9196–9203
Madhav S, Ahamad A, Singh P, Mishra PK et al (2018) A review of textile industry: Wet processing, environmental impacts, and effluent treatment methods. Environ Qual Manage 27:31–41
Mahmud I, Kaiser S (2020) Recent progress in waterless textile dyeing. J Text Sci Eng 10:1–4
Miśkiewicz P, Frydrych I, Cichocka A et al (2020) Application of Physical Vapor Deposition in Textile Industry. Autex Research J 22:1–7
Missoum K, Belgacem MN, Bras J et al (2013) Nanofibrillated cellulose surface modification: a review. Materials (Basel) 6:1745–1766
Nema SK, Jhala PB (2015) Plasma technologies for textile and apparel. CRC Press, New Delhi
Nova-Institute announces new market study on bioplastics with latest data for 2018 (2018). https://renewable-carbon.eu/news/nova-institute-announces-new-market-study-on-bioplastics-with-latest-data-for-2018/. Accessed 18 Apr 2024

Rakshit AK (2000) Reliance industries introduces recron dope-dyed blackpolyester fibers. Int Fiber J 15:72–74

Saus W, Knittel D Schollmeyer E et al (1993) Dyeing of textiles in supercritical carbon dioxide. Text. Res. J. 63:135

Sawhney A, Condon B, Singh K, Pang S, Li G, Hui D et al (2008) Modern applications of nanotechnology in textiles. Text Res J 78:731

Schmidt A, Bach E, Schollmeyer E et al (2003) The dyeing of natural fibres with reactive disperse dyes in supercritical carbon dioxide. Dyes Pigm 56:27–35

Shahidi S, Ghoranneviss M, Moazzenchi B, Rashidi A, Dorranian D et al (2013) Study of surface modification of wool fabrics using low temperature plasma. In: Proceedings of the 3rd international conference on the frontiers of plasma physics and technology, Islamic Azad University, Iran

Sicardi S, Manna L, Banchero M et al (2000) Comparison of dye diffusion in Poly (ethylene terephthalate) films in the presence of a supercritical or aqueous solvent. Ind Eng Chem Res 39:4707

Sisilia (2024) Inside the masterbatch manufacturing process: behind fascinating color and performance. In: Masterbatch Global. https://masterbatchglobal.com/articles/inside-the-masterbatch-manufacturing-process. Accessed 24 Oct 2024

Srivastava JK, Narasimhan KV, Vaidya AA et al (1987) Mass-coloration of polyester. Textile Asia 8:142–147

Srivastava JK, Ramesh R et al (2023) Mass coloration of polyester fiber—emerging opportunities for green Sustainable dyeing technology. Eur Chem Bull 1182–1191

Tabata I, Lyu J, Cho S, Tominaga T, Hori T et al (2001) Relationship between the solubility of disperse dyes and the equilibrium dye adsorption in supercritical fluid dyeing. Color Technol 117:346

Tang B, Zhang M, Hou X, Li J, Sun L, Wang X et al (2012) Coloration of cotton fibers with anisotropic silver nanoparticles. Ind Eng Chem Res 51:12807

Tang YL, Jin S, Lee CH, Law HS, Yu J, Wang Y, Kan C et al (2023) Reverse micellar dyeing of cotton fabric with reactive dye using biodegradable non-ionic surfactant as nanoscale carrier: an optimisation study by one-factor-at-one-time approach. Polymers 15:1–23

Tomasino C, Cuomo JJ, Smith CB, Oehrlein G et al (1995) Plasma treatments of textiles. J Coat Fabr 25:115–127

Varma AK, Ghosh S et al (2020) Supercritical carbon dioxide: a glimpse from the modern era of green chemistry. In: Inamuddin AA (eds) Advanced nanotechnology and application of supercritical fluids. Nanotechnology in the life sciences. Springer, Cham

Vogue (2024) Waterless dyeing techniques new innovation in textile processing. In: Vogue Institute. https://www.voguefashioninstitute.com/waterless-dyeing-techniques-new-innovation-in-textile-processing. Accessed 24 Oct 2024

Wang Y, Tang Y, Lee C, Kan C et al (2019a) Reverse micellar dyeing of wool fabric with reactive dyes. Fibers Polym 20:2367–2375

Wang Q, Zhou W, Du S, Xiao P, Zhao Y, Yang X, Zhang M, Chang Y, Cui S et al (2019b) Application of foam dyeing technology on ultra-fine polyamide filament fabrics with acid dye. Text Res J 89:4808–4816

Wertheimer MR, Schreiber HP et al (1981) Surface property modification of aromatic polyamides by microwave plasmas. J Appl Polym Sci 26:2087–2096

Wu T, Han B et al (2019) Supercritical carbon dioxide (CO2) as green solvent. In: Han B, Wu T et al (eds) Green chemistry and chemical engineering. Encyclopedia of sustainability science and technology Series, Springer, New York.

Yücetürk M, Seydibeyoglu MO et al (2024) Bio-based polyethylene masterbatch preparation and investigation of the effect of process conditions on pigment dispersion. Color Technol 1–11

Yusuf M, Shahid M (2022) Emerging technologies for textile coloration. CRC Press, Boca Raton

Zaidy SS, Vacchi FI, Umbuzeiro GA et al (2019) Approach to waterless dyeing of textile substrates–use of atmospheric plasma. J Ind Eng Chem 58:18478–18487
Zhu D, Wan Z, Zhao X, Liao S et al (2022) Foaming indigo: an efficient technology for yarn dyeing. Dyes Pigm 197:109862

Chapter 4
Emerging Trends in Solvent-Based Textile Dyeing

Adrian Kreuz, Priscila Aparecida Horn, Laís Feltrin Sidou, Omar A. El Seoud, and Jürgen Andreaus

Abstract As a part of fiber wet processing, dyeing is considered one of the most polluting stages in the textile industry, consuming high amounts of fresh water and chemicals and generating large volumes of effluents loaded with salts, residual dyestuff, and other dyeing auxiliaries. Different approaches have been made to reduce both water consumption and wastewater output. Replacing water as a dyeing medium is a logical solution. Due to the development of new solvents, this strategy has gained increasing attention. Starting with a general insight into problems and challenges of textile dyeing, this chapter gives an updated review on dyeing using non-aqueous solvents from chlorinated solvents to the latest trends involving neoteric solvents such as ionic liquids and deep eutectic solvents, amongst others.

Keywords Non-aqueous dyeing · Solvent dyeing · Ionic liquids · Deep eutectic solvents · Silicon solvents · Alcohols · Reverse micellar systems

A. Kreuz · P. A. Horn · L. F. Sidou · J. Andreaus (✉)
Department of Chemistry, Regional University of Blumenau, Blumenau, Brazil
e-mail: jandr@furb.br

A. Kreuz
e-mail: adriankreuz@yahoo.com.br

P. A. Horn
e-mail: phorn@furb.br

L. F. Sidou
e-mail: lfsidou@rbox.me

L. F. Sidou
Department of Museology, Conservation and Restoration, Federal University of Pelotas, Pelotas, Brazil

O. A. E. Seoud
Institute of Chemistry, University of São Paulo, São Paulo, Brazil
e-mail: elseoud.usp@gmail.com

S. Maiti et al. (eds.), *Sustainable Coloration Techniques in Textiles*,
https://doi.org/10.1007/978-981-96-4975-4_4

4.1 Introduction

The global textile market size was valued at USD 1,837.27 billion in 2023, and an increase in this value is estimated at a compound annual growth rate (CAGR) of 7.4% in terms of revenue from 2024 to 2030 (Grand View Research 2024). While total textile fiber production reached 113,800 metric tons in 2022, 87,600 metric tons (77%) are due to chemical fibers, such as polyesters, polyamides, and manmade cellulosic fibers (Statista 2024). Among natural fibers, cotton is the most expressive, thanks to its properties such as high resistance, body humidity absorption, and color retention, being responsible for over 38% share of overall revenue in 2023 (Grand View Research 2024). However, the most consumed textile fiber is synthetic, namely polyethylene terephthalate, which is mostly simply referred to as polyester.

As well as the enormous economic importance of the textile sector, the ecological impact is also considerable. In the European Union, the estimated textile per capita consumption requires *ca.* 400 m^2 of land, 9 m^3 of water, and 391 kg of raw materials, and results in a carbon footprint of 270 kg of CO_2. In this way, the textile sector was the third largest source of water degradation and land use in 2020 (European Parliament 2024). While each European buys per year almost 26 kg of new textiles and discharges 11 kg of textile waste, only 1% of used clothes are recycled into new clothes (Gersmann 2024).

Textile properties are mostly defined by the intrinsic properties of the fiber polymer and during chemical wet processing, which is responsible for most of the negative environmental impact of the textile industry because of the consumption of elevated amounts of water, energy, and chemicals and the generation of high volumes of polluted effluents.

Depending on the textile´s structure, its wet processing may include desizing, scouring, bleaching, mercerization, dyeing, and finishing steps, and as suggested by its name, makes use of a solvent, which is traditionally water (Burkinshaw 2016a, b, c). Under certain circumstances, which will not be further discussed here, textile coloring can be carried out without the use of any solvent, e.g., as in mass dyeing of textile fibers, where the dye or pigment is added to the molten fiber polymer during melt spinning, or during transfer printing, where the dye is transferred from colored paper to the textile by dye sublimation at elevated temperature. Mass dyeing of textile fibers is usually limited to a few colors only, mainly because in large production facilities, the change from one color to another is time-consuming and costly, while transfer printing is limited to synthetic fibers, mainly polyester, and requires the preparation of the transfer paper, which is also expensive (Broadbent, Dyers et al. 2001).

Important functions of the solvent are (Burkinshaw 2016a, b, c):

1. **Wetting and swelling of fibers** to allow transport by diffusion and fixation of the dyes inside the fibers;
2. **Dissolving or dispersing** dyes and auxiliaries, to guarantee homogenous solutions or uniform dispersion (leveling) during wet processing and after washing to

remove unfixed dyes, auxiliaries, and impurities from the fibers after the dyeing or finishing process;

3. **Plasticization of fiber polymers**: solvent molecules contribute to break or weaken intermolecular bonds between fiber polymer chains, decreasing thus the glass transition temperature, T_G, which results in changes in fiber properties such as rigidity, hydrophobicity, and modulus, and permits lower dyeing temperatures;
4. Medium for heat and mass transfer during wet processing.

At the end of the wet processing of textiles, large volumes of water contaminated with starch, waxes, pectin, chemicals such as oxidizing/reducing agents, salts, heavy metals, dyes, resins, and sulfur and phosphate-rich species high in COD (Chemical Oxygen Demand), and BOD (Biological Oxygen Demand) are generated (Senthil Kumar and Grace Pavithra 2019). For comparison, the amount of water involved in the production of a single cotton T-shirt, 2700 L, is enough to meet a person's drinking needs for 2.5 years (European Parliament 2024).

Considering the growing demand for natural resources, and in line with the increasingly alarming climate changes, world leaders and the scientific community have made efforts to seek more sustainable production processes (United Nations 2023). Some of the 17 objectives of sustainable development of the United Nations are related to sustainable economic growth and attention to climate change (Biermann et al. 2022; United Nations 2023). Also, according to the principles of Green Chemistry, industrial activities involving chemical processes are guided to eliminate the use of solvents and auxiliaries whenever possible; minimize the generation of hazardous substances; and should prioritize processes with energy efficiency (Anastas and Warner 1998; Poliakoff and Licence 2007).

Considerable efforts have been made to make environmentally friendly methodologies viable, such as using agro-industrial waste, incorporating the textile industry into the concept of biorefineries; or using enzyme-based technologies, reducing the use of chemicals and operating under milder conditions, as well as generating less waste (Shahid et al. 2025). From a broad perspective, employing technologies that replace or minimize water consumption in textile processes can reduce the pressure on aquatic ecosystems, mitigate the effects of climate change, as well as reduce production costs, and increase production efficiency.

4.2 Basics and Challenges of Textile Dyeing

Textiles are much more than merely protecting materials against cold, heat, dirt, etc., and their aesthetic appeal is closely related to color and coloration. The dyeing stage involves several physico-chemical processes (Burkinshaw 2016a, b, c). Dyeability depends on the interaction between fabric and dye, and the dyeing mechanisms, as well as the classes of dyes used. The latter are related to the specific characteristics of each fiber, such as its polarity and the presence of certain functional groups (Broadbent et al. 2001). A fundamental parameter in the dyeing process is the solubility of the dye and its partition coefficient between the solvent and the fiber (Burkinshaw 2016a, b, c). The dyeing process involves the transfer of dye molecules from bulk solution to the textile fiber, initially by an adsorption process. The dye must be sufficiently soluble in the solvent/fluid phase to avoid aggregation, but it should be less soluble than in the fiber; otherwise, dyeing will be inefficient. If its affinity (or the chemical potential μ) for the solvent is greater than for the fiber, the dye tends to remain in solution rather than moving into the fiber (Ad 2022). Less dye-to-fiber adsorption and lower dye-fiber fixation result in less intense colors, reduced washing fastness, and higher dye loads in the discharged dyeing effluents.

On the other hand, dyes with low solubility in the solvent tend to form aggregates, which may lead to the breakage of dye dispersions, cause precipitates, and hinder transport and penetration into textile fibers. The undesired results are uneven dyeing and the formation of spots on the textile material. Optimal solvent solubility allows the dye to move efficiently from the dyeing bath to the fiber, permitting its adsorption on the fiber until saturation and resulting in good fixation, intense colors, and satisfying fastness (Ad 2022).

Cotton, bast fibers (flax, hemp, ramie, and jute), and man-made cellulosic fibers (viscose rayon, modal or High-Wet Modulus rayon, lyocell) are mostly dyed with reactive dyes, in aqueous solution, in the presence of auxiliary chemicals such as inorganic salts (NaCl, Na_2SO_4), alkali, amongst others. The function of electrolytes is to reduce electrostatic repulsion between anionic dye molecules and the negatively charged cellulose fiber surface. The addition of salts (such as NaCl) provides Na^+ cations that are chemically adsorbed onto the negatively charged cellulose surface, facilitating binding of the reactive dye and increasing dye bath exhaustion. Reactive dyes bind covalently to cellulose –OH groups via nucleophilic substitution or addition reactions, that are favored by high temperatures and alkaline conditions (Broadbent et al. 2001).

Alkalis, e.g., Na_2CO_3 or NaOH are needed for pH adjustment and as catalyst for dye-fiber bonding, and besides, depending on the dye type used (e.g. vinyl sulphonic dyes), they are essential for dye activation. Other auxiliaries such as surfactants, dispersants or anti-crease agents, amongst others, may also be added.

However, alkaline pH (above 10) and high temperature favor dye hydrolysis, a secondary, unwanted reaction of reactive dyes with water, which reduces the fixation yield and requires post-washing to remove unfixed dye molecules. Reactive dyes display high washing fastness due to covalent bonding; however hydrolyzed dye molecules that remain adsorbed to the fiber are responsible for lower washing or rubbing fastness and need to be removed by appropriate post-washing.

Process selection may reduce the consumption of water and inorganic salts during dyeing. Application techniques that permit to reduce the amount of water and electrolytes used in dyeing are working at low liquor ratios, that is the amount of liquid used per kg of textile (L/kg), such as impregnation followed by squeezing using a padding machine, e.g. cold pad-batch, pad-steam, pad-roll, pad-dry, vapor-phase dyeing (thermosol), or techniques with process intensification using ultrasound.

Among the main challenges for more sustainable textile dyeing are:

- Reduced water (or solvent) consumption, or solvent-free dyeing
- Reduced use of chemicals, including salt and dyeing auxiliaries
- Lower energy consumption (cold dyeing)
- Less effluent generation (or: solvent-free dyeing)
- Less polluted effluents
- Reduced dye hydrolysis in reactive dyeing
- Increased dye exhaustion and fixation

Nonaqueous solvents, in addition to their environmental appeal, may have the advantage that reactive dyes undergo less hydrolysis in the absence of water; and therefore, reduced dye loss. However, the reaction with the selected solvent (solvolysis) must be evaluated.

The requirements for ideal textile dyeing media depend on the specific textile material, the polymer structure of the fiber, and the available dyes or pigments for this material. The properties of a universal ideal dyeing bath are as follows:

1. Inert under dyeing conditions; nonreactive
2. Nontoxic
3. Recoverable and reuseable
4. Low boiling point
5. Low viscosity
6. Low heat capacity
7. Sufficient solubility for the dyes
8. Good balance between solubility for the dye and the dye solubility in the fiber
9. Minimal, or non-existent hydrolysis (solvolysis) of reactive dyes
10. Reduces the T_G of fiber polymer to permit lower dyeing temperatures
11. Permit dyeing with reactive dyes with lower load of electrolyte and at nearly neutral pH
12. Reduces the need for other dyeing auxiliaries

4.3 Chlorinated Organic Solvents

The first efforts to replace water in dyeing processes used chlorinated organic solvents, in particular, perchloroethylene (PCE) or tetrachloroethylene, which was already used in dry cleaning (Burkinshaw 2016a, b, c; Tang and Kan 2020; Alebeid et al. 2021; Ceballos, Fellows et al. 2021; Pawar et al. 2022; National Center for Biotechnology Information 2024). Its advantages include relatively low production cost, stability, non-flammability, lower energy requirements for vaporization, faster drying of the impregnated fabric due to the lower boiling point of the solvent, and better fiber wettability (thanks to its low surface tension). PCE also swells synthetic polymers relatively easily. In the case of polyester (PET), for example, the solvent decreases the glass transition temperature from 85 to 56 °C, enabling, in principle, faster dyeing with disperse dyes (Gebert 1971).

However, due to the high solubility of disperse dyes in PCE, and considering that the partition coefficient between the solvent and the fiber determines the dyeability (Burkinshaw 2016a, b, c), low dye transfer to the polyester fiber during dyeing in this solvent is observed. The color fixation increases strongly in the presence of water: The addition of just 1% water increases color yield by 430%. This effect is attributed to the decrease in the solubility of the dye in the liquid phase; as well as to the formation of H-bond interactions between dye and PES polymer chains (Gebert 1971). It has also been suggested that water plays a role in reducing the glass transition temperature of the polymer, in dyeing studies on polyester fiber in PCE (Roberts and Solanki 2008).

Despite the efficiency of dyeing synthetic fibers in this medium, a significant fraction of the solvent is absorbed by the substrate during the dyeing process, leading to solvent losses, lower lightfastness of dyed polyester, as well as a risk to the consumer's health, considering the toxicity and potentially carcinogenic nature, which represents risks to both humans and nature (Ceballos, Fellows et al. 2021; National Center for Biotechnology Information 2024).

Especially for disperse dyes and dyes that do not form covalent bonds with the textile fiber, a general rule may be formulated as follows: Dyes, which penetrate easily into the fiber polymer structure, also leave or go off easily. The presence of a solvent that favors the diffusion of the dye into the fiber also contributes to the inverse movement, so it is very important to remove residual solvents after dyeing.

A system based on EtOH/CCl_4/water in the proportion 55:40:5 (v/v) was used in the dyeing of cotton yarns previously treated with Na_2CO_3 (Wang, Xia et al. 2020). Six reactive dyes of the chlorotriazine and vinyl-sulfone classes were evaluated. Higher dye particle sizes and lower fiber zeta potential are observed in the EtOH/CCl_4/water system compared to the water medium; which may be related, respectively, to the involvement of hydrogen bonds between ethanol and water, leaving fewer water molecules available for dissolving the dyes; and to the lower dielectric constants and surface tension of the EtOH/CCl_4/water mixture. As a result, dye aggregation is facilitated, without the need for salts, and without compromising the homogeneity of dyeing. Low water content leads to less hydrolysis of dyes. It is worth noting that total fixation (>80%) in the first 10 min is similar to the conventional method in 60 min. This faster dyeing is related to better fiber wettability of the solvent mixture. The authors argue that the higher cost of the method compared to the conventional method is offset by its more environmentally friendly, salt-free nature and reduced wastewater discharge. Solvents can be recovered at a later stage.

A similar ternary system was used for wool dyeing with three reactive dyes (Wang et al. 2021), in this case EtOH/CCl_4/water in the proportion 45:50:5 (v/v). Higher K/S, and improved mechanical properties compared to conventional dyed samples were obtained by this method.

4.4 Paraffin

Liquid paraffin is a colorless, non-volatile liquid derived from petroleum and composed mainly of alkanes (IFA), which is used in cosmetics and in medical applications (Sharif et al. 2001). This solvent has been investigated as an alternative for dyeing polyester and cotton textiles.

In the case of polyester, the dyeing affinity of dispersed dyes in liquid paraffin is much lower than in aqueous medium, due to the greater solubility of the dye in paraffin (An et al. 2022). Therefore, like what was observed for chlorinated solvents and reported for silicon solvents (see Sect. 4.5), it is necessary to add small amounts of water to reduce the solubility of dyes favoring its tendency to migrate to the fiber surface and improve dyeing. Furthermore, the plasticizing effect of water reduces the glass transition temperature (Tg) of polyester (An et al. 2021).

Cotton dyeing with reactive dyes has shown good results compared to the conventional method (An et al. 2020). In this case, small amounts of water were needed to pre-dissolve the dyes. With 120% on weight of fabric (owf) of water, a dye uptake of almost 100% and a dye fixation up to 94% were obtained. Water beyond this amount had adverse effects such as low levelness, and hydrolysis of the dye. Na_2CO_3 is necessary to increase the reactivity of cotton cellulose, but no other electrolyte is needed. For comparison, dyeing in aqueous medium containing NaCl results in ca. 40% dye uptake and less than 40% total fixation.

In another study, cotton dyeing with reactive dyes in five different solvents was compared, namely siloxane, paraffin, isooctane, M-oil (white oil), and water (Wang et al. 2018). The non-aqueous media yielded similar results to each other, with more than 94% dye uptake and 80–90% dye fixation, while aqueous dyeing resulted in ca. 60% dye uptake and about 50% dye fixation. In this case, water was also used (130% owf) in the pre-dissolution of reactive dyes. The adsorption kinetics of the dyes in non-aqueous solvents was also faster compared to water dyeing, which was correlated with the lower surface tension of non-aqueous solvents and the better fiber wetting in these solvents.

4.5 Silicons

Another alternative solvent tested for textile dyeing is the organosilicon compound decamethylcyclopentasiloxane (D5) (Fig. 4.1), a colorless, odorless, non-polar (dielectric constant ε: 2.50), slightly viscous and volatile (bp: 213 °C), water immiscible liquid, with low surface tension (Haynes 2016).

Its use includes formulations in diverse cosmetics (Montemayor et al. 2013), antiperspirants, skin care products, and hair care products (Rousselle 2017). Due to its low polarity, D5 does not dissolve water-soluble dyes, such as reactive dyes, satisfactorily, so it is necessary to pre-dissolve dyes in a small amount of water and suspend the aqueous solution in D5 or paraffin, generating an emulsion or suspension

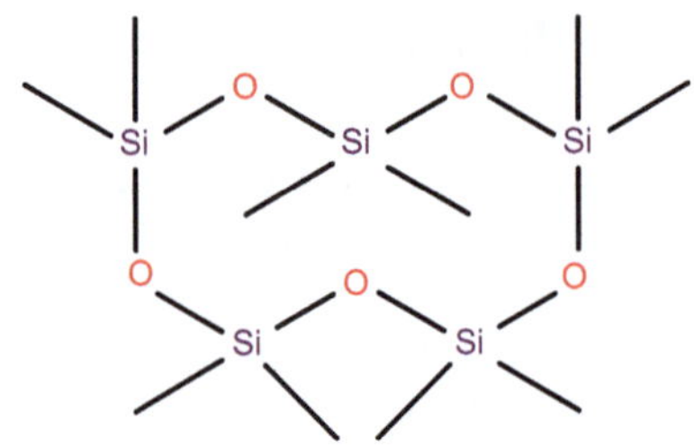

Fig. 4.1 Molecular structure of decamethylcyclopentasiloxane (D5)

(Feng et al. 2022; Li et al. 2022). In this emulsion system, because of the low solubility in the D5 system and high affinity for cotton fiber, reactive dyes are forced to diffuse into the fiber, leading to more than double the dye uptake achieved in aqueous dyeing, and significantly shorter absorption times. It was suggested, that a microscopical phase separation occurs at the fiber surface leading to a water film, that favors the migration of the dye from the liquid phase to the fiber (Pei et al. 2019).

The beneficial effect of adding small amounts of water to D5 in PET dyeing was also recognized as being caused by the synergistic effect between the two solvents on fiber swelling, allowing better dye diffusion and fixation. The addition of solvents such as ethanol or acetone also causes a (slight) increase in dye absorption. Optimum dyeing conditions with D5 are 60 min at 130°C, in the presence of 2.5% water and 1.5% ethanol (Li et al. 2011).

Exhaustion of Reactive red 195 in silk dyeing in D5 was practically quantitative (> 99%), with an increase of about 20% in K/S when compared to aqueous dyeing. Optimum operating conditions were 70 °C for 75 min in D5 containing 0.4% (owf) $NaHCO_3$. The fabrics also displayed better tensile strength than those dyed in water (Luo et al. 2022).

Dyeing in D5 can usually be carried out without electrolyte, unlike aqueous dyeing. However, when an electrolyte such as Na_2SO_4 (13% owf) was added, an increase in the adsorption rate of C.I. Reactive Red 120 on cotton was observed (+130%), which was attributed to the decrease in the repulsive forces between dye and fiber and the resulting increase in the dye diffusion rate. Although it accelerates dyeing, increasing further the electrolyte concentration had a negative effect on the uniformity of the dyeing, and a minimal (positive) change in the color depth of the dyed fabric and the final uptake of dye (Sun et al. 2023).

In another dyeing study with C.I. Reactive Black 5 in D5, supported by Molecular Dynamics (MD) simulation studies, it was concluded that the addition of Na_2SO_4 causes dye molecules to aggregate in the form of trimers and oligomers, since the interaction between sodium ions and the sulfonate groups of the dye reduces the electronic repulsion of the molecules. This aggregation decreases the diffusion coefficient of the molecule. The salt also causes an increase in Van der Waals energy between the cotton fibers and the dye, therefore increasing the attraction between these species (Hu et al. 2023).

Wool dyeing with Monascus pigments in D5 rendered an increase in K/S in comparison to water-dyed wool, although numerical data have not been provided.

The optimum dyeing conditions in D5 were 50 min at 80 °C and pH 3 (Feng et al. 2022).

Compared to traditional dyeing solvents, the cost of D5 as a solvent is relatively high and the economic viability of dyeing with this solvent depends on the efficiency of the dyeing process and the recycling step. D5 is considered relatively safe, with low acute toxicity, but chronic, long-term exposure may be associated with adverse effects, such as impacts on the endocrine system and fertility. In addition, it is persistent in the environment and can bioaccumulate in aquatic organisms, which may lead to potential long-term ecological effects (Fairbrother et al. 2015; Gobas et al. 2015). More data on the safety of this solvent and its recyclability are needed.

4.6 Alcohols

Alcohols are considered green solvents, but in the past, they have been solely used as dye-dissolving aids and dyeing auxiliaries such as benzyl alcohol in wool dyeing (Beal et al. 1960; Peters et al. 1960). Only more recently has their use as solvents been investigated in dyeing of cotton with reactive dyes (Xia et al. 2021; Periyasamy et al. 2024; Periyasamy and Negi 2024).

Xia et al. (2021) proposed a procedure to dye raw cotton fabrics in an EtOH–H_2O mixture (95% v/v ethanol) with different reactive dyes containing the vinyl sulphonic and/or monochlorotriazine reactive groups. Initially, the greige cotton fabrics were preswollen for 10 min in alkaline aqueous ethanol (90% alcohol; 15 g L^{-1} of Na_2CO_3) at 90 °C. Dyeing bath exhaustion, total dye fixation, and K/S values were significantly better than in the conventional dyeing procedure, and residual dye concentration in the dyeing bath was almost zero. Besides recycling of the swelling solution (up to five times) and recovery of ethanol, the process was also assessed successfully on an industrial scale, circumventing scouring or bleaching stages, and reducing chemicals to ethanol and sodium carbonate only. The main advantages of this procedure were attributed to the lower surface tension and better wetting properties of ethanol, reduced secondary hydrolysis reaction of dyes, and differences in dye aggregation.

Ethanol, isopropyl alcohol, and n-propanol were explored as dyeing media for cotton with eight reactive dyes (Periyasamy et al. 2024). To increase the reactivity ("activate") of the cotton hydroxyl groups, the cotton fabric was previously treated with Na_2CO_3 by padding. For the initial solubilization of the dye, 4.5% water was required, which also reduces the flammability of the organic vapors in the medium. As a result, in general (except for highly ionic dyes, that are poorly soluble in organic media), there is an increase in the exhaustion and fixation of dyes, as well as in color parameters; without, however, affecting the wash-, rub-, and light-color fastness properties of the textile. For example, in the case of the Reactive Red 35 (RR35) dye, increases of 249%, 114%, and 369% in K/S are observed in ethanol, isopropanol, and n-propanol, respectively, when compared with dyeing in an aqueous bath.

These results are related to the physical-chemical properties of the solvents. Alcohols provide better wettability than water, due to their lower surface tension, allowing for better accessibility of dyes to the interior of the fibers, generating strong bonds. Because of the lower dielectric constant of alcohols, the negative charge of the dyes is less pronounced, and the electronic repulsion between dye and fiber is lower, resulting in successful dye fixation, even in the absence of electrolytes. Additionally, solvolysis, the reaction of the dye with the solvent, the equivalent to dye hydrolysis in aqueous dyeing, is less likely in the organic medium, because of the lower dielectric constant of these solvents, which diminishes the nucleophilic character of the alcohols. However, in an organic solvent, the tendency of dye aggregation persists, which compromises levelling in some cases.

In another study, the same research group pretreated cotton with different sodium alkoxides to increase the nucleophilicity of its –OH groups and thereafter carried out dyeing in ethanol medium (Periyasamy and Negi 2024). This method was compared with the pre-treatment of cotton with Na_2CO_3, dyed in ethanol, and with the conventional aqueous method, in the presence of NaCl and Na_2CO_3. For most dyes, the results indicated increased dye uptakes and greater K/S values. For example, dyeing with reactive dye C.I. Reactive Red 239 by the conventional method resulted in a K/S of 3.4, while a considerable increase with the alkoxide-activated method was achieved, yielding a K/S of 8.9 for Na_2CO_3/EtOH, and 16.2 for C_2H_5ONa/EtOH.

The morphology of the alkoxide-treated fiber is not significantly damaged by the treatment, and its fastness properties, air permeability, and crease recovery angle are similar to those of fabrics dyed using an aqueous bath. Dye exhaustion and fixation rates are high, which reduces the dye concentration in the final effluent. On the other hand, in the case of water-soluble dyes, less soluble in ethanol, the results in organic media are worse than in conventional aqueous media (Periyasamy and Negi 2024).

4.7 Dimethylsulfoxide (DMSO)

Dimethylsulfoxide (DMSO) is an aprotic dipolar solvent, which is known for its excellent cellulose swelling properties because of disrupting and reducing intra- and intermolecular hydrogen bonds, and it was reported as cosolvent in dissolving cellulose (Shigemasa et al. 1990; Kohler and Heinze 2007; Ferreira et al. 2019).

Because of its high dipole moment ($\mu = 13.5 \cdot 10^{-30}$ C·m = 2.96 D) (Haynes 2016) and dielectric constant ($\varepsilon_r = 46.5$), and due to the presence of lone electron pairs, it is a good electron-pair donor (EPD) (Reichardt and Welton 2011). As a result, it effectively solvates cations, while anions remain as "naked ions" displaying increased reactivity in nucleophilic substitution reactions (S_N2). Furthermore, the negative charge developed in the transition state of the S_N2 reaction is stabilized, decreasing activation energy and, consequently, increasing the reaction rate (Anslyn and Dougherty 2006). Considering that some reactive dyes (e.g., containing the chlorotriazine group) react with the hydroxyl groups of cellulose fibers via nucleophilic substitution, it is expected that DMSO and other aprotic dipolar solvents favor dyeing with these dyes.

Consequently, DMSO has been successfully applied in dyeing of natural and synthetic fibers, often in combination with other components, taking advantage of synergistic effects. For example, dyeing poly-m-phenylene isophthalamide (PMIA), a synthetic fiber with high resistance to heat and chemicals, was improved by using DMSO in conjunction with electrolytes, e.g., NaCl and Na_2SO_4. While DMSO increased dye uptake, the addition of electrolytes further enhanced the process by reducing electrostatic repulsion between the dye and the fiber (Zhuo et al. 2025).

In another approach, a multicomponent solvent system comprising ethanol, urea, DMSO, and supercritical carbon dioxide ($scCO_2$) was employed to dye cotton with reactive dyes. This method involved a multi-step process, including pretreatment with KOH and urea, dye dissolution in DMSO, and fixation in $scCO_2$. This approach minimized dye hydrolysis and offered significant energy and material cost savings compared to traditional aqueous dyeing methods (Zhao et al. 2022).

A novel strategy utilized perchloroethylene (PCE) and DMSO, two solvents with temperature-dependent miscibility, to dye cotton with reactive dyes. At elevated temperatures, the solvents mix, allowing for dye diffusion and fixation. Upon cooling, the solvents separate, simplifying the washing process. This method demonstrated high dye fixation rates and reduced the need for extensive washing (Deng et al. 2019).

In nonaqueous solvents, surfactants form the so-called reverse micelles where the polar or ionic groups form the interior of the colloidal aggregate, see Sect. 4.11. A reverse micellar system based on N-ethyl perfluorooctyl sulfonimide (N-EtFOSA), together with an ionic liquid, 1,1,3,3-tetramethylguanidinium ($[(CH_3)_2N]_2C = NH_2^+$) acetate (TMGA) in $scCO_2$ was tested in the dyeing of linen fabrics. The addition of DMSO to this system improved dye fixation by reducing viscosity and enhancing dye diffusion into the fibers (Ma et al. 2023). Dyeing wool with reactive dyes in a mixture of DMSO and dimethyl carbonate resulted in higher dye fixation compared to conventional aqueous methods. This approach also minimized fiber damage and eliminated the need for pH adjustment (Chen et al. 2015). Finally, DMSO has been utilized to produce polymeric dyes (pigments) using regenerated cellulose and reactive dyes. These pigments were then used to dye various fabrics, demonstrating significant dye savings compared to traditional aqueous dyeing methods (Zhao et al. 2019).

Aprotic dipolar solvents, particularly DMSO, offer promising avenues for improving textile dyeing processes. They enhance dye uptake and fixation, reduce dye hydrolysis, and provide opportunities for more sustainable and efficient dyeing methods. However, further research is necessary to address challenges such as solvent recovery and potential environmental concerns associated with certain solvents.

4.8 Ionic Liquids

Ionic liquids (IL), also known as Room Temperature Ionic Liquids (RTIL) or "molten salts", are electrolytes, generally composed of bulky organic cations and organic or inorganic anions, with a low melting point, typically below 100 °C. The univalent and asymmetric ions make tight crystalline packing difficult which reduces the melting point. Such solvents display negligible vapor pressure, good thermal stability, non-flammability, wide electrochemical windows and a wide range of solubility and miscibility, which are desirable properties for a good solvent. As a special feature their properties are easily "customizable" based on the appropriate combination of the cation/anion pair, being also denominated as "designer solvents". (Wasserscheid and Welton 2008; MacFarlane et al. 2017).

Among the applications of this class of solvents are their use as catalysts (McNeice et al. 2021); solvents for syntheses (Wasserscheid and Welton 2008), for extraction and purification of inorganic ions (Atanassova 2021) proteins (Nunes et al. 2021) and natural products (Gong et al. 2023); for absorption of gases such as CO_2 and SO_2 (Scaglione et al. 2024) and as electrolytes for batteries (Ha et al. 2021).

Among the characteristics that limit their applicability on a large scale, at least in the short term, are high costs, high viscosity, and the lack of knowledge regarding its toxicity and biodegradability (Wasserscheid and Welton 2008; MacFarlane et al. 2017).

In the textile area, some academic studies have been carried out, but practical application is still limited (Meksi and Moussa 2017). Some ILs are quite efficient in swelling and dissolving cellulose (Rinaldi 2011; El Seoud et al. 2019; Cao et al. 2024). In some cases, fibers such as cotton (Özdemir et al. 2022), wool (Jiang et al. 2021a, b; Jiang et al. 2021a, b), ramie (Yuan et al. 2013) have been treated or modified with an ionic liquid prior to dyeing, thus increasing the affinity between the fiber and the dye. For example, cotton fabrics were treated with imidazolium (1-ethyl-2,3-dimethylimidazolium ethylsulfate; EDMIM[ESO_4]) and ammonium (methyl-tri-n-butylammonium methylsulfate; MTBA[MSO_4]) ILs, before dyeing with reactive dye (Özdemir et al. 2022). In this case, the IL is deposited on the fiber, forming a coating, which increases the fiber's affinity for the dye, as evidenced by higher K/S values.

Various ILs have been studied as solvent media in dyeing of different textile fibers (Fig. 4.2). The possibly first reports on dyeing textiles in ILs may be attributed to the group of Schollmeyer (Knittel and Schollmeyer et al. 2006; Knittel and Schollmeyer 2007), followed by a patent filed by the group of Earl and Seddon (Earle and Seddon 2009a, b; Earle and Seddon 2009a, b). Despite great promises and interesting ideas (no auxiliaries, dyeing without salt, shorter 10-min dyeing times, etc.), nothing has been proved so far in more detail or taken to another level.

Bianchini et al. (2015) dyed wool, polyester, and cotton with disperse Red 13 in aqueous medium, with 2 g/L 1-(2-hydroxyethyl)-3-methylimidazolium chloride (HEMIM[Cl]) as the only additive. The procedure was carried out in an open vessel at 95 °C, milder conditions than generally chosen for dyeing polyester. Despite the use of water as dyeing medium, the reported merit of this study is that the presence of the IL leads to total exhaustion of the dye, enabling recycling of the dyeing bath, which was only composed of water, dye, and IL. However, the publication does not include any quantitative data on the increased color yield and dye exhaustion, and the good color fastness to crocking (4 to 5 color change and transfer) was not compared to traditional dyeing with water.

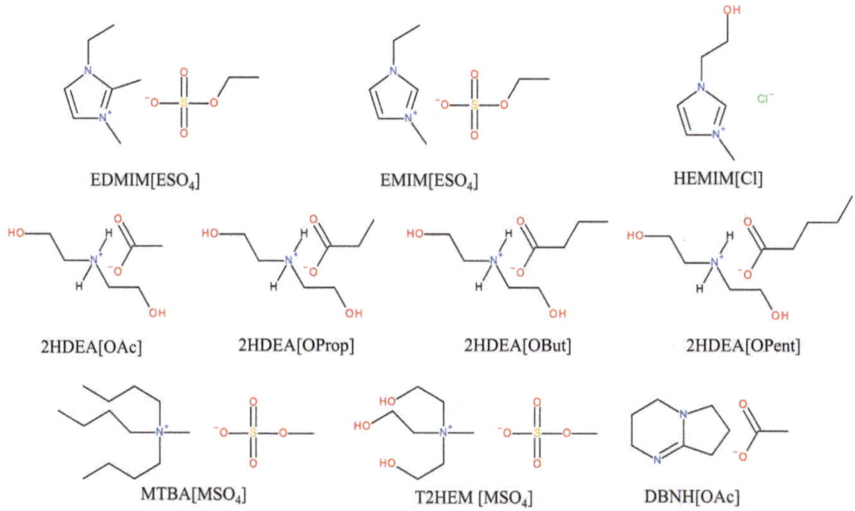

EDMIM[ESO$_4$]	1-ethyl-2,3-dimethylimidazolium ethylsulfate
EMIM[ESO$_4$]	1-ethyl-3-methylimidazolium ethylsulfate
HEMIM[Cl]	1-(2-hydroxyethyl)-3-methylimidazolium chloride
2HDEA[OAc]	2-hydroxy diethylammonium acetate
2HDEA[OProp]	2-hydroxy diethylammonium propionate
2HDEA[OBut]	2-hydroxy diethylammonium butanoate
2HDEA[OPent]	2-hydroxy diethylammonium pentanoate
MTBA[MSO$_4$]	methyl-tri-n-butylammonium methylsulfate
T2HEM [MSO$_4$]	Tris(2-hydroxyethyl)methylammonium methylsulfate
DBNH[OAc]	1,5-diazabicyclo[4.3.0]non-5-ene acetate

Fig. 4.2 Structure of different ionic liquids evaluated in dyeing of textile materials

Andrade et al. (2017) investigated the use of 17 protic ionic liquids (PILs), a class of ILs that can be synthesized at a much lower cost than imidazolium ILs, in cotton dyeing without the addition of auxiliaries or alkali. According to the authors, it is possible to reuse the dye bath. However, a closer analysis of the results revealed some contradictions (Andreaus and Sidou 2019). The very small K/S values (below 0.07), and a* and b* parameters (CIELAB), indicate little or insignificant fixation of the dye. Furthermore, the results obtained were not compared to conventional, aqueous reactive dyeing of cotton using auxiliaries. In the case of the reaction between cotton and a reactive dye of the sulfato ethylsulfone class, alkali is necessary to activate the sulfato ethylsulfone residue, generating a vinylsulfone group. Since the authors did not use an alkali in the procedure, in addition to the low values of K/S and a* and b*, it is concluded that it is unlikely that any significant fixation occurred.

Some PILs from a homologous series, (2-hydroxy diethylammoniumacetate (2HDEA[OAc]), 2-hydroxy diethylammonium propionate (2HDEA[OProp]), 2-hydroxy diethylammonium butanoate (2HDEA[OBut]) and 2-hydroxy diethylammonium pentanoate (2HDEA[OPent])) were tested in dyeing of precationized cellulose diacetate, cotton, polyamide, acrylic and wool fibers (Barros et al. 2022). The fibers were pretreated with a cationizing agent, which increased the affinity for the reactive dye, Drimaren Yellow CL-2R. It was found that K/S values increased as the anion size of the PIL increased, which may be related to the greater contact between IL and fiber, facilitating penetration of the dye into the fiber. Furthermore, density and surface tension of the PILs decrease with increasing length of the anion, facilitating the wettability of the textile material. Excellent rubbing and perspiration fastness were observed.

Opwis et al. (2017) showed that it is possible to dye cotton and PET blends in an open non-pressurized system in a single step; mixing in the same dyeing liquor the reactive dye for cotton and the disperse dye for PET. Besides avoiding the use of dyeing auxiliaries, it was also possible to reuse the dye bath for several cycles (10), without significant loss in color intensity. Several ILs were tested, but the one that presented the most advantageous results was Tris(2-hydroxyethyl)methylammonium methylsulfate (T2HEM [MSO_4]). In comparison to conventional aqueous PET dyeing, higher temperatures of 140–160 °C, were necessary, because lower temperatures resulted in insufficient color yields. No significant fiber damage and good fastness were observed.

The same group investigated dyeing of amorphous and crystalline m-aramid fabrics (Nomex 450), a synthetic highly resistant polymer, in ILs without any additional auxiliaries using cationic and anionic dyes. Out of 7 tested ILs, 1-ethyl-3-methylimidazolium ethylsulfate (EMIM[ESO_4]) was selected because of good color yields and without compromising fiber properties. While for amorphous m-aramid fibers at 120–140 °C with all tested dyes, excellent color shades were achieved, for crystalline m-aramid fibers, depending on dye stability, dyeing at 180 °C was necessary to obtain higher color yields. The authors suggested, especially for the cationic dyes, dyeing temperatures in the range of 120–140 °C. Furthermore, excellent results in terms of fastness to washing, light (generally a challenge in aqueous dyed aramid fibers), and friction, indicating deep fiber penetration, were obtained (Opwis et al. 2020). Dyeing of aramid fibers by conventional methods is in general problematic, due to its highly crystalline structure and the presence of strong hydrogen bonds (Azam et al. 2019).

Using a spin-dyeing process (Lyocell-type Ioncell®), cellulose fibers dyed with a series of vat dyes were carried out (Nygren et al. 2024). Cellulose is dissolved in an ionic liquid, 1,5-diazabicyclo[4.3.0]non-5-ene acetate (DBNH[OAc]), and the dye is incorporated into this solution. The cellulose fiber is then spun by means of coagulation in an aqueous bath, generating a fabric with the dye already incorporated. As a result of the immobilization of the vat dye inside the fiber, in addition to its low solubility in water, textiles with high quality, excellent light (grade 6 to 8), wash and rubbing fastness (both 4/5 to 5) are produced, in addition to the fact that the need

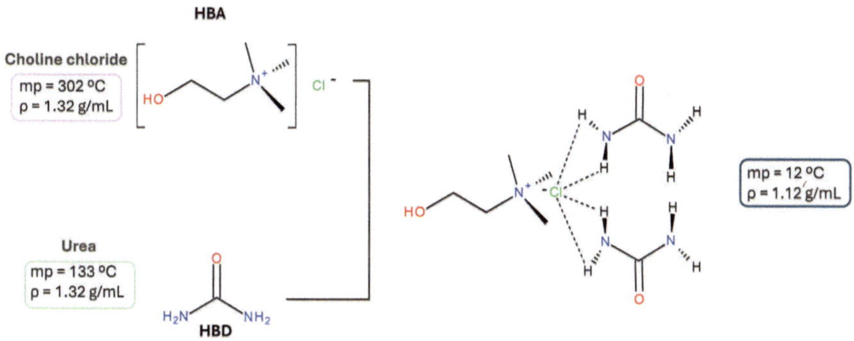

Fig. 4.3 The preparation of a deep eutectic solvent (DES) from choline chloride as hydrogen bonding acceptor (HBA) and urea as hydrogen bonding donor (HBD)

for intensive washing to remove excess dye is reduced. However, the authors did not specify whether the IL is recycled in this specific spin-dyeing process.

Its use in the textile sector, although advantageous and innovative, is limited mainly by the high cost.

4.9 Deep Eutectic Solvents

Deep Eutectic Solvents (DESs) are solvents formed by mixing two, occasionally three components, that present dramatically lower freezing points (or melting point) than the isolated components (Abbott et al. 2003). The occurrence of hydrogen bonding between the components, a hydrogen bond donor (HBD) and a hydrogen bond acceptor (HBA), is involved in lowering the freezing point and is fundamental for the solubilization capacity (Figs. 4.3 and 4.4 and Table 4.1). Moreover, systems are highly conductive, confirming that the ionic species are dissociated and can diffuse through the medium (Marcus 2019).

DESs are promising for applications in several areas, such as synthesis, biocatalysis, and extractions (Marcus 2019; Prabhune and Dey 2023). As well as ILs and supercritical carbon dioxide (sc-CO_2), DESs are neoteric solvents, that have been gaining prominence for their "green" characteristics: they are ecologically friendly and safe, non-toxic (or less toxic than ordinary organic solvents), biocompatible and biodegradable. They are also non-flammable, have negligible vapor pressure, and are prepared easily, simply by physically mixing the components (i.e. there is no chemical reaction between them). Precursors are low-cost materials, mostly from natural origin (NADES—Natural deep eutectic solvents) and are tunable with respect to their physical properties, by varying the HBA/HBD combination (Marcus 2019; Prabhune and Dey 2023). A recent review of the physical properties of several DES was compiled by Omar and Sadeghi (2023).

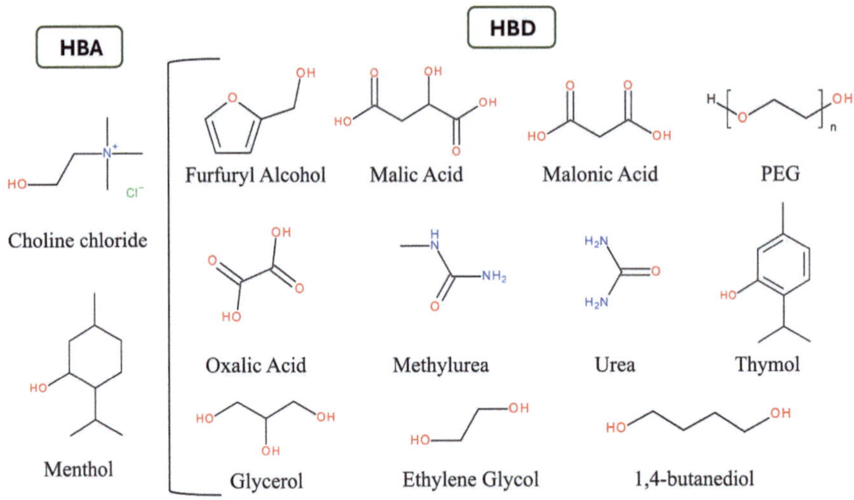

Fig. 4.4 Chemical structures of the constituents of deep eutectic solvents used in textile fiber dyeing, which are divided into hydrogen bond acceptors (HBA) and hydrogen bond donors (HBD)

More recently these solvents have attracted the attention of textile chemists and engineers and so far, only a few investigations into their application in textile dyeing have been carried out. Pawar et al. (2019), and Pawar and Adivarekar (2020) were among the first to publish on the use of DESs-based dyeing processes for polyester textiles, however advantages over water-based processes are not clearly evidenced. While Pawar and Adivarekar (2020) used choline chloride:urea 1:1, Pawar et al. (2019) used a ternary mixture of choline chloride:urea:glycerin 1:1:1. The same group reported later on dyeing of silk assisted by choline chloride:urea 1:1 as swelling agent that proved to be better than water (Pawar et al. 2021).

In a more recent study, polyester was dyed with the disperse dyes C.I. Disperse Blue 79 and C.I. Disperse Red 343 in DES prepared from choline chloride as HBA and ethylene glycol as HBD (ChCl:EG), and results were compared to aqueous dyeing using dispersants (Zheng et al. 2023). Solubilities of both dyes in ChCl:EG is considerable between 100–110 °C, which means there is no need for dispersing agents. They are chemically stable, with no UV/Vis spectral differences being observed after a few hours of contact with DES. As for polyester dyeing, K/S were improved by 15%–42% in comparison to the aqueous method, which was attributed to better dye dissolution and particle size reduction, and therefore, better permeation into the amorphous structure of the polymer. Uniform dyeing (evenly dye distribution) and good dry and wet rubbing fastness were obtained. (Zheng et al. 2023).

Cellulose nanofibers (CNFs) were dyed with two reactive anionic dyes, CI Reactive Red 195, an azo dye, and C.I. Reactive Blue 19, an anthraquinone dye, in choline chloride: ethylene glicol (ChCl:EG) at a molar ratio 1:2, without the addition of salt (Mehdi et al. 2022). Optimum dyeing conditions for 3% owf of dye were 60 min at 70 °C, which produced color yields 28% higher for C.I. Reactive Blue 19 and

Table 4.1 Selected deep eutectic solvent systems that have been investigated in textile dyeing

DES	DES abbreviated	Molar ratios	Fiber	Reference
Choline chloride: Ethylene glicol	ChCl:EG	1:2 1:1 1:2	PET cotton PICT	Zheng et al. (2023), Jiang et al. (2024), Mehdi et al. (2022)
Choline chloride: Furfuryl alcohol	ChCl: FA	1:2	PICT	Mehdi et al. (2022)
Thymol: Menthol	Thy:Men	1:1	cotton	Wei et al. (2021a, b)
Choline chloride: Oxalic acid	ChCl:OA	1:1	PLA	Ying et al. (2022)
Choline chloride: Malonic acid	ChCl:MaloAc	1:1 to 1:4	PET	Cao et al. (2022)
Choline chloride: Malic acid	ChCl:MalAc	1:1 to 1:4	PET	Cao et al. (2022)
Choline chloride: Urea	ChCl:Ur	1:1 to 1:4 1:1 1:1	PET PET silk	Cao et al. (2022), Pawar and Adivarekar (2020), (Pawar et al. 2021)
Choline chloride: Urea: Glycerol	ChCl:Ur: Gly	1:1:1	PET	Pawar et al. (2019)
Choline chloride: Methylurea	ChCl:methUr	1:1 to 1:4	PET	Cao et al. (2022)
Choline chloride: Polyethylene glycol	ChCl:PEG	1:1 to 1:4	PET	Cao et al. (2022)
Choline chloride: 1,4-butanediol	ChCl:Butdiol	1:1 to 1:4	PET	Cao et al. (2022)
Choline chloride: Glycerol	ChCl:Gly	1:1 to 1:4	PET	Cao et al. (2022)

PET polyethylene terephthalate (polyester)
PLA polylactic acid
PICT poly(1,4-cyclohexane dimethylene isosorbide terephthalate)

31% higher for C.I. Reactive Red 195 when compared to conventional processing in aqueous medium. The DES treatment made CNFs more hydrophilic according to water contact angle measurements, and led to CNF with greater tensile strength, probably due to an increase in the crystalline phase and the amount of –OH groups (Mehdi et al. 2022).

Nanofibers made from poly(1,4-cyclohexane dimethylene isosorbide terephthalate) (PICT), a bio-based polymer, were dyed with C.I. Disperse red 167 in two DESs, namely, choline chloride: furfuryl alcohol (ChCl:FA) at 1:2 molar ratio; and choline chloride: ethylene glycol (ChCl:EG) at 1:2 molar ratio, using a batch dyeing and ultrasonic dyeing method. The PICT nanofibers dyed by the ultrasound method presented an approximately 25% higher color strength compared to fibers dyed by the batch dyeing method. ChCl:EG produced the highest color strength. The results were attributed to better disaggregation of the dyes, leading to improved dye transfer to the fiber surface. In addition to the improved color strength, the ultrasonically dyed nanofibers had increased tensile strength compared to neat PICT and PICT dyed by the batch method, and slightly better washing fastness, but greater fiber shrinkage was observed (Hussain et al. 2021).

ChCl:EG (molar raio 1:1) was also used both as solvent and mordant in the dyeing of wool and cotton using the natural pigment from *Platycodon Grandiflorus* root (Jiang et al. 2024). Materials dyed in the presence of the DES, especially wool, displayed higher color strength than materials dyed without any mordant and materials with Al^{3+} as mordant. According to the authors, ChCl:EG binds to the structure of cotton and wool, improving the fixation of the natural dye. Materials dyed with natural dye in the presence of DES also presented antibacterial, anti-UV, antioxidant activity, and flame retardancy, with good fastness (Jiang et al. 2024). Some of the arguments raised by the authors are not sufficiently supported by the presented data, especially with respect to DES-fiber and DES-dye interactions and the function of the DES as mordant.

DESs can be used to modify the polymer surface to increase its dye uptake. Poly(lactic acid) (PLA) was "reconstructed"/functionalized with a DES composed of choline chloride and oxalic acid (ChCl:OA) in a molar ratio 1:1. The modification with ChCl:OA converts the hydrophobic PLA into a new material, DES-PLA, which is hydrophilic. The resulting material (DES-PLA) was dyed with three disperse dyes (C.I.Disperse Red 167, C.I.Disperse Orange 30, and C.I.Disperse Blue 79), and its properties were compared with the precursor polymer PLA (Ying et al. 2022). As the bioderived PLA has limited thermal stability, dyeing should be ideally carried out at relatively low temperature. Color strength increases as the temperature increases, however, in aqueous dyeing, color strength reaches only significant values at 110 °C. When dyeing DES-PLA, the color strength is much higher than PLA, allowing dyeing at lower temperatures. For example, the K/S of DES-PLA dyed with C.I.Disperse Red 167 at 90 °C is close to the K/S of original PLA dyed at 110 °C. In addition to preserving the polymer, less energy is consumed. Such distinguished dyeing behavior was attributed to chemical modification of the surface through the introduction of polar groups, as well as the induction of roughness that increases the surface area of the polymer. The resulting material was suggested as a candidate to replace PET (Ying et al. 2022).

Cao and co-workers explored the use of DESs to dye PET fabrics with anthraquinone dyes (Cao et al. 2022). DESs were designed from the combination of choline chloride (ChCl) as HBA; and various HBD, namely malonic acid, malic acid, urea, methylurea, polyethylene glycol, 1,4-butanediol, and glycerol. The dyes used were 1,4-dihydroxyanthraquinone (DHA), 1-amino-4-hydroxy-anthraquinone (AHA), 1,4-bis (methylamino)-anthraquinone (BMA), and 1-hydroxy-4-p-toluidino-anthraquinone (HTA). Results showed that the adsorption of dyes on the fiber depends on several factors, including the type of DES, the HBA:HBD ratio, the dye concentration, the temperature and the dyeing time. The HBA: HBD molar ratio influences the physicochemical properties of the DES, impacting dye solubility and interaction with the fiber. No clear trend in the influence of HBA: HBD ratio on dye adsorption and dyeing was found, suggesting a balance between dye solubility and the ability of

DES to penetrate the fibers. Water has a relevant and complex influence on dyeing. Adding water decreases viscosity, facilitating dye diffusion and adsorption on the PET fiber. It may also influence the polarity of the medium, decreasing the DES: dye interaction, and facilitating dye fixation. The ideal water concentration varies according to the dye used. For DHA and especially BMA, the K/S increased considerably with the addition of 30 v/v % water. For HTA and AHA, the optimal amount of water was 20 v/v %, but K/S decreased with a water concentration above this point. The negative effect in this case may be due to the weakening of the interaction between DES and PET, reducing the ability of DES to penetrate the fiber. Tensile strength of fabrics decreased after dyeing with AHA and BMA, probably due to some disturbance of the crystalline structure. When dyeing with DHA, on the contrary, this property increased, possibly due to the establishment of crosslinking hydrogen bonds between functional groups of the dye and the polymeric chains of PET, which strengthened the cohesion and increased the resistance of the material. The dye bath could be reused three times, simply by adding more dye, until the ideal concentration was obtained for the next operating cycle. DESs were recovered by adding water, precipitating the dye, which is then adsorbed onto the polyester fabric (Cao et al. 2022). Unfortunately, any comparison in this work between DES dyeing and conventional dyeing in water with dispersants is missing. It should furthermore be considered that acidic dyes, with carboxy, sulfonic or hydroxyl groups, may be able to interact or react with the basic component of some DES, interfering with dyeing.

A hydrophobic deep eutectic solvent (HDES) consisting of thymol and menthol (Thy-Men) was used in an attempt for salt-free dyeing of cotton with reactive dyes (Wei et al. 2021a, b). In this case, small fractions of water were needed to dissolve the dyes and to pre-swell the cotton (in the form of aqueous Na_2CO_3), which was done by a dip-squeeze procedure, similar to padding to achieve an adequate pick-up. About 1:1 water:cotton is necessary in the dye bath to favor dye diffusion through the fabrics, ensuring high color yield and levelness. A W/O (water-in-oil) emulsion system was formed between water and the DES. By adjusting the water content (1:1 water:fabric), pick-up, and concentration of alkali solution (60 g/L), excellent levelness, satisfactory colorfastness, and color strength were obtained with the Thy-Men solvent. Additionally, the dyed fabrics showed good colorfastness to crocking and washing, comparable to the aqueous method. Tone, chromaticity, colorfastness, as well as physical properties such as hydrophilicity, tensile strength, and elongation, besides sensory properties such as resilience, softness, and smoothness, were not significantly altered by the DES treatment. The Thy-Men dyeing medium could be reused for 5 dyeing cycles, demonstrated good color consistency, implying reduced effluent generation (Wei et al. 2021a, b).

4.10 Supercritical Carbon Dioxide (Sc-CO_2)

A supercritical fluid is a substance at a temperature and pressure condition above its critical point, i.e., in which liquid and gas are indistinguishable. For CO_2, this critical point is 31 °C and 73 atm, which are relatively easily achievable conditions, considering that the pressure of CO_2 in a cylinder is approximately 50 atm. In the supercritical state, the density and viscosity are much lower than in ordinary liquids. The diffusion of solutes is faster in supercritical fluids than in ordinary liquids (Reid et al. 1987; Levine 2009). Furthermore, the relatively low viscosity and low interfacial tension of these fluids lead to a much higher diffusivity in(to) fibrous materials than that of common liquids, favoring mass transport of (hydrophobic) dyes through the textile fiber.

Interestingly, the physical properties of the fluid near the critical point vary significantly as a function of pressure and temperature, so it is possible to "tune" such properties by varying p and T (Levine 2009). Supercritical carbon dioxide, sc-CO_2, is a nonpolar, non-Hydrogen Bonding Donor solvent; it is a good solvent for low molecular mass nonpolar substances, but a poor solvent for polar solutes (Reichardt and Welton 2011). However, the addition of cosolvents (e.g., methanol, ethanol, acetone) can aid in increasing the solubility of polar substances (Wu and Han 2019).

Sc-CO_2 presents itself as an interesting alternative to organic solvents in industrial processes because it is non-flammable, environmentally safe, cheaper, easily available, and easily recoverable, being considered a green solvent (Wu and Han 2019). The solvent is exploited for extraction purposes, such as in the production of decaffeinated coffee (Zabot 2020), extraction of vanillin (Fujii et al. 2018), and essential oils (Yousefi et al. 2019). Its advantage is that because the critical point is close to room temperature, thermal degradation is reduced. It also finds use in dry cleaning (Sutanto et al. 2014).

sc-CO_2 can be removed after the reaction, by depressurization, evaporating into air, or it can be reused by condensation into a recovery vessel. In this way, the dye not fixed in the textile material can also be recovered (Alebeid et al. 2021).

Regarding the dyeing of textile fibers, the method is applicable to hydrophobic fibers dyed with non-polar dyes (e.g., PES dyeing with disperse dyes). In this case, sc-CO_2 acts as a solvent for the dye, and as a plasticizer for the synthetic fiber. The fluid increases the mobility of the polymer chains, facilitating the diffusion of the dye within the polymer fibers (Burkinshaw 2016a, b, c).

Dyeing in sc-CO_2 medium is applied commercially, for example by the companies Hkrita and eCO2Dye (Hkrita; eCO2Dye). Dyes typically used for dyeing synthetic fibers, such as polyester, disperse dyes, are not soluble in water by themselves; they require the addition of dispersing agents/surfactants to stabilize them in this medium. sc-CO_2, as well as nonpolar solvents, on the other hand, dissolve this class of dyes without necessarily adding such dispersing agents.

However, dyeing polar fibers such as cotton with polar dyes (e.g., direct anionic, acidic, and reactive dyes) is not feasible, because, in addition to not adequately solubilizing the dye, sc-CO_2 does not swell the fibers and is not capable of breaking the hydrogen bonds of the polymer, reducing any diffusion within the fiber (Saus et al. 1993; Mohamed et al. 2013). The topic has been revised (Bach et al. 2002; Liao and Chang 2012; Goñi et al. 2021; de Oliveira et al. 2024).

The solubilization of disperse dyes in sc-CO_2 occurs without the need for dispersing agents, unlike dyeing with this type of dye in water. Additionally, the method allows for shorter dyeing times, lower energy and water consumption, and lower effluent production. In short, it is more environmentally friendly (Burkinshaw 2016a, b, c).

The rates of dyeing PET with disperse dyes increase with increasing pressure, and especially with increasing temperature; thanks to the increased solubility of the dye in the polymer (Tušek et al. 2000; Burkinshaw 2016a, b, c). The effect of temperature can be understood as a result of the increase in free volume in the polymer and the greater kinetic energy of the dye molecules (Burkinshaw 2016a, b, c).

In polyester dyeing with hydrazono propanenitrile dyes (dyes that can generate fabrics with antibacterial properties) in sc-CO_2, it was observed that the optimum operating conditions are 120 °C and 15 MPa, and dye concentrations from 2 – 6% owf. The obtained color yield, washing, rubbing, and light fastness were excellent (Abou Elmaaty et al. 2014a, b); the light fastness obtained was better than those of the aqueous method (Abou Elmaaty et al. 2014a, b). Similar strategies were used to dye polypropylene (Elmaaty et al. 2018; Abou Elmaaty et al. 2019), poly(ethylene terephthalate) (PET), and nylon (Elmaaty et al. 2019), also with good results.

To overcome the difficulties encountered in dyeing polar fibers such as cotton, silk, and wool in sc-CO_2, three strategies may be adopted (Burkinshaw 2016a, b, c; Broadbent et al. 2023):

- Chemical modification of existing disperse dye molecules; with a reactive group that covalently binds to the substrate (cotton/wool/silk). In other words, converting a disperse dye into a reactive dye;
- Covalently modify the substrate (cotton/wool/silk) with carbon residues (non-polar), making the fiber surface hydrophobic, increasing the affinity between the textile material and the non-polar dye;
- Add auxiliaries (co-solvents) to sc-CO_2, improving the dissolution, transport, and fixation of the (polar) dye on the fiber.

Obviously, the additional synthesis step, in strategies (1) and (2), represents additional costs, involving the consumption of chemicals, time, and energy, in addition to the generation of waste. These additional costs are perhaps justified when accompanied by better fabric properties, such as better resistance to washing, crease, easy care, etc. (Broadbent et al. 2023).

An example of a modified natural fiber is cellulose acetate; Broadbent and coworkers studied the dyeing of this fiber and polyester in sc-CO_2 and compared it with the traditional method in aqueous medium (Broadbent et al. 2023). Increased colour strength, uniformity, and fastness were observed in the sc-CO_2 medium, in comparison with the aqueous method.

For strategy (3), a small amount of a miscible co-solvent (modifier) such as ethanol, methanol, or acetone, for example, is used in order to increase the solubility of the dye as well as improve dyeing (dye uptake) of natural, polar fibers in sc-CO_2 (Wu and Han 2019).

For example, Van der Kraan and co-workers (van der Kraan et al. 2007) investigated the dyeing of silk and wool, and also synthetic fibers such as polyester and nylon, with disperse reactive dyes (vinylsulphone and dichlorotriazine reactive groups) in sc-CO_2 with varying amounts of water/humidity, obtaining good dyeing results.

For nylon, silk, and wool fibers, coloration and fixation increase positively when both the fiber and sc-CO_2 are saturated with water vapor. In the case of polyester, however, these parameters were independent of relative humidity. With vinylsulphone dye, fixation values of 94, 85 and 90%, respectively for nylon, silk, and wool are reached in 2 h. With respect to dyeing with dichlorotriazine in sc-CO_2 saturated with water, the fixations of polyester, silk, and wool were 93, 88 and 79%. The role of water was recognized as a swelling agent of the fiber (van der Kraan et al. 2007), but also probably increasing the reactivity of the dye and the fiber, by some polarity effect.

An improvement in the absorption and fixation of disperse dyes by cotton in the presence of moisture was also observed by Long et al. (2012). The authors also found that the efficiency of dye absorption and fixation increased remarkably as temperature and pressure increased.

Long et al. dyed cotton with reactive disperse dyes and used the catalyst triethylene diamine (TEDA) for the fixation step (Long et al. 2012). The addition of the catalyst was necessary to increase the nucleophilicity of cotton hydroxyl groups, and reacting more readily with the dye. Even so, optimized K/S values are relatively low, such as 0.374, with dyeing at 140 °C, 6.0 MPa, and 60 min of reaction, in the presence of the catalyst. Additionally, the fabric displayed poor wash fastness (rated 2–3).

In another example (Banchero et al. 2010), higher solubilities, accompanied by better dye uptake, were observed for polyethylene terephthalate yarn dyeing with C.I. Disperse Blue 79, C.I. Disperse Orange 3 and C.I. Solvent Brown 1 dyes in sc-CO_2 containing 5.5% ethanol.

The addition of the co-solvent increases the solubility of the dyes in the fluid (from 300% up to 2400%), as well as dye uptake by the polymer, polyethylene terephthalate yarn (from 10% up to 120%). These quantities also increase as a function of increasing temperature and pressure.

This effect was explained as a result of the reduction of intermolecular interactions between polymer chains, thanks to their interactions with ethanol, increasing mobility, and consequently better swelling and better absorption of the dye. The expressive effect of the modifier enables efficient dyeing at lower pressures, making the method milder (Banchero et al. 2010).

Within the "Green Chemistry" context, in addition to reducing water consumption, the use of raw materials of biological and renewable origin, also attracts growing interest. Curcumin, a natural dye, was applied to polyester, in scCO_2, without fiber pretreatment and in the absence of auxiliary chemicals and mordants; conditions that are essential in conventional dyeing processes of natural dyes. Bright yellow shades with excellent color strength, levelness, and fastness properties were obtained, and the optimal operating conditions were 120 °C, 25 MPa, and 1 h of dyeing time. High K/S values were obtained with low curcumin concentrations (0.1–1% owf). (Abate et al. 2019). This study is a considerable advance, considering that natural dyes generally lack the necessary/desirable properties for textile material, namely levelness and fastness.

The lower consumption of water, energy, and chemicals such as electrolytes and dispersants, in addition to the generation of a reduced quantity of effluents, and the possibility of recovering the solvent, makes the technique attractive, compared to the aqueous method. Disadvantages include the high operational costs, with an equipment that must withstand high pressures and temperatures; in addition to an applicability that is limited to dyeing hydrophobic fibers with non-polar dyes. Different cases, such as polar fibers, require adaptations, such as chemical modification of the dye or fiber, or the addition of co-solvents.

4.11 Reverse Micellar Systems

Reverse micelles or inverted micelles are spherical, thermodynamically stable, self-assembled nanostructures composed of surfactants, formed in non-polar media when its concentration in solution is higher than the critical micelle concentration (CMC). They differ from normal micelles in that the polar, hydrophilic headgroups of the surfactant point inwards, while the nonpolar, hydrophobic tails point outwards and are in contact with the organic phase through dispersion forces (Hiemenz 1997; Correa et al. 2012) (Fig. 4.5).

A small amount of water is added to the non-polar solvent in the presence of surfactant, creating an internal polar microenvironment, generally referred to as "water pool". It is interesting to note that the water characteristics in smaller reverse

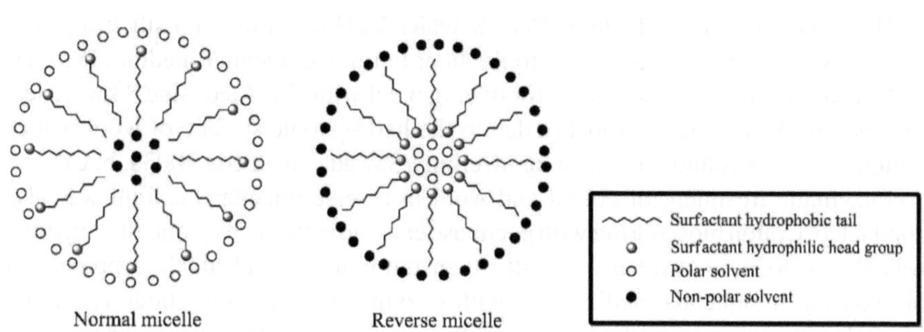

Fig. 4.5 Normal and reverse micelles

micelles contrast with the properties of "bulk" water, for example, increased micro-viscosity and slower solvation dynamics are observed for water confined in such a microenvironment (Correa et al. 2012). This system can be used to solubilize polar substances in non-polar media and is applied in drug delivery systems (Jorgensen et al. 2023), selective separation and purification of biomolecules (Shin 2018), catalysis (Arivalagan and Zhao 2015), organic synthesis (Correa et al. 2004), and preparation of nanoparticles (Wei and Matsui 2014).

Likewise, polar/water-soluble dyes can be dispersed in a non-polar medium, through reverse micelles, enabling the dyeing of textile fibers. Dyes are encapsulated by the reverse micelles and carried to and deposited on the textile fibers, subsequently reacting with the reactive groups of the fiber, e.g. cellulose (Tang et al. 2024).

The first investigations using reverse micellar system in textile dyeing processes were based on an anionic surfactant, sodium bis-2-ethylhexylsulphosuccinate (Aerosol-OT, AOT). Excellent results have been reported for dyeing cotton (Sawada and Ueda 2003a, b, c; Sawada and Ueda 2003a, b, c), wool, and silk fabrics (Sawada and Ueda 2003a, b, c) with direct, acid, and reactive dyes, even in the absence of auxiliaries. Electrolytes are not necessary in this system, but effective dye fixation requires the use of alkali or heating (Sawada and Ueda 2003a, b, c). The high exhaustion of the dye in this method was attributed to the very low bath ratio (water–fabric ratio) compared to conventional aqueous dyeing process (Sawada and Ueda 2003a, b, c).

Dyeing of animal fibers turned out to be more complex to understand: Dyeing silk in reverse micellar medium with C.I. Acid Orange 7 was less efficient than in water, whereas C.I. Acid Blue 83 dyed silk successfully. This difference in dyeing behavior was attributed to the dye's size and related to the competition between C.I. Acid Orange 7, a relatively small dye molecule, and the surfactant for adsorption on the fiber surface, while with C.I. Acid Blue 83, a comparatively larger dye, the same did not occur (Sawada and Ueda 2003a, b, c). Apparently, the dye adsorption process (and therefore, dyeing) is influenced by the charge of the surfactant molecules. Reverse micelles derived from ionic surfactants affect the polarity of the water-pool, generating uneven microenvironments, leading to non-reproducible results (Tang and Kan 2020). Non-ionic surfactants, such as Triton X-100, produce better results.

However, direct dye C.I. Direct Red 28, which had little affinity for silk in aqueous dyeing, was successfully adsorbed to the fiber in reverse micellar medium (Sawada and Ueda 2003a, b, c). The results for dyeing wool with C.I. Acid Blue 83 were less satisfactory than for silk, probably due to the hydrophobic surface of wool, which requires a larger volume of water for swelling (Sawada and Ueda 2003a, b, c).

Enzymatic treatment of cotton and wool in reverse micellar medium was also conducted. Cotton bioscouring with pectinase enzyme in the reverse micellar medium was shown to be equivalent or better than the conventional alkaline process or bioscouring in aqueous media, even with enzyme concentration about 100 times lower than the aqueous method. The modification of wool with protease, on the other hand, was less efficient than the aqueous method, which was explained as a result of inhibition by the products of the reaction itself. Shrinkage and tear strength were similar for reverse micelle medium and water. Moreover, the same authors succeeded in dyeing cotton and polyester in a reverse micellar system with C.I. Disperse Violet 1 and C.I. Reactive Red 2, respectively, which means that blends of these fibers can be dyed simultaneously (Sawada and Ueda 2001).

To avoid synthetic surfactants derived from fossil sources, an alkyl polyglucoside (APG)-based surfactant (decyl-β-D-glucopyranoside) synthesized from renewable raw materials/ starch derivatives, was utilized in dyeing cotton (Lee et al. 2021; Tang et al. 2022). Several reactive dyes were used in a reverse micellar system of decyl-β-D-glucopyranoside (TritonTM CG-110), with n-octanol as co-surfactant to reinforce the structure of reverse micelle and in D5 as solvent. Na_2CO_3 was added to regulate the pH, but no further addition of electrolyte was necessary. Still, water was needed to swell cotton fiber pores, and to provide the polar microenvironment inside the micelle (Lee et al. 2021; Tang et al. 2022). Higher color yield (K/S) and lower color reflectance were observed, attesting better dye uptake and fixation than in the water-based conventional method (Lee et al. 2021; Tang et al. 2022). Reactive dyes in reverse micelles are more resistant to hydrolysis, since they are confined (Lee et al. 2021).

It has been demonstrated that micelle stability is a crucial factor in dyeing with this system. Size and stability of micelles are in turn linked to the hydrophilic-lipophilic balance (HLB) of non-ionic surfactant (i.e., the relationship or balance between the hydrophilic to the lipophilic portion of the surfactants) (Tang et al. 2018). It was observed that reflectance percentage increases and K/S sum value decreases with an increase in HLB value.

Recently, Tang and collaborators (Tang et al. 2024) presented a protocol for dyeing cotton materials with reactive, salt-free, and alkali-free dyes, namely Levafix Red CA, Levafix Navy Blue E-BNA, and Levafix Yellow E-3RL. They used a reverse micellar system with limonene as a bio-solvent, rhamnolipid as bio-surfactant, and n-octanol as co-surfactant. Rhamnolipid is a microbial bio-surfactant with low ecotoxicity and high biodegradability, derived from the gram-negative bacteria *Pseudomonas aeruginosa*. Limonene, in turn, is a monoterpene hydrocarbon derived from citrus fruits and is non-toxic. For dyeing, they used two different one-bath approaches, a one-step procedure with dye addition at the initial stage of dyeing, and a two-step process, where the dye was added in the intermediate stage of dyeing, which they compared to conventional dyeing in water, with NaCl and Na_2CO_3. Dyeing in reverse micelle medium proved superior to conventional water-based method, which was attributed to reduced ionization effects between dyes and fibers and improved swelling of cotton fibers in non-aqueous media. The one-step procedure presented better color yields, probably due to a more efficient encapsulation of the dye, but poorer levelness than the two-step method. Both methods yielded darker shades (lower reflectance) than aqueous dyeing, and excellent washing and crocking fastness, comparable to water dyeing, and no significant fiber damage which usually occurs under harsh conditions.

The authors also argued that the lack of need to add any alkalizing agent was due to a reaction between the surfactant and the co-surfactant n-octanol, which might have increased the pH by forming some basic species, which was confirmed by measuring a medium pH of ca. 11. Howsoever, measuring the pH of non-aqueous media is not trivial and should not be based on the pH scale from 0 to 14, which might leave some doubts on the pH measurement, and about the argument to explain the good performance in the absence of any added base.

Reverse micellar media based on dialkyl maleic acid ester surfactants and reactive dyes were found to deliver similar dyeing quality (leveling, fastness) to conventional aqueous dyeing, with savings of 60% of electrolytes, 60% of water, and 40% of steam (Xie et al. 2011). In the reverse micellar method, it was possible to use a lower liquor ratio, 1:5 (fabric:water), compared to the aqueous method (ratio 1:15). To sustain the consequently higher dye concentration, the use of surfactant is essential to avoid aggregation, which would interfere with the homogeneity of dyeing (leveling) (Xie et al. 2011).

4.12 Final Considerations and Perspectives

Water is the solvent of probably all biological processes, so its substitution as a solvent in textile dyeing is an audacious attempt. As shown in various parts of this review, the use of water as a solvent is essential or even indispensable in various stages of textile processing. This may be the case in redox reactions, such as bleaching, or in ionic reactions, like reactive dyeing, where reactions require a medium with a high dielectric constant, or in polyester dyeing as a plasticizer of the fiber polymer, or as polarizing, swelling, dissolving or washing agent. It seems unlikely that water will be completely replaced in the textile production chain. Each textile fiber available on the market, natural, synthetic, hydrophilic, or hydrophobic, has its own characteristics and affinity for a certain class of dyes and, therefore, requires specific processing conditions and a specific solvent. From this point of view, a "universal dyeing medium" different from water that would suit all textile fibers seems unrealistic. However, as shown in this review, different solvents can be combined, and solvent combinations can be designed that obtain advantages over single solvents or fluids due to synergistic effects. Several examples of solvent combinations have been described in this chapter with very promising results. The applicability of non-aqueous solvents or their combinations depends on factors such as environmental friendliness (e.g. toxicity), the intrinsic characteristics of the solvents, such as their physical properties, their capacity to solubilize a broad range of dyes, their interaction with textile fibers, compatibility with existing equipment, machinery and processes, in addition to the cost of the solvents and their thermal and chemical stability for recovery and recyclability.

Procedures developed and described so far aimed at reducing the consumption of water, electrolytes, auxiliary chemicals, and dyes accompanied by low energy consumption and a reduction in waste generation, which complies with the principles of Green Chemistry. Solvent recovery and recycling are mandatory; otherwise, new waste will be generated, raising questions about the environmental appeal of the proposed method.

The main advantages of non-aqueous solvents revealed so far are related to their intrinsic properties, such as lower surface tension, non-nucleophilicity, good fiber swelling properties that favor dyeing and reduce secondary hydrolysis reactions. On the other hand, high viscosities, as in the case of some ILs and DESs, reduce the rate of dye diffusion in the fiber. However, the use of higher temperatures or addition of water or molecular solvents may circumvent this problem. Uncomplete dye exhaustion and fixation will inevitably require post-washing procedures to remove unfixed dyes and ensure excellent fastness to washing and crocking.

Some of the presented approaches are apparently simple and straightforward, with excellent dyeing results, such as the use of aqueous ethanol, while other proposals seem quite complex to be implemented on an industrial scale in the near future. DESs and ILs are intriguing solvents, and their possibilities need to be further investigated. Replacing (at least in part) water by other dyeing media may lead to increased operating costs. Governmental incentives (e.g. lower taxes) may be necessary to make such processes economically viable.

In summary, only significant advantages over aqueous dyeing will justify the implementation of new processes with new solvents.

Acknowledgements The authors thank Conselho Nacional de Desenvolvimento Científico e Tecnológico (CNPq) and Fundação de Amparo à Pesquisa e Inovação do Estado de Santa Catarina (FAPESC) for grants of A. Kreuz and P.A. Horn (Project FAPESC/contract 2023TR000249), CNPq for grants of J. Andreaus (CNPq 308536/2020-7; CNPq 303633/2023-9) and O.A. El Seoud (CNPq 307022/2014-5), and Coordenação de Aperfeiçoamento de Pessoal de Nível Superior (CAPES) for the MSc. scholarship of L.F. Sidou. O.A. El Seoud would also like to thank FAPESP (grant 2024/13668-4) and J. Andreaus is grateful for the financial support of FAPESC through various projects (contracts: 2021TR1500, 2022TR000166, 2023TR000249). Furthermore, authors acknowledge the support of FURB—Universidade Regional de Blumenau.

Declarations Conflict of interest: The authors declare that they have no conflict of interest.

Abbreviations

D5	Decamethylcyclopentasiloxan
DES	Deep Eutectic Solvents
HBA	Hydrogen bond acceptor
HBD	Hydrogen bond donor
IL	Ionic Liquids
owf	On weight the fabric
PES	Polyester
$scCO_2$	Supercritical carbon dioxide

References

Abate MT, Ferri A, Guan J, Chen G, Nierstrasz V (2019) Colouration and bio-activation of polyester fabric with curcumin in supercritical CO_2: part I—Investigating colouration properties. J Supercrit Fluids 152:104548. https://doi.org/10.1016/j.supflu.2019.104548

Abbott AP, Capper G, Davies DL, Rasheed RK, Tambyrajah V (2003) Novel solvent properties of choline chloride/urea mixtures. Chem Commun (Camb). https://doi.org/10.1039/b210714g:70-71.10.1039/b210714g

Abou Elmaaty T, El-Taweel F, Abd El-Aziz E, Yuesf M, Okubayashi S (2014a) Facile bifunctional dyeing of polyester fabrics with new antibacterial β-oxoalkanenitriles disperse dyes. Int J Sci Eng Res 5:703–706

Abou Elmaaty T, Ma J, El-Taweel F, Abd El-Aziz E, Okubayashi S (2014b) Facile bifunctional dyeing of polyester under supercritical carbon dioxide medium with new antibacterial Hydrazono Propanenitrile dyes. Ind Eng Chem Res 53:15566–15570. https://doi.org/10.1021/ie502088r

Abou Elmaaty T, Sofan M, Elsisi H, Kosbar T, Negm E, Hirogaki K, Tabata I, Hori T (2019) Optimization of an eco-friendly dyeing process in both laboratory scale and pilot scale supercritical carbon dioxide unit for polypropylene fabrics with special new disperse dyes. J CO_2 Util 33:365–371. https://doi.org/10.1016/j.jcou.2019.06.018

Ad W (2022) Equilibrium of dyeing: a review. Adv Res TextE Eng 7. https://doi.org/10.26420/advrestexteng.2022.1073

Alebeid OK, Hassan EAM, Pei L (2021) Recent advances in non-aqueous dyeing systems. In: sustainable practices in the textile industry, pp 43–73

An Y, Ma J, Zhu Z, Hu MG, Shao J (2020) Study on a water-saving and salt-free reactive dyeing of cotton fabrics in non-aqueous medium of liquid paraffin system. J TextE Inst 111:1538–1545. https://doi.org/10.1080/00405000.2020.1711552

An Y, Miao J, Fan J, Li M, Hu MG, Shao M, Shao J (2021) High-efficiency dispersant-free polyester dyeing using D5 non-aqueous medium. Dye Pigment 190. https://doi.org/10.1016/j.dyepig.2021.109303

An Y, Miao J, Wang L, Fan J, Li M, Hu MG, Shao M, Shao J (2022) Dyeing of polyester fabrics based on a liquid paraffin system: thermodynamic, kinetic, and molecular dynamics simulation studies. J TextE Inst 114:1423–1430. https://doi.org/10.1080/00405000.2022.2131314

Anastas PT, Warner JC (1998) Green chemistry: theory and practice. Oxford University Press

Andrade RS, Torres D, Ribeiro FR, Chiari-Andreo BG, Oshiro Junior JA, Iglesias M (2017) Sustainable cotton dyeing in nonaqueous medium applying protic ionic liquids. ACS Sustain Chem Eng 5:8756–8765. https://doi.org/10.1021/acssuschemeng.7b01555

Andreaus J, Sidou LF (2019) Comment on sustainable cotton dyeing in nonaqueous medium applying protic ionic liquids. ACS Sustain Chem Eng 7:7999–8000. https://doi.org/10.1021/acssuschemeng.9b01112

Anslyn EV, Dougherty DA (2006) Modern physical organic chemistry: university science books

Arivalagan PR, Zhao Y (2015) Interfacial catalysis of aldol reactions by prolinamide surfactants in reverse micelles. Org Biomol Chem 13:770–775. https://doi.org/10.1039/C4OB02074J

Atanassova M (2021) Solvent extraction chemistry in ionic liquids: an overview of f-ions. J Mol Liq 343. https://doi.org/10.1016/j.molliq.2021.117530

Azam F, Iqbal K, Safdar F, Hussain T, Ashraf M, Rehman A (2019) An investigation into the effect of different parameters on the dyeing of high-performance m-Aramid fiber and its optimization. Cloth Text Res J 38:90–103. https://doi.org/10.1177/0887302x19870872

Bach E, Cleve E, Schollmeyer E (2002) Past, present and future of supercritical fluid dyeing technology—an overview. Rev Prog Color Relat Top 32:88–102. https://doi.org/10.1111/j.1478-4408.2002.tb00253.x

Banchero M, Manna L, Ferri A (2010) Effect of the addition of a modifier in the supercritical dyeing of polyester. Color Technol 126:171–175. https://doi.org/10.1111/j.1478-4408.2010.00243.x

Barros S, Andrade R, Torres D, Chiari-Andréo B, Veloso G, Gonzalez C, Iglesias M (2022) Eco-friendly technology for reactive dyeing of cationized fabrics: protic ionic liquids as innovative media. Cellul Chem Technol 56:403–425. https://doi.org/10.35812/CelluloseChemTechnol.2022.56.36

Beal W, Dickinson K, Bellhouse E (1960) The dyeing of wool by solvent-assisted processes. J Soc Dyers Colour 76(6):333–341. https://doi.org/10.1111/cote.1960.76.issue-6, https://doi.org/10.1111/j.1478-4408.1960.tb02380.x

Bianchini R, Cevasco G, Chiappe C, Pomelli CS, MaJ Rodríguez Douton, (2015) Ionic liquids can significantly improve textile dyeing: an innovative application assuring economic and environmental benefits. ACS Sustain Chem Eng 3:2303–2308. https://doi.org/10.1021/acssuschemeng.5b00578

Biermann F, Hickmann T, Sénit C-A, Beisheim M, Bernstein S, Chasek P, Grob L, Kim RE, Kotzé LJ, Nilsson M, Ordóñez Llanos A, Okereke C, Pradhan P, Raven R, Sun Y, Vijge MJ, van Vuuren D, Wicke B (2022) Scientific evidence on the political impact of the sustainable development goals. Nat Sustain 5:795–800. https://doi.org/10.1038/s41893-022-00909-5

Broadbent PJ, Carr CM, Lewis DM, Rigout ML, Siewers EJ, Shojai Kaveh N (2023) Supercritical carbon dioxide (SC-CO_2) dyeing of cellulose acetate: an opportunity for a "greener" circular textile economy. Color Technol 139:475–488. https://doi.org/10.1111/cote.12690

Broadbent AD, So Dyers, Colourists (2001) Basic principles of textile coloration. Society of Dyers and Colourists

Burkinshaw SM (2016) Dyeing theory. In: Burkinshaw SM (ed) Physico-chemical aspects of textile coloration, pp 209–247

Burkinshaw SM (2016) Non-aqueous dyeing. In: Burkinshaw SM (ed) Physico-chemical aspects of textile coloration, pp 571–599

Burkinshaw SM (2016) The role of water in aqueous dyeing. In: Burkinshaw SM (ed) Physico-chemical aspects of textile coloration, pp 81–152

Cao R, Long Y, Li T, Lv W, Wu H, Wang B, Song Y, Gao H, Nie Y (2024) Swelling and dissolution behaviors of cellulose in phosphate-based ionic liquids-H_2O binary systems: in-situ observation and molecular dynamics simulation. J Mol Liq 415:126315. https://doi.org/10.1016/j.molliq.2024.126315

Cao D, Liu M, Su Y-C, Yang Z, Bi W, Chen DDY (2022) Adsorption behavior of anthraquinones in deep eutectic solvent on polyester fiber and its application. Sustain Chem Pharm 27. https://doi.org/10.1016/j.scp.2022.100680

Ceballos DM, Fellows KM, Evans AE, Janulewicz PA, Lee EG, Whittaker SG (2021) Perchloroethylene and dry cleaning: it's time to move the industry to safer alternatives. Front Public Health 9:638082. https://doi.org/10.3389/fpubh.2021.638082

National Center for Biotechnology Information P (2024) PubChem compound summary for CID 31373, Tetrachloroethylene. https://pubchem.ncbi.nlm.nih.gov/compound/Tetrachloroethylene. Accessed 11 Nov 2024

Chen L, Wang B, Chen J, Ruan X, Yang Y (2015) Characterization of dimethyl sulfoxide-treated wool and enhancement of reactive wool dyeing in non-aqueous medium. Text Res J 86:533–542. https://doi.org/10.1177/0040517515591784

Correa NM, Zorzan DH, D'Anteo L, Lasta E, Chiarini M, Cerichelli G (2004) Reverse micellar aggregates: effect on ketone reduction. 2. surfactant role. J Org Chem 69:8231–8238. https://doi.org/10.1021/jo049172v

Correa NM, Silber JJ, Riter RE, Levinger NE (2012) Nonaqueous polar solvents in reverse micelle systems. Chem Rev 112:4569–4602. https://doi.org/10.1021/cr200254q

de Oliveira CRS, de Oliveira PV, Pellenz L, de Aguiar CRL, da Silva Júnior AH (2024) Supercritical fluid technology as a sustainable alternative method for textile dyeing: an approach on waste, energy, and CO_2 emission reduction. J Environ Sci 140:123–145. https://doi.org/10.1016/j.jes.2023.06.007

Deng Y, Xu M, Zhang Y, Zhou G, Li N, Qiu X (2019) Non-water dyeing process of reactive dyes in two organic solvents with temperature-dependent miscibility. Text Res J 89:3882–3889. https://doi.org/10.1177/0040517518819840
Earle MJ, Seddon KR (2009) Dyeing fiber for fabric material or textiles, involves contacting the fiber with dye solution comprising ionic liquid. Univ Queens Belfast
Earle MJ, Seddon KR (2009) Dyeing process in ionic liquid solvents. The Queen's University Of Belfast
eCO2Dye eCO2Dye—Waterless Textile Dyeing. http://eco2dye.com/about-us.html. Accessed 30 Dec 2024
El Seoud OA, Kostag M, Jedvert K, Malek NI (2019) Cellulose in ionic liquids and alkaline solutions: advances in the mechanisms of biopolymer dissolution and regeneration. Polymers (Basel) 11:1917. https://doi.org/10.3390/polym11121917
Elmaaty TA, El-Taweel F, Elsisi H, Okubayashi S (2018) Water free dyeing of polypropylene fabric under supercritical carbon dioxide and comparison with its aqueous analogue. J Supercrit Fluids 139:114–121. https://doi.org/10.1016/j.supflu.2018.05.018
Elmaaty TA, Sofan M, Kosbar T, Elsisi H, Negm I (2019) Green approach to dye PET and nylon 6 fabrics with novel pyrazole disperse dyes under supercritical carbon dioxide and its aqueous analogue. Fibers Polym 20:2510–2521. https://doi.org/10.1007/s12221-019-1208-7
European Parliament EP (2024) The impact of textile production and waste on the environment
Fairbrother A, Burton GA, Klaine SJ, Powell DE, Staples CA, Mihaich EM, Woodburn KB, Gobas FAPC (2015) Characterization of ecological risks from environmental releases of decamethylcyclopentasiloxane (D5). Environ Toxicol Chem 34:2715–2722. https://doi.org/10.1002/etc.3041
Feng L, Ren L, Zhao Y, Dong C, Zhang H (2022) Sustainable dyeing of wool yarn with Monascus pigments in non-aqueous medium. J TextE Inst 114:1550–1556. https://doi.org/10.1080/00405000.2022.2144666
Ferreira DC, Oliveira ML, Bioni TA, Nawaz H, King AWT, Kilpelainen I, Hummel M, Sixta H, El Seoud OA (2019) Binary mixtures of ionic liquids-DMSO as solvents for the dissolution and derivatization of cellulose: effects of alkyl and alkoxy side chains. Carbohydr Polym 212:206–214. https://doi.org/10.1016/j.carbpol.2019.02.024
Fujii T, Matsuo Y, Kawasaki S-i (2018) Rapid continuous supercritical CO_2 extraction and separation of organic compounds from liquid solutions. Ind Eng Chem Res 57:5717–5721. https://doi.org/10.1021/acs.iecr.8b00812
Gebert K (1971) The dyeing of polyester textile fabric in perchloroethylene by the exhaust process. J Soc Dyers Colour 87:509–513. https://doi.org/10.1111/j.1478-4408.1971.tb02997.x
Gersmann H (2024) Die Tonne ist künftig tabu—was die neuen EU-Vorgaben für alte Textilien bedeuten. Frankurter Rundschau
Gobas FAPC, Powell DE, Woodburn KB, Springer T, Huggett DB (2015) Bioaccumulation of decamethylpentacyclosiloxane (D5): a review. Environ Toxicol Chem 34:2703–2714. https://doi.org/10.1002/etc.3242
Gong J, Liang C, Majeed Z, Tian M, Zhao C, Luo M, Li C (2023) Advances of imidazolium ionic liquids for the extraction of phytochemicals from plants. Separations 10. https://doi.org/10.3390/separations10030151
Goñi ML, Gañán NA, Martini RE (2021) Supercritical CO_2-assisted dyeing and functionalization of polymeric materials: A review of recent advances (2015–2020). J CO_2 Util 54:101760. https://doi.org/10.1016/j.jcou.2021.101760
Grand View Research GVR (2024) Textile market size, share & trends analysis report by raw material (wool, chemical, silk), by product (natural fibers, polyester), by application, by region, and segment forecasts, 2024–2030. https://www.grandviewresearch.com/industry-analysis/textile-market. Accessed 30 Dec 2024
Ha TA, Li H, Wang X, O'Dell LA, Forsyth M, Pozo-Gonzalo C, Howlett PC (2021) Functional binders based on polymeric ionic liquids for sodium oxygen batteries using ionic liquid electrolytes. ACS Appl Energy Mater 4:434–444. https://doi.org/10.1021/acsaem.0c02352

Haynes WM (2016) CRC handbook of chemistry and physics. CRC Press
Hiemenz PC, Rajagopalan R (1997) Colloidal structures in surfactant solutions: association colloids. In: Hiemenz PC, Rajagopalan R (eds) Principles of colloid and surface chemistry, revised and expanded, 3rd edn. CRC Press
Hu Q, Pei L, Shen J, Sun S, Wang J (2023) Level dyeing property and aggregation morphology of reactive black 5 on cotton fabric in a salt-free and less-water dyeing system. Text Res J 94:598–608. https://doi.org/10.1177/00405175231214191
Hussain N, Hussain S, Mehdi M, Khatri M, Ullah S, Khatri Z, Van Langenhove L, Kim IS (2021) Introducing deep eutectic solvents as a water-free dyeing medium for poly (1,4-cYclohexane dimethylene isosorbide terephthalate) PICT Nanofibers. Polymers (Basel) 13. https://doi.org/10.3390/polym13162594
IFA IfAdDGU—Paraffinöl. https://gestis.dguv.de/data?name=091110#01221. Accessed 05 Dec 2024
Jiang Z, Zhang Y, Wang Q, Wang P, Yu Y, Zhou M, Li E (2021a) Enhancing surface performance of wool using reduced ionic liquid. J TextE Inst 113:983–992. https://doi.org/10.1080/00405000.2021.1912952
Jiang Z, Zhang Y, Wang Q, Wang P, Yu Y, Zhou M, Li E (2021b) Thiol-based ionic liquid: an efficient approach for improving hydrophilic performance of wool. J Nat Fibers 19:9729–9740. https://doi.org/10.1080/15440478.2021.1993403
Jiang Z, Zheng G, Cui Y, Wang W, Shang X, Wei Y, Zhou M, Wang P, Yu Y, Wang Q (2024) Natural deep eutectic solvent: a novel and green mordant for the natural dye. Chem Eng J 481. https://doi.org/10.1016/j.cej.2023.148319
Jorgensen AM, Steinbring C, Stengel D, To D, Schmid P, Bernkop-Schnurch A (2023) Self-emulsifying drug delivery systems (SEDDS) containing reverse micelles: advanced oral formulations for therapeutic peptides. Adv Healthc Mater 12:e2302034. https://doi.org/10.1002/adhm.202302034
Knittel D, Schollmeyer E (2007) Ionic liquids for textile finishing, part 1: dyeing of textiles. Melliand Textilberichte/International Textile Reports 88:E14
Knittel D, Schollmeyer E, Knittel DP (2006) Färbeverfahren für Synthese- und Naturfasermaterial mit Hilfe ionischer Flüssigkeiten (Process to dye man-made or natural fibres in which the dye is applied as an 'ionic fluid'). Deutsches Textilforschungszentrum Nord-West (Dete-Non-Standard)
Kohler S, Heinze T (2007) New solvents for cellulose: dimethyl sulfoxide/ammonium fluorides. Macromol Biosci 7:307–314. https://doi.org/10.1002/mabi.200600197
Lee CH, Tang YL, Wang Y, Kan C-w (2021) Dyeing of cotton fabric in decamethylcyclopentasiloxane using alkyl polyglucoside-based reverse micelle as reactive dye carrier. Fibers Polym 23:107–118. https://doi.org/10.1007/s12221-021-0382-6
Levine IN (2009) Physical chemistry. McGraw-Hill Education
Li SZ, Liu JQ, Li YQ, Li LM (2011) Study on the dyeing of PET fiber with disperse dyes in D5 media. Adv Mater Res 331:334–337. https://doi.org/10.4028/www.scientific.net/AMR.331.334
Li M, Fan J, An Y, Shao M, Miao J, Wang L, Hu MG, Shao J (2022) Thermodynamic characteristics of cotton dyeing with reactive dyes in non-aqueous media. J TextE Inst 114:952–958. https://doi.org/10.1080/00405000.2022.2105030
Liao S-K, Chang P-S (2012) Special issue—supercritical fluids literatures on dyeing technique of supercritical fluid carbon dioxide. Am J Anal Chem 03:923–930. https://doi.org/10.4236/ajac.2012.312A122
Long J-J, Xiao G-D, Xu H-M, Wang L, Cui C-L, Liu J, Yang M-Y, Wang K, Chen C, Ren Y-M, Luan T, Ding Z-F (2012) Dyeing of cotton fabric with a reactive disperse dye in supercritical carbon dioxide. J Supercrit Fluids 69:13–20. https://doi.org/10.1016/j.supflu.2012.05.002
Luo Y, Zhu L, Pei L, Cai Z, Wang J (2022) Sustainable silk dyeing in a silicon non-aqueous dyeing system with reactive red 195: salt-free and less alkali. Text Res J 92:4921–4932. https://doi.org/10.1177/00405175221113080

Ma Y, Zheng H, Xiong X, Cai T, Zheng F, Zheng L (2023) Dyeing of linen fabrics in supercritical CO_2—using a reverse micellar system with ionic liquid domains. J Nat Fibers 20. https://doi.org/10.1080/15440478.2023.2222555

MacFarlane DR, Kar M, Pringle JM (2017) Fundamentals of Ionic Liquids: from chemistry to applications. Wiley

Marcus Y (2019) Deep eutectic solvents. Springer Cham

McNeice P, Marr PC, Marr AC (2021) Basic ionic liquids for catalysis: the road to greater stability. Catal Sci Technol 11:726–741. https://doi.org/10.1039/d0cy02274h

Hkrita Non-aqueous Solvent Medium (NASM) Dyeing for Textiles. https://www.hkrita.com/en/our-innovation-tech/projects/non-aqueous-solvent-medium-nasm-2. Accessed 04 Nov 2024

Mehdi M, Hussain N, Khatri M, Hakro RA, Ghaffar A, Khatri Z, Kim IS, Zhang K-Q (2022) Salts and water-free dyeing of cellulose nanofibers using novel green deep eutectic solvents: isotherm, kinetics, and thermodynamic studies. J Appl Polym Sci 139:52279. https://doi.org/10.1002/app.52279

Meksi N, Moussa A (2017) A review of progress in the ecological application of ionic liquids in textile processes. J Clean Prod 161:105–126. https://doi.org/10.1016/j.jclepro.2017.05.066

Mohamed AL, Er-Rafik M, Moller M (2013) Supercritical carbon dioxide assisted silicon based finishing of cellulosic fabric: a novel approach. Carbohyd Polym 98:1095–1107. https://doi.org/10.1016/j.carbpol.2013.06.027

Montemayor BP, Price BB, van Egmond RA (2013) Accounting for intended use application in characterizing the contributions of cyclopentasiloxane (D5) to aquatic loadings following personal care product use: antiperspirants, skin care products and hair care products. Chemosphere 93:735–740. https://doi.org/10.1016/j.chemosphere.2012.10.043

Nunes JCF, Almeida MR, Faria JL, Silva CG, Neves MC, Freire MG, Tavares APM (2021) Overview on protein extraction and purification using ionic-liquid-based processes. J Solution Chem 51:243–278. https://doi.org/10.1007/s10953-021-01062-x

Nygren N, Schlapp-Hackl I, Heimala S, Sederholm H, Rissanen M, Hummel M (2024) Spin-dyeing of cellulose fibres with vat dyes using the Ioncell process. Carbohydr Polym 346:122578. https://doi.org/10.1016/j.carbpol.2024.122578

Omar KA, Sadeghi R (2023) Database of deep eutectic solvents and their physical properties: a review. J Mol Liq 384:121899. https://doi.org/10.1016/j.molliq.2023.121899

Opwis K, Benken R, Knittel D, Gutmann JS (2017) Dyeing of PET fibers in ionic liquids. Int J New Technol Res (IJNTR) 3:101–108

Opwis K, Celik B, Benken R, Knittel D, Gutmann JS (2020) Dyeing of m-aramid fibers in ionic liquids. Polymers (Basel) 12. https://doi.org/10.3390/polym12081824

Özdemir HN, Kartal GE, Keskİn ÖY, Sekİ Y (2022) Employment of imidazolium and ammonium-based ionic liquids to enhance the dyeability of cotton fibers with reactive dyestuff. Tekstil Ve Konfeksiyon 33:269–276. https://doi.org/10.32710/tekstilvekonfeksiyon.1156626

Pawar SS, Adivarekar R (2020) A novel approach for dyeing of polyester using non-aqueous deep eutectic solvent as a dyeing medium. Pigm Resin Technol 50:1–9. https://doi.org/10.1108/prt-09-2019-0085

Pawar SS, Maiti S, Biranje S, Kulkarni K, Adivarekar RV (2019) A novel green approach for dyeing polyester using glycerine based eutectic solvent as a dyeing medium. Heliyon 5:e01606. https://doi.org/10.1016/j.heliyon.2019.e01606

Pawar SS, Athalye A, Adivarekar RV (2021) Solvent assisted dyeing of silk fabric using deep eutectic solvent as a swelling agent. Fibers Polym 22:405–411. https://doi.org/10.1007/s12221-021-0142-7

Pawar SS, Shahid M, Adivarekar RV (2022) Nonaqueous dyeing of textile materials. In: Yusuf M, Shahid M (eds) Emerging technologies for textile coloration. CRC Press, Boca Raton

Pei L, Liu J, Gu X, Wang J (2019) Adsorption kinetic and mechanism of reactive dye on cotton yarns with different wettability in siloxane non-aqueous medium. J TextE Inst 111:925–933. https://doi.org/10.1080/00405000.2019.1696642

Periyasamy AP, Negi A (2024) Alkoxide-based solvent dyeing: a feasible strategy for pollution minimization and sustainable approach for the reactive dyeing of cellulosic materials. Cellulose 31:7765–7791. https://doi.org/10.1007/s10570-024-06059-5

Periyasamy AP, Dutta S, Tehrani-Bagha AR (2024) Solvent-assisted salt-free reactive dyeing of cotton fabric. Cellulose 31:7101–7118. https://doi.org/10.1007/s10570-024-05928-3

Peters L, Stevens CB, Budding J, Burdett BC, Sykes JAW (1960) The effect of solvents in wool dyeing. J Soc Dyers Colour 76(9):543–550. https://doi.org/10.1111/cote.1960.76.issue-9, https://doi.org/10.1111/j.1478-4408.1960.tb02396.x

Poliakoff M, Licence P (2007) Green chemistry. Nature 450:810–812. https://doi.org/10.1038/450810a

Prabhune A, Dey R (2023) Green and sustainable solvents of the future: deep eutectic solvents. J Mol Liq 379:121676. https://doi.org/10.1016/j.molliq.2023.121676

Reichardt C, Welton T (2011) Solvents and solvent effects in organic chemistry. Wiley

Reid RC, Prausnitz JM, Poling BE (1987) The properties of gases and liquids. McGraw-Hill

Rinaldi R (2011) Instantaneous dissolution of cellulose in organic electrolyte solutions. Chem Commun (Camb) 47:511–513. https://doi.org/10.1039/c0cc02421j

Roberts G, Solanki R (2008) Carrier dyeing of polyester fibre part III—the role of water in solvent dyeing. J Soc Dyers Colour 97:220–223. https://doi.org/10.1111/j.1478-4408.1981.tb03583.x

Rousselle DC (2017) Opinion of the scientific committee on consumer safety (SCCS)—final version of the opinion on decamethylcyclopentasiloxane (cyclopentasiloxane, D5) in cosmetic products. Regul Toxicol Pharmacol 83:117–118. https://doi.org/10.1016/j.yrtph.2016.11.016

Saus W, Hoger S, Knittel D, Schollmeyer E (1993) Faerben aus ueberkritischem Kohlendioxid-Dispersionsfarbstoffe und Baumwollgewebe. Textilveredlung 28:38–38

Sawada K, Ueda M (2001) Enzyme processing of textiles in reverse micellar solution. J Biotechnol 89:263–269. https://doi.org/10.1016/S0168-1656(01)00310-8

Sawada K, Ueda M (2003a) Adsorption and fixation of a reactive dye on cotton in non-aqueous systems. Color Technol 119:182–186. https://doi.org/10.1111/j.1478-4408.2003.tb00170.x

Sawada K, Ueda M (2003b) Adsorption behavior of direct dye on cotton in non-aqueous media. Dyes Pigm 58:37–40. https://doi.org/10.1016/s0143-7208(03)00027-5

Sawada K, Ueda M (2003c) Dyeing of protein fiber in a reverse micellar system. Dyes Pigm 58:99–103. https://doi.org/10.1016/S0143-7208(03)00048-2

Scaglione N, Wylie L, Padua A, Costa Gomes M (2024) Improved reversible and selective SO_2 absorption by a stable phosphonium carboxylate ionic liquid. ACS Sustain Chem Eng 12:10486–10497. https://doi.org/10.1021/acssuschemeng.4c02659

Senthil Kumar P, Grace Pavithra K (2019) 2—Water and textiles. In: Muthu SS (ed) Water in textiles and fashion. Woodhead Publishing, pp 21–40

Shahid M, Mallick A, Debnath S (2025) Advances in renewable natural materials for textile sustainability. Taylor & Francis Group

Sharif F, Crushell E, Driscoll K, Bourke B (2001) Liquid paraffin: a reappraisal of its role in the treatment of constipation. Arch Dis Child 85:121. https://doi.org/10.1136/adc.85.2.121

Shigemasa Y, Kishimoto Y, Sashiwa H, Saimoto H (1990) Dissolution of cellulose in dimethyl sulfoxide effect of thiamine hydrocloride. Polym J 22:1101–1103. https://doi.org/10.1295/polymj.22.1101

Shin YD (2018) Extraction of biomolecules from aqueous solutions by reverse micelles. In: Vera JH, Wilczek-Vera G (eds) Ionic surfactants and aqueous solutions: biomolecules, metals and nanoparticles. De Gruyter, pp 155–204

Statista (2024) Production volume of chemical and textile fibers worldwide from 1975 to 2022. Accessed 02 Jan 2025

Sun S, Pei L, Chen J, Shen J, Alebeid OK, Xu J, Luo C, Zhang X, Zhang S, Wang J (2023) Adsorption of reactive red 120 in decamethyl-cyclopentasiloxane non-aqueous dyeing system. Coatings 13. https://doi.org/10.3390/coatings13030502

Sutanto S, van Roosmalen MJE, Witkamp GJ (2014) Mechanical action in CO_2 dry cleaning. J Supercrit Fluids 93:138–143. https://doi.org/10.1016/j.supflu.2013.09.019

Tang AYL, Lee CH, Wang YM, Kan CW (2018) Effect of hydrophilic-lipophilic balance (HLB) values of PEG-based non-ionic surfactant on reverse micellar dyeing of cotton fibre with reactive dyes in non-aqueous medium. Fibers Polym 19:894–904. https://doi.org/10.1007/s12221-018-8061-y

Tang AYL, Kan CW (2020) Non-aqueous dyeing of cotton fibre with reactive dyes: a review. Color Technol 136:214–223. https://doi.org/10.1111/cote.12459

Tang YLA, Lee CH, Chan CY, Wang Y, Kan C-W (2022) Alkyl polyglucoside (APG) nonionic surfactant-based reverse micellar dyeing of cotton fabric—a study of reactive dyes with different functional groups. J Nat Fibers 20. https://doi.org/10.1080/15440478.2022.2136324

Tang YLA, Yu J, Lee CH, Wang Y, Kan C-W (2024) Limonene biosolvent-based non-aqueous dyeing medium for salt-free and alkali-free rhamnolipid (RL) biosurfactant reverse micellar dyeing of cotton fabric with reactive dyes. Sustain Mater Technol 42. https://doi.org/10.1016/j.susmat.2024.e01137

Tušek L, Golob V, Knez Ž (2000) The effect of pressure and temperature on supercritical CO_2 dyeing of PET-dyeing with mixtures of dyes. Int J Polym Mater 47:657–665. https://doi.org/10.1080/00914030008031320

United Nations UN (2023) The sustainable development goals report: special edition

van der Kraan M, Fernandez Cid MV, Woerlee GF, Veugelers WJT, Witkamp GJ (2007) Dyeing of natural and synthetic textiles in supercritical carbon dioxide with disperse reactive dyes. J Supercrit Fluids 40:470–476. https://doi.org/10.1016/j.supflu.2006.07.019

Wang J, Gao Y, Zhu L, Gu X, Dou H, Pei L (2018) Dyeing property and adsorption kinetics of reactive dyes for cotton textiles in salt-free non-aqueous dyeing systems. Polymers (Basel) 10:1030. https://doi.org/10.3390/polym10091030

Wang A, Liu F, Xia L, Zhang C, Zhou S, Fu Z, Wang Y, Xu W (2021) Solvent assisted dyeing of wool fibers with reactive dyes in a ternary solvent system for protecting fibers against damage. J Ind Eng Chem 103:329–339. https://doi.org/10.1016/j.jiec.2021.07.048

Wang A, Xia L, Zhou S, Guo H, Wang Y, Xu W (2020) Solvent assisted reactive dyeing of cotton with high exhaustion in the absence of salt. J Clean Prod 267. https://doi.org/10.1016/j.jclepro.2020.122062

Wasserscheid P, Welton T (2008) Ionic liquids in synthesis. Wiley

Wei Y, Jiang Z, Wang Q, Wang P, Yu Y (2021) A salt-free and water-saving approach as a green alternative to conventional reactive dyeing of cotton. Sustain Chem Pharm 24. https://doi.org/10.1016/j.scp.2021.100536

Wei Z, Matsui H (2014) Rational strategy for shaped nanomaterial synthesis in reverse micelle reactors. Nat Commun 5:3870. https://doi.org/10.1038/ncomms4870

Wu T, Han B (2019) Supercritical carbon dioxide (CO_2) as green solvent. In: Han B, Wu T (eds) Green chemistry and chemical engineering. Springer, New York, New York, NY, pp 173–197

Xia L, Wang A, Wang Y, Zhang C, Wang Y, Zhou S, Fu Z, Zhao H, Ding C, Xu W (2021) Eco-friendly dyeing of raw cotton fibres in an ethanol–water mixture without scouring and bleaching pretreatments. Green Chem 23:796–807. https://doi.org/10.1039/D0GC02839H

Xie K, Cheng F, Zhao W, Xu L (2011) Micelle dyeing with low liquor ratio for reactive dyes using dialkyl maleic acid ester surfactants. J Clean Prod 19:332–336. https://doi.org/10.1016/j.jclepro.2010.10.006

Ying L, Zhao H, Li C, Yang H, Hu C, Wang Z (2022) Surface reconstruction and low-temperature dyeing performances of a poly(lactic acid) filament pretreated with a choline chloride and oxalic acid deep eutectic solvent. Macromolecules 55:6238–6246. https://doi.org/10.1021/acs.macromol.2c01012

Yousefi M, Rahimi-Nasrabadi M, Pourmortazavi SM, Wysokowski M, Jesionowski T, Ehrlich H, Mirsadeghi S (2019) Supercritical fluid extraction of essential oils. TrAC, Trends Anal Chem 118:182–193. https://doi.org/10.1016/j.trac.2019.05.038

Yuan J, Yu Y, Wang Q, Fan X, Chen S, Wang P (2013) Modification of ramie with 1-butyl-3-methylimidazolium chloride ionic liquid. Fibers Polym 14:1254–1260. https://doi.org/10.1007/s12221-013-1254-5

Zabot GL (2020) Chapter 11—Decaffeination using supercritical carbon dioxide. In: Inamuddin, Asiri AM, Isloor AM (eds) Green sustainable process for chemical and environmental engineering and science. Elsevier, pp 255–278

Zhao J, Ding L, Sui X, Mao Z, Xu H, Zhong Y, Zhang L, Chen Z, Wang B (2019) Bio-based polymer colorants from nonaqueous reactive dyeing of regenerated cellulose for plastics and textiles. Carbohydr Polym 206:734–741. https://doi.org/10.1016/j.carbpol.2018.11.056

Zhao H, Wang H, Wang M, Bai P, Tan L, Xiong X, Zheng L (2022) A recyclable anhydrous cotton dyeing technology with low energy consumption and excellent dyeing effects by mixing supercritical carbon dioxide, ethanol, and dimethyl sulfoxide. J Clean Prod 367. https://doi.org/10.1016/j.jclepro.2022.133034

Zheng M, Sun Y, Li C, Lu Y, Dai Y, Wang Z (2023) A novel and eco-friendly approach for dyeing polyester fabrics by liquid disperse dyes treated with deep eutectic solvent. Color Technol 139:552–564. https://doi.org/10.1111/cote.12673

Zhuo Y, Wang K, Hou L, Liu J, Fan Z, Fu Y, Dong A, Zhu B (2025) Molecular dynamics study on structural modulation and dyeing property optimization of meta-aramid in a DMSO/electrolyte system. J Colloid Interf Sci 678:732–741. https://doi.org/10.1016/j.jcis.2024.06.149

[illegible]

Chapter 5
Ultrasound-Assisted Dyeing: Efficiency, Performance, and Environmental Advantages

Md. Reazuddin Repon, Ashraful Alam Aulia, Latifa Noushin, Md. Shalit Hasan, Pranto Biswas, and Meherin Sultana

Abstract Ultrasound dyeing aims to decrease the environmental impact by assisting the dyeing efficiency with low disposal of waste material and moving towards sustainable and eco-friendly dyeing process as well as dye extraction. The ultrasound method assists in maintaining higher efficiency of natural dye extraction and enhancing the exhaustion percentage with improved dyeing properties as the usage of different mordants increases the effectiveness and lower disposal of wastewater. Ultrasound-assisted dyeing improves specific characteristics, such as exhaustion rate, Lower energy consumption, Uniform dye distribution, Colour fastness, better penetration, and Lower environmental impact. The paper aims to summarize the dye extraction process from different sources through ultrasound systems and how ultrasound systems enhance the dye extraction process from natural sources. The dye exhaustion process of natural dyes is also assessed in this chapter using the ultrasound system. The following is a detailed exposition of how the dyeing properties of materials dyed with natural dyes can be boosted by employing an ultrasound environment.

Keywords Textile coloration · Environmental impact · Ultrasound · Wastewater

Md. R. Repon (✉)
ZR Research Institute for Advanced Materials, Sherpur, Bangladesh
e-mail: reazmbstu.te@gmail.com

A. A. Aulia
Department of Dyes and Chemical Engineering, Bangladesh University of Textiles, Dhaka, Bangladesh

L. Noushin · Md. S. Hasan
Department of Wet Processing Engineering, Bangladesh University of Textiles, Dhaka, Bangladesh

P. Biswas
Department of Textile Machinery Design and Maintenance, Bangladesh University of Textiles, Dhaka, Bangladesh

M. Sultana
Department of Textile Fashion and Design, Bangladesh University of Textiles, Dhaka, Bangladesh

S. Maiti et al. (eds.), *Sustainable Coloration Techniques in Textiles*,
https://doi.org/10.1007/978-981-96-4975-4_5

5.1 Introduction

Traditional dyeing methods such as vat dyeing, direct dyeing, and reactive dyeing have been prolific in the textile industry for quite some time. These methods include methods where the fabrics allow the dye molecules to diffuse into the material to get the needed dyeing level (Hassan and Saifullah 2023). However, the traditional form of dyeing is a source of some serious environmental problems. For instance, vat dyeing uses a significant proportion of water and energy primarily for heating baths of vats and sustaining predetermined temperatures for dyeing (Lara et al. 2022). Further, the application of severe chemicals during the dyeing process leads to considerable chemical wastage and, hence, water pollution and harm to the water dwellers (Lellis et al. 2019).

Ultrasound technology provides a new technique in textile dyeing with less environmental effects and improved dyeing performance by using high-frequency sound waves (Hassan and Saifullah 2023; Islam et al. 2023). One of the primary mechanisms is acoustic cavitation, which entails the forming of bubbles in the dye solution and, after their collapse, forms micro jets that assist in improving the dye penetration on the textile fibers. As a result, there are enhanced dye penetration, dye color uniformity, and dye exhaustion rates (Vajnhandl and Majcen Le Marechal 2005; Repon et al. 2023). Furthermore, ultrasound-assisted dyeing also greatly minimizes the use of high temperatures and long processing periods, thus leading to low energy and water consumption.

This chapter aims to describe the applications of ultrasound technology in the dyeing of textiles, especially about efficiency, performance as well as sustainability. It will also try to explain how the application of ultrasound in dyeing will be instrumental in the conservation of the environment through the reduction of the formulation of chemical wastes, energy, and water, among others. In addition, it will also evaluate the applications of ultrasound during the processes of dye extraction and its capability of changing the vision of the textile industry on sustainability.

5.2 Ultrasound-Assisted Textile Dyeing

Ultrasound-assisted dyeing is a novel method that uses High-Frequency sound waves to augment dyeing processes. This method is suitable for both synthetic and natural dyes with improved uptake, penetration, and overall dyeing process (Khalil et al. 2023; Rai et al. 2023). The present work on the use of ultrasound in textile dyeing is of importance in reducing disturbance to the environment through the reduction of rigorous chemicals as well as conservation of water and energy.

5.2.1 Ultrasound-Assisted Textile Dyeing with Synthetic Dyes

Some synthetic dyes include acid dyes, vat dyes, reactive dyes, direct dyes, basic dyes, sulfur dyes, and disperse dyes, which are largely used in the textile industry due to their bright colors, various shades, and good color fastness (Islam et al. 2022). However, dyeing with those dyes traditionally has problems related to energy consumption, water pollution, and chemical waste (Lellis et al. 2019). This is considered a solution through enhancing the dye's uptake and penetration, enhancing dyeing processes.

Ultrasound also takes dye into its solution and improves the penetration as it forms cavitation bubbles in the dye solution. These bubbles burst rapidly and generate micro jets that aid in the uniform dispersion of the dye molecules on the textile fibers, resulting in enhanced dye fixation as well as improved color strength (Harifi and Montazer 2015; Yang et al. 2024). The ultrasonic waves also shorten the dyeing time by increasing the rate of dye–fiber interaction; these, in turn, also enhance the dye exhaustion rates.

5.2.1.1 Ultrasound-Assisted Dyeing with Acid Dyes

Acid dyes are anionic dyes whose affinities are good for protein fibers like wool, silk, and nylon in aqueous baths. These dyes are inclined towards the fibers-based contact between the dye molecules and the amino groups of the fibers (Musa 2013). The ultrasound-assisted dyeing with acid dyes results in enhanced dye fixation and color strength with improved penetration of dye molecules in the fiber structure, making uniform dye distribution and better fastness properties. This has been done by research proving that the use of ultrasound in the dyeing process augments the uptake of dye by as much as 30%, dicing the time used in the dyeing process in half and decreasing the amount of energy used by 40% (Ferrero and Periolatto 2012; Tissera et al. 2016).

5.2.1.2 Ultrasound-Assisted Dyeing with Vat Dyes

Vat dyes are insoluble in water and are mainly employed for the dyeing of cellulose fiber, for instance, cotton fabrics. They need to be reduced and processed to make them soluble in leuco form, which is then able to penetrate the fiber and then oxidize back to insoluble form (Broadbent 2001). The use of ultrasound also facilitates the dispersion of vat dyes and helps in achieving a better distribution of the dyes since the dyes tend to cluster together and thus need to be processed under high temperatures until they are in small particles that can easily penetrate the fibers (Vouters et al. 2004; Adeel et al. 2020).

5.2.1.3 Ultrasound-Assisted Dyeing with Reactive Dyes

Reactive dyes actually come into chemical action with the fiber's molecules; this makes them have good wash fastness and give a good depth of shade. These dyes are widely applied to dye cotton and other cellulosic fibres (Hunger 2008; Repon et al. 2024). Using ultrasound in reactive dyeing improves the dye-fiber bonding process since the reactivity of dye and fiber molecules is improved, leading to the formation of strong covalent bonds (Bian et al. 2022). This leads to an increase in fixation rate and a decrease in dyeing time, which in turn makes it an efficient process.

5.2.1.4 Ultrasound-Assisted Dyeing with Direct Dyes

Direct dyes, as their name suggests, are dyes that are soluble in water and can dye cellulosic fibers without the former for mordant. They are often employed in the dyeing of cotton and rayon and other cellulosic fiber systems (Rahman et al. 2019). Another advantage of ultrasound in the dyeing process is that it assists the direct dyes to have better penetration and distribution in the fiber by using the microcurrent to carry the dye molecules, which results in better dyeing and color strength (Udrescu et al. 2014; Hassan and Bhagvandas 2017).

5.2.1.5 Ultrasound-Assisted Dyeing with Basic Dyes

Basic dyes are blue-black in shade and are among the cationic types of dyes that are widely used in the dyeing of synthetic fabrics, particularly those made from acrylic (Eren 2012). These dyes easily bind to synthetic fibers carrying a negative charge, leading to highly colorful and beautiful shades. The application of ultrasound technology raises the exhaustion rates and dyeing uniformity of the basic dyes by increasing the interaction of the dye with the fiber and decreasing the wastage of the dye (Hassan and Saifullah 2018).

5.2.1.6 Ultrasound-Assisted Dyeing with Sulfur Dyes

Sulfur dyes are insoluble in water dyes, and they are used to heat cellulose fibers and use darker shades. They are associated with a very strong smell and a negative environmental impact, particularly because of sulfur-containing compounds (Asfaram et al. 2015). Ultrasound also heavily facilitates dyeing with sulfur dyes by imitating the solubilization of the dye and, therefore, the penetration of the dye onto the fiber and color development. This lowers the need to apply a lot of dye and prevents the emission of sulfur compounds into the atmosphere (Khandate 2012; Hosseinnezhad et al. 2022).

Table 5.1 Summary of ultrasound-assisted textile dyeing performance with different synthetic dyes

Dyes	Textile fiber	Types of sonicator	Dyeing performance	Refs.
Acid dyes	Wool, silk, nylon	Ultrasonic bath	Improved dye fixation, enhanced color strength	Ferrero and Periolatto (2012)
Vat dyes	Cotton	Probe sonicator	Better dispersion, improved color uniformity	Vouters et al. (2004)
Reactive dyes	Cotton, viscose	Ultrasonic bath	Enhanced dye-fiber bonding, reduced dyeing time	Habib et al. (2023)
Direct dyes	Cotton, rayon	Probe sonicator	Enhanced penetration, uniform dye distribution	Hassan and Saifullah (2023)
Basic dyes	Acrylic	Ultrasonic bath	Improved uniformity, higher exhaustion rates	Hassan and Saifullah (2018)
Sulfur dyes	Cotton	Probe sonicator	Enhanced solubilization, better penetration	Ma et al. (2020)
Disperse dyes	Polyester	Ultrasonic bath	Improved dispersion, better color fastness	Udrescu et al. (2014)

5.2.1.7 Ultrasound-Assisted Dyeing with Disperse Dyes

Disperse dyes are non-ionic dyes, and hydrophobic synthetic fibers such as polyester are suitable for fiber-to-dye systems. These dyes are not soluble in water and hence need a dispersing agent in order to get an even dispersion (Lee and Kim 2001; Repon et al. 2023). Ultrasound technology improves the wetting of the surfaces and the diffusion of disperse dyes in the dye bath and assertively improves dye penetration and color fastness. This method also allows organometallic intermediates with high temperatures to not undergo side reactions, thus saving energy and decreasing the byproducts that harm the environment (Burkinshaw and Jeong 2012; Udrescu et al. 2014) (Table 5.1).

5.2.2 Ultrasound-Assisted Textile Dyeing with Natural Dyes

Natural dyes are extracted from plants, animals, and minerals and are slowly included in the fashion industry due to being friendly to the environment and having little effect on the ecosystem. Natural dyes are eco-friendly or less hazardous to the environment because they can be washed out and do not have the synthetic toxic components of chemical dyes (Sheikh et al. 2016). Nonetheless, using natural dyes can be difficult because of poor color fastness and lower exhaustion. Some disadvantages are associated with natural dyeing, including low percentage yield, long time for dyeing, and low color development. Ultrasound technology overcomes these challenges through better dye extraction rate and dye-fiber affinity uptake (Sivakumar et al. 2009; Qi et al. 2023). The use of ultrasound makes it easy for the plant cells to be disrupted with a larger amount of dye compounds being released into the dye bath, hence enhancing the dye. Ultrasound dyeing has been reported to offer up to a 50% increase in dye diffusion rates and % dye extinction by enhancing vibrancy and color fastness (Banna et al. 2023). This method also helps to minimize dyeing at high temperatures, energy consumption, and the impacts that it has on the environment. The use of ultrasound increases the effectiveness of natural dyes because it helps to increase dye penetration and fixation, leading to better dye build-up and dye fastness characteristics (Gonzalez et al. 2019). It is also less damaging to the environment since Ultrasound also cuts down on the utilization of mordants which in several conventional natural dyeing approaches is vital in fixing the dye to the fiber (Hao et al. 2012) (Table 5.2).

Ultrasound-assisted textile dyeing is a major innovation in the dyeing process regarding dye uptake, pollution minimization, and dyeing effectiveness for both synthetic and natural dyes. This means that the ultrasound dyeing process is an efficient way of dyeing fabrics while avoiding some of the environmental practices, such as the resettlement of water by conventional dyeing procedures (Das et al. 2022).

Table 5.2 Summary of ultrasound-assisted textile dyeing performance with different natural dyes

Types of natural dyes	Textile fiber	Types of sonicator	Dyeing performance	Refs.
Indigo	Cotton	Ultrasonic bath	Enhanced dye uptake, better color fastness	Baig et al. (2021), Wang et al. (2023)
Madder	Wool	Probe sonicator	Improved extraction, uniform dye distribution	Helmy (2020)
Turmeric	Silk	Ultrasonic bath	Better color strength, reduced dyeing time	Ma et al. (2020), Habib et al. (2024)
Henna	Cotton	Probe sonicator	Enhanced penetration, improved fixation	Ahmad and Mitran (2019)
Saffron	Wool	Ultrasonic bath	Increased dye yield, better color consistency	Ma et al. (2020)

5.3 Environmental Advantages of Ultrasound-Assisted Textile Dyeing

The use of ultrasound-assisted textile dyeing can be noted as one of the innovative technologies that have remarked the textile dyeing industry due to the minimal environmental drawbacks compared with that of the traditional dyeing process. This solution employs ultrasonic waves to improve dyeing processes, which results in significant reductions in water, energy, chemicals, and utilization of waste.

5.3.1 Reduction in Water and Energy Consumption

The textile dyeing process has been known to consume water and energy, and many rinses are performed to get the required level of dye fixation and CI value. Bothered by the high water and energy usage of conventional dyeing processes, Ultrasound-assisted dyeing provides a sustainable solution (Shang 2013). Traditional dyeing processes entail water use in other states, such as washing, dyeing, and rinsing. For instance, it requires approximately 100–150 L of water to process 1 kg of cotton fabric (Aivazidou and Tsolakis 2018; Islam et al. 2023). On the other hand, ultrasound-assisted dyeing minimizes water use by half, improves the dye uptake to the fiber, and cuts down the number of rinse cycles required. This is done through cavitation, and the mechanism involves the formation of microbubbles that can cause enhanced heat and pressure region at the vicinity, enhancing dyes' solubility and their penetration into the fibers. Energy utilization is another important aspect, which can be marked as the advantage of the ultrasound-assisted dyeing process. In the conventional dyeing techniques, it is mandatory to heat substantive dyes at higher temperatures (60–100 °C) for longer durations to permit dye substantivity and exhaustion (Kamel et al. 2007; Islam et al. 2022). The overall treatment with ultrasound, though, facilitates dyeing at reduced temperature, commonly ranging between 30 and 40 °C. Therefore, energy consumption is reduced to 40% (Hassan and Saifullah 2023). This reduction is advantageous in fibers such as polyester, which require high thermal energy to disperse dyes. The working of ultrasound technology results in the formation of acoustic cavitation that supports the growth of the mass transfer rate between the dye and the fabric (Harifi and Montazer 2015). Local micro-streaming and shock waves of cavitation improve the kinetic energy of dye molecules and enhance penetration at a lower temperature and for shorter periods through the fiber. For example, the dyeing time of cotton using the reactive dye can be reduced from 60 to 30 min with the help of ultrasound, which shows a saving of 50% in the dyeing time (Li et al. 2020).

5.3.2 *Minimization of Chemical Usage and Waste Generation*

Another environmental issue that goes hand in hand with textile dyeing, particularly when it comes to choosing the type of dye to use, is the issue to do with chemicals and effluent pollution (Körlü and Bahtiyari 2021). Ultrasound-assisted dyeing solves this problem by reducing the amount of chemicals necessary for dyeing fixations and enhancing the dye penetration rates.

In conventional dyeing techniques, large proportions of dye materials, including salts, alkalis, and fixatives, are used to fix the dye in the textile material. Nevertheless, ultrasound-assisted dyeing cuts down the usage of these chemicals by improving the dyeing and fixation rates. For instance, Ultrasound has decreased salt consumption by about 60% during the reactive dyeing of cotton with the same or even better color yield and fastness properties (Harifi and Montazer 2015). This reduction in the amount of chemicals used also reduces the cost of raw materials, chemical production, and transportation. This method also ensures that the concentration of unreacted dyes and chemicals in effluent is minimized, reducing effluent pollution. This loss of dyes is rather high, ranging between 20 and 30%, and the effluent streams contain high concentrations of pollutants, creating a major environmental concern (Tripathi et al. 2023). The application of ultrasounds enhances Figure eight exhaustion ratios, a measure of dye perhaps as high as 95% for specific dyes like the reactive and direct dyes. This high dye exhaustion minimizes the amount of dye and chemical residues in the wastewater, hence limiting the extent of wastewater treatment and dangers of water pollution.

5.3.3 *Improved Environmental Footprint and Sustainability*

The application of ultrasound-assisted dyeing technology helps to reduce dramatically the contribution of textile dyeing to polluting the environment. This improvement aligns with the increasing need for environmentally friendly production in the textile business. Ultrasound-assisted dyeing reduces the environmental footprint in several key areas, including using water, energy, and chemicals and producing waste. By reducing water and energy consumption to 40–50% and chemical needs, this system reduces greenhouse gas emissions, protects natural resources, and minimizes the role of textiles in adverse environmental effects (Alsukaibi 2022). Further, it has been observed that the new process of dyeing made using ultrasound technology takes shorter production cycles, putting less emphasis on many batches and thus working to reduce carbon footprint. For textile manufacturers, the features of the proposed technology are not limited to environmental improvements. The outcome of lowering water intake, energy, and chemicals is the reduction in expenses and the enhancement of production and product quality (Hasanbeigi and Price 2015). In addition, it becomes easier to meet high environmental compliance standards and, therefore, avoid the chances of incurring hefty fines. It also improves the company's image on issues of corporate social responsibility.

5.3.4 *Case Studies and Examples*

A number of textile companies have effectively applied ultrasound-assisted dyeing, showing that this process is effective in terms of environmental impact and cost. The increased employment of an ultrasound-assisted dyeing method by a textile mill in India has seen a 40% cut in the use of water and 30% in energy during the dyeing process; at the same time, chemical usage has been cut in half (Jajpura and Nayak 2020). Another example is the dyeing factory in China that used ultrasound technology in polyester and attained a 60% reduction in the generation of wastes and 70% increase in the uptake of dyes (Zhao et al. 2022). The above examples show that sustainability in the textile industry and economic returns can be achieved through ultrasound-assisted dyeing. In terms of resource conservation and an overall reduction in pollution, ultrasound-assisted textile dyeing is highly efficient and thus can be considered a sustainable method compared to the traditional methods of dyeing. It can also be instrumental in introducing sustainable practices incorporated into textile manufacturing worldwide.

5.4 Conclusion

The use of ultrasound in textile dyeing proves to be advantageous over conventional techniques in all aspects, such as efficiency and efficacy, and it contributes to the conservation of the environment. It also helps improve dye uptake and fixation, reduce dyeing time, and lower the temperatures necessary for dyeing, thus saving great energy and water. It also ensures that chemical utilization are limited to positively impact the environment and the effluent pollution of textile dyeing processes. These benefits shed light on the role of ultrasound technology in changing textile dyeing practices with improved sustainability. A large-scale application of the ultrasound-assisted dyeing process in textiles is possible. As the issue of sustainability is gaining importance for manufacturers and consumers all over the world, implementing such technologies could be in line with sustainable development objectives. It is imperative that manufacturers embrace ultrasound-assisted dyeing techniques to meet the increased regulatory compliance on environmental standards regarding corporate social responsibility and yet make a bottom line in cost by delivering high-quality products. More studies are required to select the best ultrasound parameters for the specific dyeing and fabrics for the best results and environmental impact. Further research on the impacts of applying ultrasound technology on different kinds of textiles and examining its compatibility with other green practices might offer significant information for developing advanced, environmentally friendly dyeing textiles.

Declaration The authors declare that they have no known competing financial interests or personal relationships that could have appeared to influence the work reported in this chapter. Generative AI and AI-assisted technologies have been used in the writing process to improve the readability and language of the work only. After using this tool, the authors carefully reviewed and revised the content to ensure accuracy and quality. The authors take full responsibility for the final version of this publication.

References

Adeel S, Saeed M, Abdullah A et al (2020) Ultrasonic assisted improved dyeing of cellulosic fabric using Vat Blue 4. J Nat Fibers 17:1–8. https://doi.org/10.1080/15440478.2018.1458683

Ahmad SW, Mitran EC (2019) An efficient ultrasonic and microwave assisted extraction of organic Henna dye for dyeing of synthetic polyester fabric for superior color strength properties. Ind Textila J 70:303–308

Aivazidou E, Tsolakis N (2018) Water footprint management in the fashion supply chain: a review of emerging trends and research challenges. In: Water in textiles and fashion: consumption, footprint, and life cycle assessment. Elsevier Ltd., pp 77–94

Alsukaibi AKD (2022) Various approaches for the detoxification of toxic dyes in wastewater. Processes 10:1–27. https://doi.org/10.3390/pr10101968

Asfaram A, Ghaedi M, Hajati S et al (2015) Simultaneous ultrasound-assisted ternary adsorption of dyes onto copper-doped zinc sulfide nanoparticles loaded on activated carbon: optimization by response surface methodology. Spectrochim Acta - Part A Mol Biomol Spectrosc 145:203–212. https://doi.org/10.1016/j.saa.2015.03.006

Baig U, Khatri A, Ali S et al (2021) Ultrasound-assisted dyeing of cotton fabric with natural dye extracted from Marigold flower. J Text Inst 112:801–808. https://doi.org/10.1080/00405000.2020.1779907

Banna BU, Mia R, Hasan MM et al (2023) Ultrasonic-assisted sustainable extraction and dyeing of organic cotton fabric using natural dyes from Dillenia indica leaf. Heliyon 9:e18702. https://doi.org/10.1016/j.heliyon.2023.e18702

Bian X, Fan S, Xia G et al (2022) Sustainable ultrasonic dyeing of unscoured and unbleached cotton sliver using low liquor ratio. J Clean Prod 379:134853. https://doi.org/10.1016/j.jclepro.2022.134853

Broadbent AD (2001) Basic principles of textile coloration. Society of Dyers and Colorists, Bradford, UK

Burkinshaw SM, Jeong DS (2012) The dyeing of poly(lactic acid) fibres with disperse dyes using ultrasound: part 1—initial studies. Dye Pigment 92:1025–1030. https://doi.org/10.1016/j.dyepig.2011.06.020

Das P, Nayak PK, krishnan Kesavan R (2022) Ultrasound assisted extraction of food colorants: principle, mechanism, extraction technique and applications: a review on recent progress. Food Chem Adv 1:100144. https://doi.org/10.1016/j.focha.2022.100144

Eren Z (2012) Ultrasound as a basic and auxiliary process for dye remediation: a review. J Environ Manag 104:127–141. https://doi.org/10.1016/j.jenvman.2012.03.028

Ferrero F, Periolatto M (2012) Ultrasound for low temperature dyeing of wool with acid dye. Ultrason Sonochem 19:601–606. https://doi.org/10.1016/j.ultsonch.2011.10.006

Gonzalez V, Wood R, Lee J et al (2019) Ultrasound-enhanced hair dye application for natural dyeing formulations. Ultrason Sonochem 52:294–304. https://doi.org/10.1016/j.ultsonch.2018.11.028

Habib N, Adeel S, Ali A et al (2023) Ultrasonic-assisted sustainable pollution free advanced method for isolation of colouring material from Amba Haldi (Curcuma aromatica) for wool dyeing. Sci Prog 106:1–20. https://doi.org/10.1177/00368504231188610
Habib N, Adeel S, Ibrahim M et al (2024) Eco-friendly ultrasonic-assisted appraisal of herbal-based yellow natural colorant for silk dyeing. Energy Sci Eng 12:5061–5077. https://doi.org/10.1002/ese3.1928
Hao L, Wang R, Liu J, Liu R (2012) Ultrasound-assisted adsorption of anionic nanoscale pigment on cationised cotton fabrics. Carbohydr Polym 90:1420–1427. https://doi.org/10.1016/j.carbpol.2012.07.010
Harifi T, Montazer M (2015) A review on textile sonoprocessing: a special focus on sonosynthesis of nanomaterials on textile substrates. Ultrason Sonochem 23:1–10. https://doi.org/10.1016/j.ultsonch.2014.08.022
Hasanbeigi A, Price L (2015) A technical review of emerging technologies for energy and water efficiency and pollution reduction in the textile industry. J Clean Prod 95:30–44. https://doi.org/10.1016/j.jclepro.2015.02.079
Hassan MM, Bhagvandas M (2017) Sustainable ultrasound-assisted ultralow liquor ratio dyeing of wool textiles with an acid dye. ACS Sustain Chem Eng 5:973–981. https://doi.org/10.1021/acssuschemeng.6b02293
Hassan MM, Saifullah K (2023) Ultrasound-assisted sustainable and energy efficient pre-treatments, dyeing, and finishing of textiles—a comprehensive review. Sustain Chem Pharm 33:101109. https://doi.org/10.1016/j.scp.2023.101109
Hassan MM, Saifullah K (2018) Ultrasound-assisted pre-treatment and dyeing of jute fabrics with reactive and basic dyes. Ultrason Sonochem 40:488–496. https://doi.org/10.1016/j.ultsonch.2017.07.037
Helmy H (2020) Extraction approaches of natural dyes for textile coloration. J Text Color Polym Sci 17:65–76. https://doi.org/10.21608/jtcps.2020.30990.1037
Hosseinnezhad M, Gharanjig K, Adeel S et al (2022) The effect of ultrasound on environmentally extraction and dyeing of wool yarns. J Eng Fiber Fabr 17:1–10. https://doi.org/10.1177/15589250221104471
Hunger K (2008) Industrial dyes: chemistry, properties, applications
Islam T, Islam T, Islam T, Repon R (2022) Synthetic dyes for textile colouration: process, factors and environmental impact. Text Leather Rev 5:327–373. https://doi.org/10.31881/TLR.2022.27
Islam T, Repon MR, Islam T et al (2023) Impact of textile dyes on health and ecosystem: a review of structure, causes, and potential solutions. Environ Sci Pollut Res 30:9207–9242. https://doi.org/10.1007/s11356-022-24398-3
Jajpura L, Nayak R (2020) Ultrasound applications in textiles and apparels. In: Nayak R (ed) Sustainable technologies for fashion and textiles. Woodhead Publishing, pp 143–161
Kamel MM, El-Shishtawy RM, Youssef BM, Mashaly H (2007) Ultrasonic assisted dyeing. IV. Dyeing of cationised cotton with lac natural dye. Dye Pigment 73:279–284. https://doi.org/10.1016/j.dyepig.2005.12.010
Khalil E, Sarkar J, Rahman MM et al (2023) Advanced technology in textile dyeing. In: Rahman MM, Mashud M, Rahman MM (eds) Advanced technology in textiles. Springer, Singapore, pp 97–138
Khandate G (2012) Ultrasound assisted oxidation process for the removal of aromatic contamination from effluents: a review. Univers J Environ Reasearch Technol 2:458–464
Körlü A, Bahtiyari Mİ (2021) Ultrasound-based wet processes in textile industry. In: Innovative and emerging technologies for textile dyeing and finishing. Wiley, pp 1–25
Lara L, Cabral I, Cunha J (2022) Ecological approaches to textile dyeing: a review. Sustainability 14:1–17. https://doi.org/10.3390/su14148353
Lee KW, Kim JP (2001) Effect of ultrasound on disperse dye particle size. Text Res J 71:395–398. https://doi.org/10.1177/004051750107100505

Lellis B, Fávaro-Polonio CZ, Pamphile JA, Polonio JC (2019) Effects of textile dyes on health and the environment and bioremediation potential of living organisms. Biotechnol Res Innov 3:275–290. https://doi.org/10.1016/j.biori.2019.09.001

Li Q, Ni L, Wang J et al (2020) Establishing an ultrasound-assisted activated peroxide system for efficient and sustainable scouring-bleaching of cotton/spandex fabric. Ultrason Sonochem 68:105220. https://doi.org/10.1016/j.ultsonch.2020.105220

Ma X, Wei Y, Wang S et al (2020) Sustainable ultrasound-assisted ultralow liquor ratio dyeing of cotton fabric with natural turmeric dye. Text Res J 90:685–694. https://doi.org/10.1177/0040517519878793

Musa H (2013) Studies on the dyeing of wool and nylon fabrics with some acid dyes. IOSR J Appl Chem 5:11–17. https://doi.org/10.9790/5736-0511117

Qi Y, Zhou Z, Xu R et al (2023) Research on the dyeing properties of Chinese fir using ultrasonic-assisted mulberry pigment dyeing. Forests 14:1–18. https://doi.org/10.3390/f14091832

Rahman SMA, Kumar Saha A, Ruhi RA et al (2019) Decolourization of textile azo dye direct Red 81 by bacteria from textile industry effluent. Int J Curr Microbiol Appl Sci 8:1742–1754. https://doi.org/10.20546/ijcmas.2019.804.203

Rai PK, Kant V, Sharma RK, Gupta A (2023) Process optimization for textile industry-based wastewater treatment via ultrasonic-assisted electrochemical processing. Eng Appl Artif Intell 122:106162. https://doi.org/10.1016/j.engappai.2023.106162

Repon MR, Dev B, Rahman MA et al (2024) Textile dyeing using natural mordants and dyes: a review. Springer International Publishing

Repon MR, Islam T, Islam T et al (2023) Cleaner pathway for developing bioactive textile materials using natural dyes: a review. Environ Sci Pollut Res 1–31. https://doi.org/10.1007/s11356-023-26131-0

Shang SM (2013) Process control in dyeing of textiles. In: Majumdar A, Das A, Alagirusamy R, Kothari VK (eds) Process control in textile manufacturing. Woodhead Publishing, pp 300–338

Sheikh J, Jagtap PS, Teli MD (2016) Ultrasound assisted extraction of natural dyes and natural mordants vis a vis dyeing. Fibers Polym 17:738–743. https://doi.org/10.1007/s12221-016-5031-0

Sivakumar V, Anna JL, Vijayeeswarri J, Swaminathan G (2009) Ultrasound assisted enhancement in natural dye extraction from beetroot for industrial applications and natural dyeing of leather. Ultrason Sonochem 16:782–789. https://doi.org/10.1016/j.ultsonch.2009.03.009

Tissera ND, Wijesena RN, De Silva KMN (2016) Ultrasound energy to accelerate dye uptake and dye-fiber interaction of reactive dye on knitted cotton fabric at low temperatures. Ultrason Sonochem 29:270–278. https://doi.org/10.1016/j.ultsonch.2015.10.002

Tripathi M, Singh S, Pathak S et al (2023) Recent strategies for the remediation of textile dyes from wastewater: a systematic review. Toxics 11:1–25. https://doi.org/10.3390/toxics11110940

Udrescu C, Ferrero F, Periolatto M (2014) Ultrasound-assisted dyeing of cellulose acetate. Ultrason Sonochem 21:1477–1481. https://doi.org/10.1016/j.ultsonch.2013.12.026

Vajnhandl S, Majcen Le Marechal A (2005) Ultrasound in textile dyeing and the decolouration/mineralization of textile dyes. Dye Pigment 65:89–101. https://doi.org/10.1016/j.dyepig.2004.06.012

Vouters M, Rumeau P, Tierce P, Costes S (2004) Ultrasounds: an industrial solution to optimise costs, environmental requests and quality for textile finishing. Ultrason Sonochem 11:33–38. https://doi.org/10.1016/S1350-4177(03)00113-5

Wang K, Meng Y, Luo Q et al (2023) An ultrasound-assisted nitrogen protection system for sustainable indigo dyeing of cotton fabrics. Text Res J 93:3355–3367. https://doi.org/10.1177/00405175221144779

Yang N, Jun BM, Choi JS et al (2024) Ultrasonic treatment of dye chemicals in wastewater: a review. Chemosphere 354:141676. https://doi.org/10.1016/j.chemosphere.2024.141676

Zhao H, Wang H, Wang M et al (2022) A recyclable anhydrous cotton dyeing technology with low energy consumption and excellent dyeing effects by mixing supercritical carbon dioxide, ethanol, and dimethyl sulfoxide. J Clean Prod 367:133034. https://doi.org/10.1016/j.jclepro.2022.133034

Chapter 6
Sustainable Colouration Through Electrolysis: The Electrochemical Approach to Dyeing

Amit Madhu, Neha, and Astha

Abstract This chapter discusses the advancements in electrochemical reduction as an innovative approach to sustainable dyeing in the textile industry. Traditional dyeing processes, often reliant on toxic chemicals, excessive water, and high energy consumption, are contrasted with electrochemical methods, which provide an eco-friendly alternative, particularly for vat and sulfur dyes in cellulosic textiles. The chapter explores the principles of direct and mediator-enhanced reduction, including electrocatalytic hydrogenation, where dyes react with in-situ hydrogen generated during water electrolysis. By addressing issues such as dyeability, fastness, and environmental sustainability, this chapter highlights the potential of electrochemical dyeing while acknowledging challenges like scalability, cost, and industrial adoption. It calls for further research into technical, economic, and environmental dimensions to realize the transformative potential of this technology for sustainable textile coloring.

Keywords Coloration · Eco-friendly · Electrochemistry · Sustainability · Textile

6.1 Introduction

Conventional textile dyeing processes are widely recognized for their significant environmental impact. The extensive use of chemicals, water, and energy in dyeing operations causes significant pollution. The principal causes of pollution in conventional textile dyeing are using dyes and auxiliaries, the discharge of untreated or partially treated wastewater, and the energy-intensive nature of dyeing operations. The traditional vat and sulfur dyeing processes substantially contribute to pollution in the textile industry, owing to the usage of vast amounts of toxic chemicals, water,

A. Madhu (✉) · Neha · Astha
Textile Chemistry Department, The Technological Institute of Textile and Sciences, Bhiwani, India
e-mail: amit.madhu5@gmail.com

S. Maiti et al. (eds.), *Sustainable Coloration Techniques in Textiles*,
https://doi.org/10.1007/978-981-96-4975-4_6

and energy. Both vat and sulfur dyes are insoluble in water and must be chemically reduced to their soluble leuco counterparts before dyeing. Typically, reducing agents like sodium dithionite ($Na_2S_2O_4$) for vat dyes and sodium sulfide (Na_2S) for sulfur dyes are used. Sodium dithionite is unstable and decomposes quickly, releasing sulfites and other hazardous byproducts that contribute considerably to water pollution. In the process of sulfur dyeing, sodium sulfide is employed, which is corrosive and toxic. This results in high levels of sulfide contamination in effluents, which can emit hydrogen sulfide (H_2S), a hazardous gas. The effluents from these dyeing processes are often highly alkaline (pH 10–12), contributing to the disruption of aquatic ecosystems and posing challenges for wastewater treatment (Chakraborty 2014; Ahmed and El-Shishtawy 2010).

In addition, vat and sulfur dyeing need a lot of water for washing and cleaning, which creates a lot of wastewater that has loose dyes, salts, heavy metals, and other coloring aids in it. These effluents often have a high chemical oxygen demand (COD) and biological oxygen demand (BOD), causing oxygen depletion in receiving waters and endangering aquatic life. Moreover, the solid trash from wastewater treatment is frequently loaded with dangerous chemicals that make it hard to get rid of and bad for the environment. The high temperature and energy requirements for dye fixation in conventional processes also contribute to carbon emissions, adding to air pollution and the overall environmental footprint of these dyeing techniques (Ahmed and El-Shishtawy 2010; Božič and Kokol 2008).

The electrochemical reduction of vat and sulfur dyes represents a sustainable alternative to conventional dyeing processes that rely on hazardous reducing agents. This chapter investigates the function of electrochemical reduction in textile dyeing, its potential to reduce environmental impact, and current technological breakthroughs that allow for industrial implementation. Focus is placed on the principles of electrochemical reduction, various methods employed, and the integration of electrolysis in dyeing processes for enhanced sustainability. Key advancements in electrochemical reactors and their role in optimizing energy and chemical efficiency are discussed.

6.2 Sustainable Alternatives to Vat and Sulfur Dyeing

Vat dyes are a class of water-insoluble dyes that require reduction to their soluble leuco form for textile dyeing. Sulfur dyes, similarly, require reduction to be solubilized. Both dyes are widely used for dyeing cellulosic fibers, particularly cotton, because of their excellent wash fastness and light fastness properties. Traditionally, sodium dithionite ($Na_2S_2O_4$) is used for this reduction, but it poses several environmental risks, including its instability, high reactivity, and contribution to water pollution. Alternative reduction techniques for vat and sulfur dyes have been developed to replace hazardous reducing agents like sodium dithionite ($Na_2S_2O_4$) for vat dyes and sodium sulfide (Na_2S) for sulfur dyes. These alternatives focus on reducing environmental pollution, enhancing sustainability, and improving overall process efficiency (Ahmed and El-Shishtawy 2010; Božič and Kokol 2008; Madhu et al. 2012).

The eco-friendly approach involves using organic reducing agents such as glucose, fructose, and sucrose. These sugars act as electron donors, reducing vat and sulfur dyes to their soluble leuco forms. Organic reducing agents offer several benefits, including biodegradability and lower toxicity, leading to significantly cleaner wastewater than traditional reducing agents. These agents are also inexpensive and widely available. However, the reduction efficiency of organic reducing agents is lower, requiring larger quantities for effective dye reduction. The reaction kinetics are also slower, extending process times. Despite these challenges, organic reducing agents are gaining traction in small- to medium-scale textile operations, though large-scale adoption remains limited due to concerns about efficiency (Chakraborty 2014; Madhu et al. 2012).

Sodium borohydride ($NaBH_4$) presents another viable alternative, offering high reduction power and faster reaction times than glucose or other organic reducing agents. Sodium borohydride is stable and can efficiently reduce vat and sulfur dyes, requiring smaller quantities than sodium dithionite or sodium sulphide. The by-products of sodium borohydride reduction, such as hydrogen gas and borates, are non-toxic, leading to significantly cleaner effluents. However, the cost of sodium borohydride is considerably higher than traditional reducing agents, and it must be handled carefully due to its reactive nature. Despite these limitations, sodium borohydride is used in specialized dyeing processes, prioritizing cleaner effluents and faster dye reduction (Chakraborty 2014; Božič and Kokol 2008; Roessler and Jin 2003).

Another alternative involves the use of hydrogen peroxide (H_2O_2) to assist in the reduction of vat and sulfur dyes. The oxidative decomposition of hydrogen peroxide generates nascent oxygen, which indirectly facilitates the reduction of dyes. This method offers several environmental benefits, including the production of water and oxygen as by-products, making it one of the cleanest reduction methods available. However, hydrogen peroxide-assisted reduction is less efficient than sodium dithionite or sodium sulphide in terms of speed and dye reduction power, which may lead to longer process times or require higher concentrations of H_2O_2. Its adoption remains limited to niche applications where environmental considerations outweigh the need for process efficiency (Ahmed and El-Shishtawy 2010; Madhu et al. 2012).

Biological reduction using microorganisms represents a highly sustainable alternative. In this method, certain bacteria reduce vat and sulfur dyes in anaerobic environments by metabolizing organic substrates, producing reducing equivalents that convert the dye into its leuco form. Biological reduction is eco-friendly, producing no harmful chemical by-products, and requires very little energy. However, the slow reaction rate and the need for precise control of conditions such as pH, temperature, and nutrient levels present significant challenges. Scaling biological reduction to industrial levels remains difficult, though pilot projects have shown promising results (Chakraborty 2014; Ahmed and El-Shishtawy 2010; Božič and Kokol 2008; Roessler and Jin 2003).

Electrochemical reduction is one of the most advanced alternatives to traditional chemical reduction methods. In this process, electrical energy is used to directly transfer electrons to dye molecules, reducing them at the cathode of an electrochemical cell. This makes the dye soluble and capable of bonding to the fiber without the use of harmful chemicals. The major advantages of electrochemical reduction include the elimination of sodium dithionite or sodium sulphide, resulting in much cleaner effluents and lower chemical oxygen demand (COD) and biological oxygen demand (BOD). Additionally, the process allows for better control of reduction through adjustable parameters like current density and voltage. However, the high initial capital costs and energy consumption required for this process pose challenges, though its industrial application, especially in vat dyeing, has shown success in reducing environmental impact (Ahmed and El-Shishtawy 2010; Roessler and Jin 2003; Bechtold and Turcanu 2009).

6.3 Sustainability Through Electrolysis

Electrolysis offers a cleaner and more controlled reduction process than traditional chemical methods. The reduction process can be fine-tuned to achieve high dyeing efficiency with minimal waste by controlling parameters such as current density, voltage, and temperature. Electrochemical reduction also allows for closed-loop systems, where water and supporting electrolytes can be reused, reducing water consumption and effluent discharge.

6.3.1 Principle of Electrochemical Reduction

Electrochemical reduction involves the transfer of electrons to the dye molecules through the use of an electric current rather than chemical reducing agents. The dye is reduced at the cathode in an electrochemical cell, gaining electrons and becoming solubilized (Bechtold and Brunner 2005; Bechtold and Turcanu 2009). There are several types of electrochemical reduction methods, each varying in its approach and mechanism. Below are the primary types of electrochemical reduction methods applied to vat, sulfur, and indigo dyes:

6.3.1.1 Direct Electrochemical Reduction

Direct electrochemical reduction involves applying an electrical potential directly to reduce dye molecules at the surface of an electrode. In its oxidized form, the dye gains electrons from the cathode, converting it to its reduced (leuco) form, which is water-soluble and can be applied to textile substrates. To preserve the dye in its solubilized state, the pH of the dye solution must be kept alkaline, and the direct

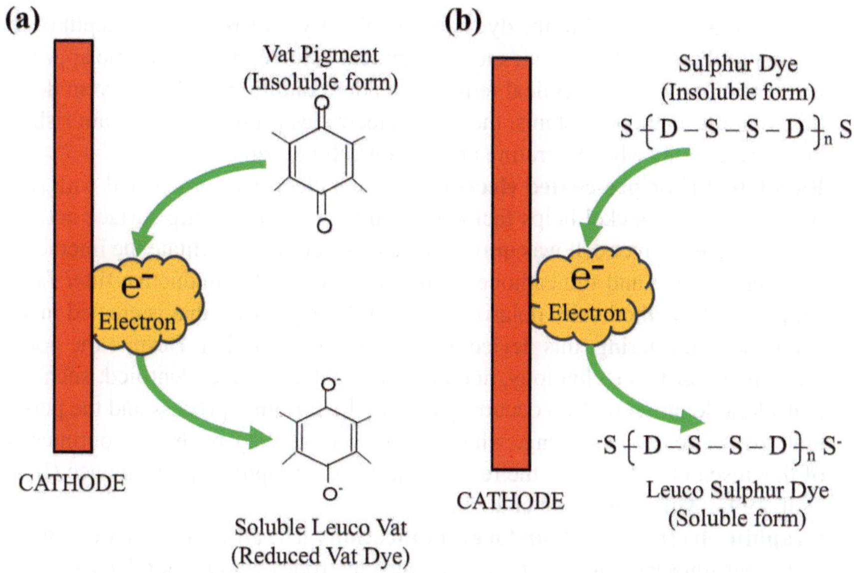

Fig. 6.1 Direct electrochemical reduction mechanism of **a** vat and **b** sulfur dyes

reduction process requires a reducing agent to begin (Holme 2002; Firoz et al. 2009). Figure 6.1 illustrates the process by which electrons are transferred from the cathode to reduce sulfur and vat dyes for application.

This process is straightforward but faces efficiency challenges. In direct electrochemical reduction, electrons must be transferred directly from the cathode surface to the surface of dispersed vat dye pigments. Vat, indigo, and sulfur dyes are water-insoluble and exist as dispersed pigments, leading to a very low likelihood of contact between the dye particles and the cathode surface (Firoz et al. 2009; Bechtold et al. 1997; Miled et al. 2014). Consequently, the rate-limiting step in direct electrochemical reduction is the electron transfer from the cathode, making direct cathodic reduction highly challenging. Furthermore, certain dyes, such as vat and indigo, require a reduction potential higher than typically achievable through direct methods (Miled et al. 2014, 2018; Roessler et al. 2001). Therefore, an alternative technique for electron transfer is needed to enhance the electron transfer rate.

The performance of direct electrochemical reduction can be improved by:

- **Multi-cathode systems**: Utilizing a multi-cathode electrolyzer increases the active surface area, enhancing electron transfer and reducing efficiency (Bechtold et al. 2009, 2008). In place of reducing agents, Bechtold et al. (2008) employed direct cathodic reduction of CI Sulfur Black 1 dye. Directly connecting a multi-cathode electrochemical cell with a dyeing machine allows instantaneous generation of a reduced dyebath and easy control of the dyeing parameters through cell current and dye-bath redox potential measurements. The investigation of the color depth using K/S and CIE lab coordinates (Bechtold et al. 2008, 2009) shows that the redox potential in the dye solution directly affects the color depth of the dyeing process. Multi-cathode electrochemical cells represent a viable approach for facilitating electrochemical reduction under conditions of low current densities. In multi-cathode systems, the initial electrodes present a significant risk for bipolar reactions when operating at elevated cell currents.

- **Raney-nickel or pre-coated electrodes**: Using electrodes pre-coated with catalysts like Raney-nickel helps facilitate reduction by increasing surface activity. The pre-coated layer cell was introduced as a solution to facilitate the interaction between the dye and the cathode. Using an electrically conductive filter fabric support and a cathodically polarized layer of Raney-nickel that is created in situ using precoat filtering, this device works based on filtration. Despite the potential benefits of this technology, notable disadvantages were identified, such as a considerable pressure drop occurring during the filtration process and the persistent risk of reactor clogging, which ultimately resulted in the discontinuation of this research. Moreover, the reactor exhibited suboptimal performance (Bond et al. 1997; Kulandainathan et al. 2008).
- **Graphite electrodes with surface modifications**: The efficiency of reduction can be further improved by modifying graphite electrodes with immobilized quinone compounds, which increase the effective surface area and improve dye interaction with the cathode (Miled et al. 2014; Roessler et al. 2002a). The effort was also coordinated to improve the chemical activity of the graphite surface, thereby increasing the reduction rate. The presence of oxygen functionalities, including quinone and hydroquinone groups, either directly on the surface of carbon black or as part of more complex compounds, is linked to the chemical activity of the graphite (carbon) surface. Therefore, by selectively creating quinine-like functionalities on the electrode surface, it is feasible to accelerate electron transport and the rate of adsorption on graphite electrodes (Roessler et al. 2002b, 2003; Roessler and Crettenand 2004). Kulandainathan et al. in their study utilized iron-deposited graphite surfaces as cathode materials for the electrochemical vat-dyeing process, employing a very low concentration of sodium dithionite. The use of electrode-posited graphite with high-surface area cathodes presents a more efficient alternative to utilizing multiple cathodes in the electrochemical cell (Kulandainathan et al. 2008; Roessler et al. 2002b).

6.3.1.2 In-Direct Electrochemical Reduction

In-direct electrochemical reduction involves using redox mediators that shuttle electrons between the electrode and the dye molecules. Mediators are compounds that can reversibly accept and donate electrons, acting as intermediaries in the electron transfer process. This method addresses the limitations of direct reduction by facilitating better electron transfer, improving overall efficiency. A redox mediator is first reduced at the cathode. This reduced mediator then transfers electrons to the dye, reducing it to its leuco form, Fig. 6.2. The oxidized mediator is then regenerated at the cathode, making the process cyclic (Ahmed and El-Shishtawy 2010; Božič and

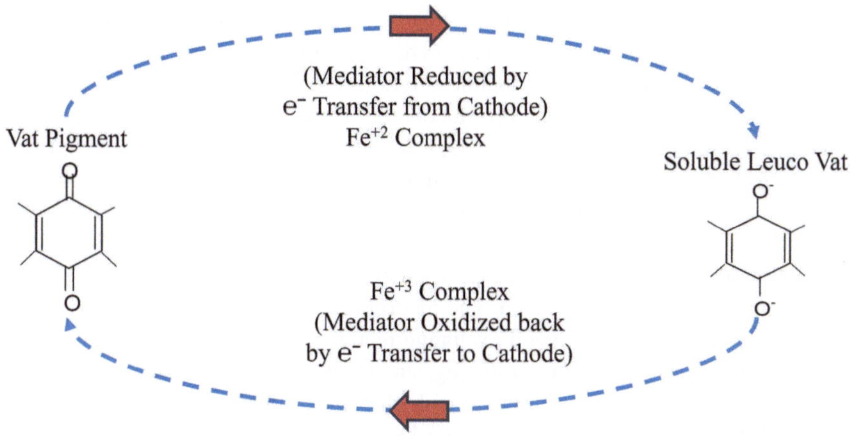

Fig. 6.2 In-direct electrochemical reduction of vat dye using Fe (Iron) based mediator

Kokol 2008; Bechtold et al. 1994). The most significant aspect of this process for enhancing sustainability is evident.

Common mediators employed for the vat and sulfur dye reduction include:

- *Substituted anthraquinones*

These compounds are organic mediators capable of reversible reduction or oxidation processes, rendering them a noteworthy class with potential applications as mediators for the indirect cathodic reduction of sulfur, indigo, and vat dyes. Anthraquininoid compounds, including anthraquinone sulfonic acids and hydroxyl anthraquinones, are significant, particularly concerning sulfur dyes, which may precipitate in the presence of iron-sulfide (iron-complex mediator). Examples include 9,10-anthraquinone-2-sulphonate, 9,10-anthraquinone-2,5-disulphonate, and dihydroxy-9–10-anthraquinone (Bechtold et al. 1994; Kulandainathan et al. 2007a). The primary drawback of these organic mediator systems is the comparatively low conversion rate at the electrodes. The limited solubility of anthraquinone-based mediators at lower pH leads to their deposition on cellulosic fibers, thereby reducing the concentration of the mediator system during recovery. Moreover, due to the coloration of many studied anthraquinones, it is essential to prevent absorption onto the fiber to avoid unwanted alterations in the dye color (Bechtold et al. 2010; DyStar 2004).

- *Iron-amine and iron-carbohydrate complexes*

These inorganic reversible redox systems have shown great promise in industrial applications. Inorganic redox systems that are most appropriate consist of salts of Fe(II)/Fe(III) complexed with ligands. Ligands coordinate with iron to form complexes, serving as mediators in the system. Ligands significantly influence the cathodic peak potential in indirect systems. The stability of coordination molecules in alkaline media and the maximum achievable redox potential are critical factors in identifying suitable ligands for Fe(II/III) complexes. Among the various ligands examined, only HEDTA (Hydroxyethylethylenediaminetriacetic acid), Bicine, Diethanolamine, and TEA (Triethanolamine) demonstrated significant stability of Fe(II) complexes in alkaline solutions (Miled et al. 2008).

Gluconic acid complexes (D-gluconate, Hepta-D-gluconate) demonstrate enhanced stability at reduced pH levels; however, the current density attained with these carbohydrate complexes is inferior to that of iron-amino complexes. A mixture of triethanolamine and D-gluconate was examined for its potential as a mediator. The utilization of a ligand mixture enables the effective integration of the benefits associated with both ligand systems, specifically the high current density of Fe(TEA) and the enhanced stability of Fe(DGL). The incorporation of Fe (DGL) and Fe(HDGL) with Fe(TEA) facilitates a ligand exchange reaction, resulting in effective vat dye reduction. In electrochemical vat dyeing, an alternative iron complex, specifically the Fe(III) + oxalate system combined with gluconate, functions as a redox complex (DyStar 2004; Miled et al. 2008) (Table 6.1).

These mediators are expensive and not entirely safe from a toxicological standpoint. The mediator must be separated from the soluble leuco dye through ultrafiltration before the dyeing process, and the concentration of the mediator in the filtrate should be increased via nano-filtration. This facilitates the recovery of the mediator, which incurs a 15% loss, along with the recycling of water and textile auxiliaries. Filtration significantly increases the costs associated with this reduction process and introduces various technical challenges, including substantial pressure drops during filtration and the persistent risk of reactor (electrochemical cell) blockage (Kulandainathan et al. 2007b; Turcanu et al. 2011). Consequently, despite the elapsed time, this mediator technique remains under development, with production trials confined to the pilot plant stage. Consequently, there remains a necessity for innovative alternatives to improve the eco-efficiency of vat dye application.

6.3.1.3 Electrolytic Hydrogenation

Electrolytic hydrogenation is an indirect electrochemical reduction method where hydrogen gas is produced in situ by electrolysis of water. The hydrogen gas generated at the cathode reacts with the adsorbed dye molecules, reducing them to their leuco forms, Fig. 6.3. This process mimics the action of traditional hydrogenation but occurs electrochemically. Catalytic hydrogenation produces leuco dye, which is then conventionally reverted to pigment form by air oxidation (Roessler and Jin 2003; Bechtold and Brunner 2005; Saïd et al. 2008).

The predominant amount of hydrogen generated through electrochemical processes does not engage with the vat dyes; rather, molecular hydrogen gas is produced, and subsequently released from the cathode. The hydrogenation process contends with the hydrogen evolution reaction, and this rivalry influences the efficiency of electrocatalytic hydrogenation (Comisso and Mengoli 2003). The issue of hydrogen ions releasing hydrogen gas rather than functioning as a reducing agent was addressed through the use of a hydrogenation metal catalyst, which facilitates the homolytic or heterolytic bond cleavage of the H–H bond as anticipated. Metal

Table 6.1 Key categories of mediator systems for indirect cathodic reductions, along with relevant examples and their appropriateness

Mediator type	Potential example	Redox potential (mV vs. Ag/AgCl)	Temperature (°C)	pH	Dyestuff	References
Anthraquinones	1,2-Dihydroxyanthraquinone; Anthraquinone-1,5-sulphonate	−500 to −700	95	11–13	Sulfur	Turcanu and Bechtold (2011)
Iron-amine complex	Fe (II/III) + TEA (Triethanolamine)	−1000	80	12–14	Vat Indigo Sulfur	Bechtold and Turcanu (2006)
Iron-carbohydrate complex	Fe (II/III) + DGL (D-Gluconate)	−750	40	11–12	Indigo	Abdelileh et al. (2020)
Binuclear iron-complex	Ca^{+} + Fe (II/III) + TEA Ca^{+} + Fe (II/III) + DGL	−1250	40	11–14	Vat Indigo Sulfur	Abdelileh et al. (2020), Bechtold and Turcanu (2004)

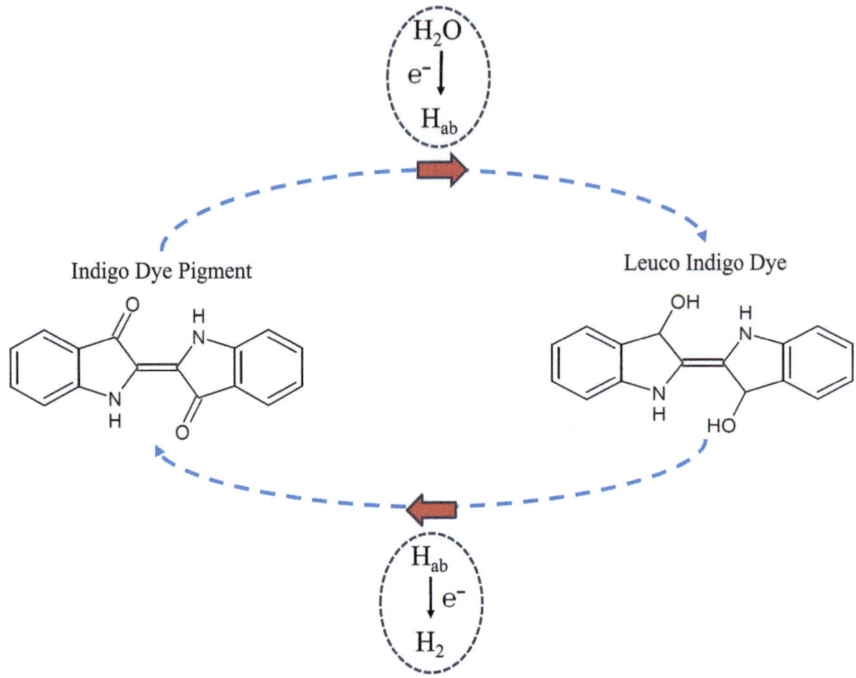

Fig. 6.3 The reduction mechanism of electrocatalytic hydrogenation

catalysts that adsorb hydrogen and produce a bulk hydride phase effectively mitigate the hazards of fire and explosion associated with gaseous hydrogen. Even when hydrogen is diluted within an inert gas carrier and loses its inflammability, the bulk hydride ensures a constant surface activity of hydrogen, thereby maintaining catalytic activity (Saïd et al. 2008; Comisso and Mengoli 2003; Roessler et al. 2002c).

Consequently, additional investigations into the electrolytic hydrogenation of vat dyes were carried out utilizing electrodes made of a thin grid that is coated with a metal layer containing fine catalyst particles. The surface of a metallic powder catalyst, such as Raney-nickel or platinum black, exhibits electrical conductivity and a low hydrogen overvoltage, suggesting a reduced likelihood of hydrogen production. The catalytic material functions as an electrode for hydrogen production while also acting as a catalyst for hydrogenation processes. Therefore, this process is distinctly different from an electron transfer process (electronation – protonation mechanism), where hydrogen participates solely in the protonation phase following the electron transfer (Turcanu et al. 2011; Yang et al. 2021).

In a divided flow cell, the industrial viability of this process was examined using a variety of metal catalyst electrode materials (Pb-black, Rh-black, Pd-black, Raney-nickel, and Raney-cobalt). Palladium and rhodium exhibit reduced catalytic activity compared to platinum. Nickel serves as the most active catalyst in Raney-type electrodes, yet its current efficiency remains at a mere 3.5%. The electro-hydrogenation

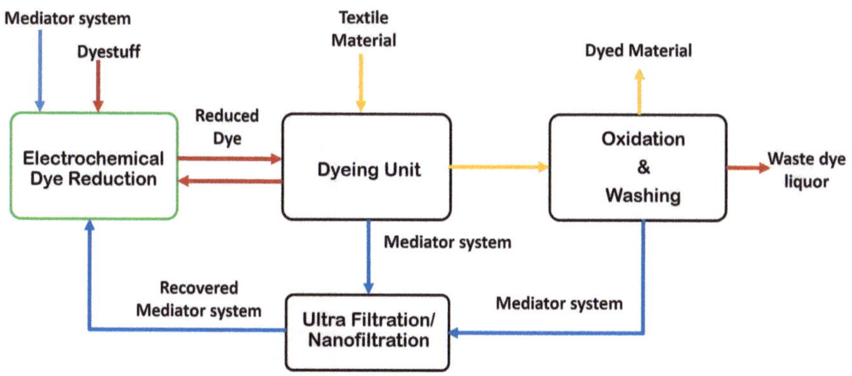

Fig. 6.4 Concept of redox-flow electrochemical reduction process for vat dyeing

rate of nickel is remarkably low, a characteristic that is atypical for a metal requiring such severe conditions. Raney-nickel was selected as the electrode material due to its interesting characteristics regarding availability, cost, and stability in alkaline media. The stability of platinum-black electrodes, recognized for their high activity, is inadequate, posing challenges for their industrial application (Kulandainathan et al. 2008; Roessler et al. 2002c; Yang et al. 2021).

6.3.1.4 Redox-Flow Electrochemical Reduction

Redox-flow electrochemical reduction is a relatively new approach that combines the principles of redox mediators with a flow-based system to enhance efficiency. In this method, the reduced and oxidized forms of redox mediators are circulated through a flow system, continuously transferring electrons to dye molecules. The mediator solution is pumped through a reactor where it is electrochemically reduced. The reduced mediator is then transferred to a dye bath, where it donates electrons to reduce the dye. The oxidized mediator is recirculated back to the electrochemical reactor for regeneration (Elgrishi et al. 2018; Huanhuan et al. 2011) Fig. 6.4.

This continuous flow system allows for greater scalability, making it suitable for large-scale textile operations. Additionally, the mediator is reused, reducing both chemical consumption and costs. Redox-flow systems are particularly beneficial in continuous dyeing operations, where high throughput is required.

6.3.1.5 Photo-Electrochemical Reduction

Photo-electrochemical reduction uses light energy in conjunction with an electrochemical process to reduce dyes. This method incorporates a photo-catalytic material that absorbs light and generates electron–hole pairs, facilitating the reduction of dye molecules. Light is absorbed by a semiconductor material, generating electron–hole

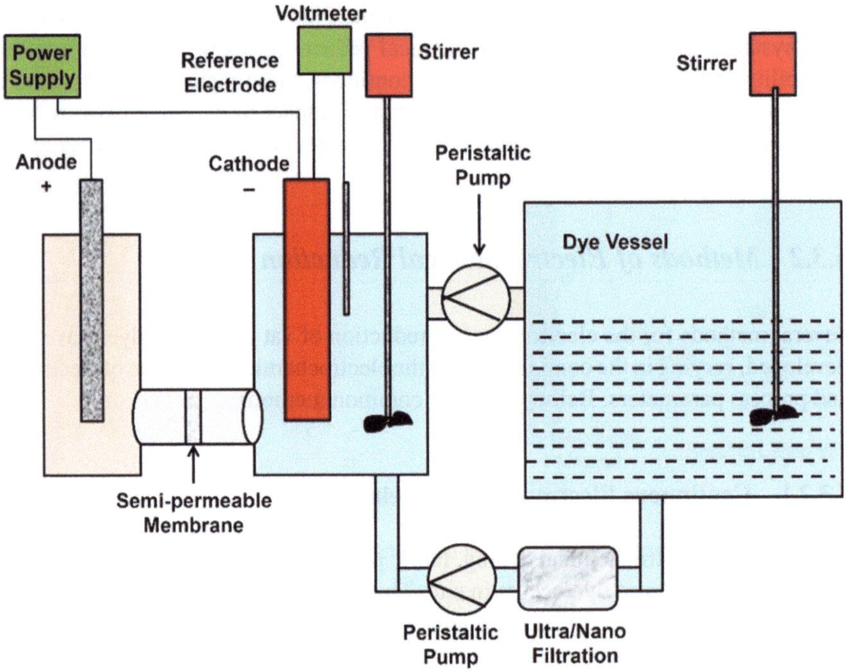

Fig. 6.6 The concept of electrochemical flow reactors for dye liquor circulation

6.4 Recent Technological Advancements

6.4.1 High-Efficiency Electrodes

Recent developments in electrode materials, such as titanium, carbon-based electrodes (graphite and carbon fiber), and conductive polymers, have improved the efficiency and selectivity of electrochemical reduction. These electrodes are more resistant to fouling and degradation, leading to longer operational lifetimes and improved process efficiency (Huanhuan et al. 2011; Kalapriya et al. 2017) (Fig. 6.6).

6.4.2 Pulsed Electrolysis

Pulsed electrolysis involves applying an intermittent electric current rather than a continuous one. This approach has demonstrated effectiveness in enhancing the reduction of vat and sulfur dyes by optimizing mass transfer and preventing electrode passivation. Additionally, pulsed electrolysis contributes to lower energy consumption and boosts the overall energy efficiency of the system (Saïd et al. 2008).

6.4.3 Hybrid Electrochemical Systems

Hybrid systems combining electrochemical reduction with other technologies, such as ultrasound or microwave-assisted electrochemical dyeing, are being developed to enhance the reduction process further. These systems allow for more efficient penetration of the reduced dye into the fiber, resulting in better color yield and reduced process time (Tan et al. 2018).

6.4.4 Process Automation and Smart Dyeing Systems

Advances in automation and process control have enabled real-time monitoring and adjustment of electrochemical dyeing parameters. Smart dyeing systems with sensors can automatically adjust current, temperature, and electrolyte concentration to maintain optimal dyeing conditions, reducing waste and improving consistency (Yi et al. 2020).

6.5 Conclusion

Electrochemical reduction of dyes can be achieved through both direct and indirect mechanisms. While direct electrochemical reduction is straightforward, it often lacks effectiveness due to the limited contact between the dispersed dye pigments and the cathode surface. The reduction potential obtained through this method can reduce sulfur dyes but is insufficient for vat dyes. However, the direct electrochemical reduction can be improved through a radical mechanism, using multi-cathode electrolyzer cells or electrodes pre-coated with materials like Raney-nickel. Modifying graphite electrodes or immobilizing quinone compounds can enhance the reduction efficiency by increasing the surface area for better dye interaction. Indirect electrochemical reduction, on the other hand, employs a mediator system that significantly boosts reduction efficiency by facilitating electron transfer. Mediators such as substituted anthraquinones, iron-amine complexes, and iron-carbohydrates are reversible redox systems that can be reused after dyeing, making this approach cost-effective. An additional method, electrolytic hydrogenation, produces hydrogen in situ through water electrolysis, where dyes adsorbed on the electrode surface react with the hydrogen to facilitate reduction. Several electrodes and catalysts have been investigated to improve this electrolytic hydrogenation process. Extensive research has been conducted both at laboratory and industrial scales to apply electrochemical reduction techniques. Compared to conventional methods, significant efforts have been made to achieve faster reduction rates, higher color yields, and better fastness properties. However, achieving high current efficiency and maintaining sufficient reduction rates remain key engineering challenges to making electrochemical reduction viable on an industrial scale. Electrochemical reduction of vat and sulfur dyes offers a promising solution for sustainable textile dyeing. The elimination of harmful chemical-reducing agents, reduced water consumption, and improved energy efficiency make this process valuable for the textile industry's shift towards eco-friendly practices. With continuous advancements in electrode materials, reactor design, and process automation, electrochemical dyeing is poised to become a mainstream solution in the industry, helping meet environmental and economic goals.

Declaration

The authors declare that they have no known competing financial interests or personal relationships that could have appeared to influence the work reported in this chapter. Generative AI and AI-assisted technologies have been used in the writing process to improve the readability and language of the work only. After using this tool, the authors carefully reviewed and revised the content to ensure accuracy and quality. The authors take full responsibility for the final version of this publication.

References

Abdelileh M, Manian AP, Rhomberg D, Ticha MB, Meksi N, Aguiló-Aguayo N, Bechtold T (2020) Calcium-iron-D-gluconate complexes for the indirect cathodic reduction of indigo in denim dyeing: A greener alternative to non-regenerable chemicals. J Clean Prod 266(1):121753. https://doi.org/10.1016/j.jclepro.2020.121753

Ahmed NSE, El-Shishtawy RM (2010) The use of new technologies in coloration of textile fibers. J Mater Sci 45:1143–1153. https://doi.org/10.1007/s10853-009-4111-6

Bechtold T, Turcanu A (2004) Fe3+–gluconate and Ca2+-Fe3+–gluconate complexes as mediators for indirect cathodic reduction of vat dyes–cyclic voltammetry and batch electrolysis experiments. J Appl Electrochem 34:1221–1227. https://doi.org/10.1007/s10800-004-1707-z

Bechtold T, Turcanu A (2006) Iron-complexes of bis(2-hydroxyethyl)-amino-compounds as mediators for the indirect reduction of dispersed vat dyes–cyclic voltammetry and spectroelectrochemical experiments. J Electroanal Chem 591(1):118–126. https://doi.org/10.1016/j.jelechem.2006.03.040

Bechtold T, Turcanu A (2009) Electrochemical reduction in vat dyeing: greener chemistry replaces traditional processes. J Clean Prod 17(18):1669–1679. https://doi.org/10.1016/j.jclepro.2009.08.004

Bechtold T, Burtscher E, Turcanu A, Bobleter O (1994) The reduction of vat dyes by indirect electrolysis. J Soc Dyers Colour 110(1):14–19. https://doi.org/10.1111/j.1478-4408.1994.tb01586.x

Bechtold T, Turcanu A, Schrott W (2008) Electrochemical reduction of CI sulphur black 1—correlation between electrochemical parameters and colour depth in exhaust dyeing. J Appl Electrochem 38:25–30. https://doi.org/10.1007/s10800-007-9390-5

Bechtold T, Fitz-Binder C, Turcanu A (2010) Electrochemical characteristics and dyeing properties of selected 9,10-anthraquinones as mediators for the indirect cathodic reduction of dyes. Dyes Pigm 87(3):194–203. https://doi.org/10.1016/j.dyepig.2010.03.026

Bechtold T, Brunner H (2005) Electrochemical processes in textile processing. In: Nunez M (ed) New developments in electrochemical research. Nova Science Publishers Inc

Bechtold T, Burtscher E, Kühnel G and Bobleter O (1997) Electrochemical reduction processes in indigo dyeing. 113(4):135–144. https://doi.org/10.1111/j.1478-4408.1997.tb01886.x

Bechtold T, Turcanu A, Brunner H and Schrott W (2009) Model calculations to optimize multi-cathode flow through electrolysers: direct cathodic reduction of C.I. Sulphur Black 1. J Appl Electrochem 39:1963–1973. https://doi.org/10.1007/s10800-009-9906-2

Bond AM, Marken F, Hill E, Compton RG, Hugel H (1997) The electrochemical reduction of indigo dissolved in organic solvents and as a solid mechanically attached to a basal plane pyrolytic graphite electrode immersed in aqueous electrolyte solution. J Chem Soc Perkin Trans 2(9):1735–1742. https://doi.org/10.1039/A701003F

Božič M, Kokol V (2008) Ecological alternatives to the reduction and oxidation processes in dyeing with vat and sulphur dyes. Dyes Pigm 76(2):299–309. https://doi.org/10.1016/j.dyepig.2006.05.041

Chakraborty JN (2014) Fundamentals and practices in colouration of textiles. Woodhead Publishing India in Textiles

Comisso N, Mengoli G (2003) Catalytic reduction of vat and sulfur dyes with hydrogen. Environ Chem Lett 1:229–232. https://doi.org/10.1007/s10311-003-0056-1

DyStar (2004) Electrochemical dyeing process from DyStar: first pilot plant for dyeing cheeses with Indigo. Pigment Resin Technol 33(3). https://doi.org/10.1108/prt.2004.12933cab.003

Elgrishi N, Rountree KJ, McCarthy BD, Rountree ES, Eisenhart TT, Dempsey JL (2018) A practical beginner's guide to cyclic voltammetry. J Chem Educ 95(2):197–206. https://doi.org/10.1021/acs.jchemed.7b00361

Firoz BK, Senthil RK, Kulandainathan AM, Noel M (2009) Ferric-oxalate-gluconate based redox-mediated electrochemical system for vat dyeing. J Appl Electrochem 39:1025–1031. https://doi.org/10.1007/s10800-008-9750-9

Holme I (2002) Recent developments in colorants for textile applications. Surface Coat Int Part B Coat Trans 85:243–264. https://doi.org/10.1007/BF02699548

Huanhuan J, Xiaofang GE, Yinghua XU, Wenwen Z, Chunan MA (2011) Indirect electrochemical reduction of indanthrene brilliant green FFB. Chin J Chem Eng 19(2):199–204. https://doi.org/10.1016/S1004-9541(11)60154-7

Kalapriya K, Prabhu HG, Nithya S (2017) Comparative study on the dyeability of cotton, silk, and polyester fabrics by conventional and electrochemical method. Rasayan J Chem 10(4):1330–1333. https://doi.org/10.7324/RJC.2017.1041923

Kalapriya K, Prabhu HG, Ramakrishnan G (2019) Electrochemical dyeing of polyester fabric using disperse dyes. Rasayan J Chem 12(1):319–323. https://doi.org/10.31788/RJC.2019.1215004

Kulandainathan MA, Muthukumaran A, Patil K, Chavan RB (2007a) Review of the process development aspects of electrochemical dyeing: its impact and commercial applications. Color Technol 123(3):143–151. https://doi.org/10.1111/j.1478-4408.2007.00082.x

Kulandainathan MA, Muthukumaran A, Patil K, Chavan RB (2007b) Potentiostatic studies on indirect electrochemical reduction of vat dyes. Dyes Pigment 73(1):47–54. https://doi.org/10.1016/j.dyepig.2005.10.007

Kulandainathan MA, Kiruthika K, Christopher G, Firoz BK, Muthukumaran A, Noel M (2008) Preparation of iron-deposited graphite surface for application as cathode material during electrochemical vat-dyeing process. Mater Chem Phys 112(2):478–484. https://doi.org/10.1016/j.matchemphys.2008.06.010

Madhu A, Singh G, Sodhi A, Malik S, Madotra Y (2012) Sulphur dyeing with non-sulfide reducing agents. J Textile Appar Technol Manag 7(4):1–5

Miled W, Said AH, Roudesli S (2014) Electrochemical reduction behavior of vat and sulfur dyes: a comparative study. Int J Appl Res Textile 2(2):1–10

Miled W, Ladhari N, Saïd AH, Roudesli S (2018) Electrochemical reduction of indigo: dyeing and fastness properties. Presentation. https://doi.org/10.13140/RG.2.2.19012.32648

Miled W, Ladhari N Saïd AH, Roudesli S (2008) Indirect electrochemical reduction of vat dyes: process optimization for indigoid, anthraquinone, and polycyclic vat dyes. Int Rev Chem Eng 20(10)

Roessler A, Crettenand D (2004) Direct electrochemical reduction of vat dyes in a fixed bed of graphite granules. Dyes Pigm 63(1):29–37. https://doi.org/10.1016/j.dyepig.2004.01.005

Roessler A, Jin X (2003) State of the art technologies and new electrochemical methods for the reduction of vat dyes. Dyes Pigm 59(3):223–235. https://doi.org/10.1016/S0143-7208(03)00108-6

Roessler A, Dossenbach O, Meyer U, Marte W, Rys P (2001) Direct electrochemical reduction of indigo. CHIMIA. 55(10):879–882

Roessler A, Crettenand D, Dossenbach O, Marte W, Rys P (2002a) Direct electrochemical reduction of indigo. Electrochim Acta 47(12):1989–1995. https://doi.org/10.1016/S0013-4686(02)00028-2

Roessler A, Dossenbach O, Rys P (2002b) Direct electrochemical reduction of indigo: process optimization and scale-up in a flow cell. J Appl Electrochem 32:647–651. https://doi.org/10.1023/A:1020198116170

Roessler A, Dossenbach O, Marte W, Rys P (2002c) Electrocatalytic hydrogenation of vat dyes. Dyes Pigment 54(2):141–146. https://doi.org/10.1016/S0143-7208(02)00035-9

Roessler A, Crettenand D, Dossenbach O, Rys P (2003) Electrochemical reduction of indigo in fixed and fluidized beds of graphite granules. J Appl Electrochem 33:901–908. https://doi.org/10.1023/A:1025876114390

Saïd AH, Miled W, Ladhari N, Roudesli S (2008) New prospects in the electrochemical dyeing of indigo. J Appl Sci 8(13):2456–2461

Tan X, Xiong W, Zhang J, Xu J, Yi H (2018) Indirect Electrochemical reduction of indigo and dyeing. Key Eng Mater 773(July):379–383. https://doi.org/10.4028/www.scientific.net/kem.773.379

Turcanu A, Bechtold T (2011) Indirect cathodic reduction of dispersed indigo by 1,2-dihydroxy-9,10-anthraquinone-3-sulphonate (Alizarin Red S). J Solid State Electrochem 15:1875–1884. https://doi.org/10.1007/s10008-010-1204-8

Turcanu A, Fitz-Binder C, Bechtold T (2011) Indirect cathodic reduction of dispersed CI Vat Blue 1 (indigo) by dihydroxy-9,10-anthraquinones in cyclic voltammetry experiments. J Electroanal Chem 654(1–2):29–37. https://doi.org/10.1016/j.jelechem.2011.01.040

Yang Z, Shen W, Chen Q, and Wang W (2021) Direct electrochemical reduction and dyeing properties of CI Vat Yellow 1 using carbon felt electrode. Dyes Pigment 184: Article No. 108835. https://doi.org/10.1016/j.dyepig.2020.108835

Yi C, Tan X, Bie B, Ma H, Yi H (2020) Practical and environment-friendly indirect electrochemical reduction of indigo and dyeing. Sci Rep 10:4927. https://doi.org/10.1038/s41598-020-61795-5

Chapter 7
Electrochemical Dyeing: A Sustainable Approach to Textile Dyeing

Tania Akther, Mohammad Shohag Babu, and Md. Rashedul Islam

Abstract Textiles become more visually appealing and more marketable through the process of dyeing. Dyeing is used not only for decorative purposes but also for functional purposes. However, conventional dyeing is a water, chemical, and energy-intensive process that pollutes water, soil, air, and other elements of the environment. The global climate and biodiversity are both adversely impacted by this pollution. Researchers are investigating alternative sustainable dyeing methods that use less water, chemicals, and energy to safeguard the environment from pollution. One of them is electrochemical dyeing. Electrochemical dyeing involves either electrical modification of the substrate to be dyed or the replacement of chemicals with electrons from electricity. This chapter covers a brief introduction to conventional dyeing as well as the history of electrochemical dyeing. It also covers problems with conventional dyeing, various electrochemical dyeing techniques for water-insoluble vat, sulfur, and indigo dyes, salt-free electrochemical dyeing of cotton, low-temperature electrochemical dyeing of polyester, and electrochemical pigment dyeing. The advantages and disadvantages of electrochemical dyeing over conventional dyeing will be addressed in the remainder of this chapter. The future of electrochemical dyeing and a summary of the conversation as a whole are covered in the conclusion.

Keywords Electrochemical dyeing · Direct · Indirect method · Mediators · Salt-free electrochemical dyeing · Low-temperature electrochemical dyeing

7.1 Introduction

The process of applying a colorant to a substrate, usually in the form of an ionic or molecular solution, to create a uniform distribution of colorant across that substrate is referred to as textile dyeing. It is the most water-intensive industrial process because

T. Akther · M. S. Babu · Md. R. Islam (✉)
Department of Wet Process Engineering, Bangladesh University of Textiles, Tejgaon, Dhaka, Bangladesh
e-mail: rashedul.islam@wpe.butex.edu.bd

S. Maiti et al. (eds.), *Sustainable Coloration Techniques in Textiles*,
https://doi.org/10.1007/978-981-96-4975-4_7

water is the most widely used medium for dyeing. The primary chemicals employed for dyeing are dyes and pigments. Over 8×10^8 kg of synthetic dyes of 1000 different types are utilized annually (Periyasamy 2024). The global market for textile dyes, valued at US$ 12,120 million in 2023, is predicted to reach US$ 17,360 million by 2030, growing at a compound annual growth rate of 5.3% between 2024 and 2030 (Choudhury 2018; Publishers 2024). Direct, azoic, vat, reactive, sulfur, acid, disperse, and basic are the most important dyes utilized to color textiles. The physical attributes and chemical constitution of the textile, the dyeing performance requirements, and the dyeing cost impact the dye choice. Besides dyes, auxiliaries are also used in dyeing for almost all dye-fiber systems; these can be basic substances like electrolytes and specialized chemical products like alkalis, acids, leveling agents, anti-creasing agents, anti-foaming agents, reducing agents, oxidizing agents, etc. Over 8000 species of chemicals are utilized in the textile dyeing sector, and most of these chemicals are thought to be harmful to people. Generally, 100% exhaustion of dyestuff and utilization of other chemical auxiliaries do not occur. Because of this inefficiency, around 2×10^8 kg of unfixed dyes, together with other auxiliaries, electrolytes, and different oxidation products, are released into effluents every year after dyeing, which causes environmental concerns.

The application of different colorants and chemicals has different impacts on the environment and humans. To overcome these problems, many researchers have studied to find a better approach by changing the chemistry and technology used in textile processing. Such a preventive action became more important worldwide during the 1990s to protect the environment. The electrochemical dyeing process is one of them. It has been 200 years since researchers began examining electrochemical methods for reducing organic materials. In 1807, Grothus published the first report on the electrochemical reduction of indigo dye. Except for one or two publications from the early twentieth century, research groups at the Swiss Federal Institute of Technology (ETH) and Leopold-Franzens-University of Innsbruck, Austria, have conducted most of the research on this (Blackburn et al. 2009). This process replaces the chemical-reducing agents used for sulfur, vat, and indigo dyeing and increases the exhaustion of reactive dye in cotton without adding salt by making cotton fabric positively charged at low temperatures. This electrochemical process has also been studied in the low-temperature dyeing of polyester fabric without carriers, pigment dyeing of wool without cationizing agents, and so on. These approaches lessen the quantity of chemicals used and the amount of effluent, each of which significantly benefits the environment. This is why the European research initiative EUREKA in June 2006 awarded an extension of the Lillehammer Prize to the project on electrochemical dyeing executed by DyStar Textilfarben GmbH in collaboration with the Institute for Textile Chemistry and Textile Physics (TID) at the University of Innsbruck in Dornbirn, Austria, and Getzner Textil AG. The prestigious project was a first step toward the industrial use of electrochemical dyeing (Ahmed and El-Shishtawy 2010; Chavan 2015).

This chapter will focus on the problems of the conventional dyeing process, electrochemical dyeing, different types of electrochemical dyeing with their advantages and disadvantages, factors affecting the electrochemical dyeing, electrochemical dyeing process of different types of fibers with different dye classes, their advantages, and disadvantages, comparison between conventional and electrochemical dyeing process and future of electrochemical dyeing process.

7.2 The Effects of Conventional Dyeing Methods on Both the Environment and People

The techniques used to dye textiles are varied and sophisticated. Every type of fiber has a distinctive set of dye classes. Wool and cotton are the two main fibers, and there are typically three to four distinct dye categories. Fewer dye classes are available for other fibers like acrylic, nylon, polyester, silk, rayon, and polypropylene. Every dye class has different effluent output volumes and properties and distinct chemistry, water, and energy needs. A major environmental concern for every dyehouse is effluent. The multicolored discharge streams known as dyehouse effluents are composed of organic materials like colorants, surfactants, starches, and other process aids, as well as non-biodegradable inorganic salts and alkalis. The discharged colors and chemicals are undesirable because they affect dissolved oxygen (DO) and photosynthetic activity due to less sunshine penetrating the water. They can also accumulate in the food chain and may cause tumors and mutations due to their carcinogenic and mutagenic qualities. Biologically non-biodegradable toxic amine-based byproducts of azo dyes can also affect the immune system and the skin. Besides hindering plant growth, seed germination, and other health hazards, dyes can also result in chromosomal abnormalities and other illnesses. Aquatic life is threatened by salts, alkalis, detergents, and different types of reducing agents because they increase total dissolved solids (TDS), biological oxygen demand (BOD), and chemical oxygen demand (COD) levels. Certain heavy metals, particularly chromium, may exist based on the chemical used for processing. In addition to being inappropriate in and of itself, they may obstruct the wastewater treatment system (Khatri and Jhatial 2015; Periyasamy 2024). The impacts of some of the dye classes on the environment and humans are mentioned in Table 7.1.

7.3 Electrochemical Dyeing

Electrochemical treatment uses electrochemistry to obtain desired results in various fields, including fuel cells, electro-synthesis, surface modification, and coatings. In this process, electrical energy is transformed into chemical energy or chemical energy into electrical energy using chemicals or electrolytes and electricity (Cao et al. 2018). During electrochemical dyeing, the flow of electrons occurs between a solid electrode and a substance. Either the electrode can take in electrons released from the chemical substance, or the electrode can transfer electrons to a chemical substance in solution

Table 7.1 Estimated degree of fixation of different types of fibers with different dyes and their effluent composition after dyeing with their hazardous impact on the environment and humans

Dye class	Fiber type	Fixation degree (%)	Used Chemicals	Pollutants in effluent	Hazard	Possible solution	References
Reactive dye	Cellulose	50–90	Salt, alkali, surfactant, defoamers, urea (in case of the pad-dry-bake process)	Salt, alkali, surfactant, defoamers, unfixed dyes, urea	Causes water and land salinization, decreases photosynthetic activity and DO, increases COD and BOD, urea increases nitrogen content of effluent	Use of organic salt, cationization of cotton, electrochemical dyeing, reverse micelle dyeing, liposome dyeing, microwave-assisted dyeing, etc	Khatri et al. (2015), Choudhury (2018), Ru et al. (2018)
Direct dye	Cellulose	70–95	Salt, alkali, dye fixing agent	Salt, unfixed dye, dye fixing agent	Cause water and land salinization	Microwave-assisted dyeing	Choudhury (2018), Bokhari et al. (2023)
Vat dye	Cellulose	80–95	Alkali, sulfur-based reducing agent (usually $Na_2S_2O_4$), oxidizing agent, detergent	Alkali, high sulfite/sulfate content due to the thermal decomposition of $Na_2S_2O_4$ in the alkaline bath, oxidative decomposition, and other decomposition, oxidizing agents, unfixed dye	Destroy waste lines (concrete pipes), the aerobic process of wastewater treatment becomes difficult, pollutes the air through the formation of H_2S, problems with nasal nuisance, high salt load	Electrochemical reduction of the dye, use of organic materials, and different natural extracts as reducing agents	Roessler and Crettenand, (2004), Kulandainathan et al. (2007b), Motaghi (2012), Xu et al. (2014), Ammayappan, Jose and Arputha Raj (2016), Patra et al. (2018), Shrwardi et al. (2023)

(continued)

Table 7.1 (continued)

Dye class	Fiber type	Fixation degree (%)	Used Chemicals	Pollutants in effluent	Hazard	Possible solution	References
Sulfur dye	Cellulose	60–90	Alkali, sulfur-based reducing agent (usually Na_2S), oxidizing agent, salt	Alkali, high sulfite/ sulfate content due to the decomposition of reducing agents, oxidizing agents, unfixed dye, salt	Pollutes water and destroys marine life, high COD, causes varying damage to the DNA of tadpoles	Electrochemical reduction of dye, use of organic materials, and different natural extracts as reducing agents	Blackburn and Harvey (2004), Motaghi (2012)
Disperse dye	Polyester	90–100	Acid, dispersing agent, carrier	Acid, dispersing agent, carrier	High temperatures of effluent cause problems for aquatic life, high BOD, and COD due to dispersing agents	Supercritical carbon dioxide dyeing, electrochemical dyeing, fiber modification, microwave-assisted dyeing, liposome-based dyeing	Al-Mousawi et al. (2013), Ge et al. (2018), Srivastava (2023)
Acid dye/ metal complex dye	Wool	90–98	Acid, levelling agent, wetting agent	Acid, levelling agent, wetting agent	High BOD and COD, high temperature of wastewater, heavy metal(metal complex dye)	Liposome-based dyeing, microwave-assisted dyeing, fiber modification	Mart\'\i et al. (2004), Kartal et al. (2020)
Basic dye/ Cationic dye	Acrylic	95–100	Acid, cationic retarding agent	Unfixed dye, acid, and cationic retarding agents	High BOD, COD	Fiber modification, Microwave-assisted dyeing	Choudhury (2018), Popescu et al. (2019)

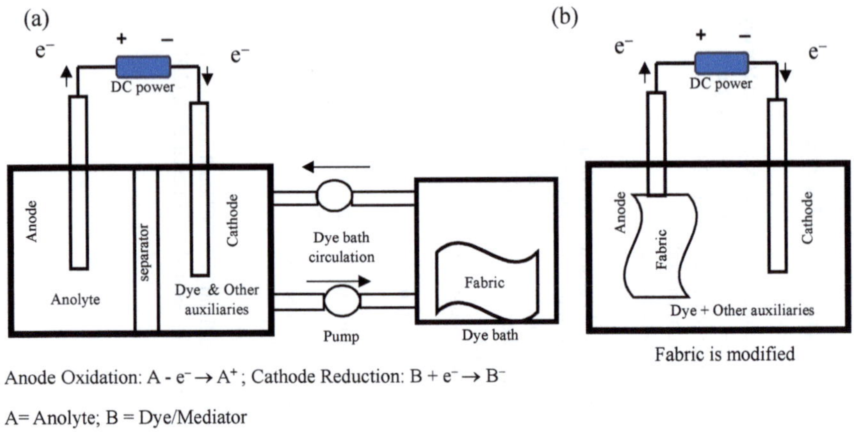

Fig. 7.1 A typical illustration of **a** Electrochemical dyeing by reduction of a dye (vat/sulfur/indigo) and **b** dyeing of fabric through electrochemical modification (Kulandainathan et al. 2007a, Hou et al. 2022)

(Blackburn et al. 2009). Figure 7.1 shows the typical illustration of electrochemical dyeing.

7.4 Methods of Dyeing of Vat, Sulfur, and Indigo Dye

Vat, indigo, and sulfur dyes show a similar dyeing process because they are water-insoluble dyes. An effective reducing agent is used for their dyeing, which reduces them and makes them water-soluble in an alkaline medium. The reduced water-soluble leuco dye is strongly attracted to cellulosic materials, and once the dyeing is completed, the leuco dye is again reoxidized to the water-insoluble oxidized form. Different types of reductants are employed to reduce vat, sulfur, and indigo dyes, some of which are listed in Table 7.2.

Among these, Na_2S and $Na_2S_2O_4$ are commonly utilized as reducing agents for sulfur and vat dye (including indigo), respectively. Nevertheless, the discharge of dye

Table 7.2 A list of reducing agents employed to reduce vat dyes (including indigo) and sulfur dyes (Blackburn and Harvey 2004; Kulandainathan et al. 2007b; Božič and Kokol 2008)

Reducing agents for vat dye (including indigo)	Reducing agents for sulfur dye
Sodium dithionite, hydroxyacetone, glutaroin, acetoin, organic sulfinic acid, glucose, thiourea dioxide, Rongalite, α-hydroxyketone, alkaline $Fe(OH)_2$, sodium borohydride, $FeSO_4$–lime method, Zn(zinc)–lime method, bisulfite-Zn(zinc)–lime method	Sodium sulfide, sodium hydrosulfide (NaHS), sodium polysulfide (Na_2S_x, where x = 1–6) thiourea dioxide, glucose, β- mercaptoethanol

solution and rinsing water containing such reducing agents results in several issues for the wastewater and increases the cost of wastewater treatment. Another problem is that these reducing agents cannot be regenerated after the dyeing process. Thus, current economic and environmental requirements are scarcely satisfied (Jiang et al. 2011; WMA et al. 2014). In the electrochemical reduction method, electrons replace the chemical reducing agents and thus eliminate the need for effluent-contaminating reducing materials. Vat, indigo, and sulfur dyes may be dyed industrially using this method in the future. For electrochemical dye reduction, the following methods can be employed (Xu et al. 2014; Chavan 2015):

1. Indirect electrochemical reduction using mediators,
2. Direct electrochemical reduction or autocatalysis electrochemical reduction, and
3. Electrochemical hydrogenation.

7.4.1 Indirect Electrochemical Dyeing

Since vat, sulfur, and indigo dyes typically remain insoluble in water and have little interaction with the solid cathode, reducing them directly on the cathode surface is not easy. To overcome this problem, Bechtold and collaborators at DyStar started exploring the indirect electrochemical reduction of dye in the 1990s (Blackburn et al. 2009). In this approach, the dye reduction happens between the electrode and the insoluble dyes using an electron carrier called a mediator rather than by direct contact between them (Božič and Kokol 2008; Jiang et al. 2011). The reaction steps of this reduction process for a dyestuff are given below (Bechtold et al. 1993, 1996):

Fe^{3+}-Ligand + e^-	↔	Fe^{2+}-Ligand	[Cathodic Process]	Step-1
Fe^{2+}-Ligand + Dye	↔	Fe^{3+}-Ligand + $Dye^{\bullet -}$	[One electron transfer]	Step-2
2 $Dye^{\bullet -}$	↔	Dye^{2-} + Dye	[Disproportionation]	Step-3
Fe^{2+}-Ligand + $Dye^{\bullet -}$	↔	Dye^{2-} + Fe^{3+}-Ligand	[Secondary reduction]	Step-4
Dye^{2-} (aq)	↔	Dye^{2-} (fiber)	[Adsorption/ exhaustion]	Step-5
Dye^{2-} (fiber)	↔	Dye (fiber) + $2e^-$	[Oxidation/ Fixation]	Step-6

Here, the Fe^{3+}-Ligand is an example of a mediator, and it is described elaborately in Sect. 7.4.1.1 of this chapter.

- Step-1: Reduction of the Fe^{3+} compounds occurs at the cathode.
- Step-2: The reduced mediator complex flows from the cathode to the dispersed dye particle, where a radical dye anion is formed by transferring one electron from the reduced compound to the dye.
- Step-3: Disproportionation stabilizes this radical anion produced in step-2 and produces one regenerated oxidized dye molecule and one entirely reduced dye molecule.
- Step-4: A further fully reduced dye anion is generated through a secondary reduction in which electrons are transferred to a radical anion produced in step-2 from a reduced mediator, because one dye molecule needs two electrons to be completely reduced (one for each keto group in the case of vat dye).
- Step-5: These reduced dyes have a strong attraction to cotton fibers. Now, dyestuff adsorption and exhaustion occur in this stage.
- Step-6: Shows that the entirely reduced dye molecule and the entirely oxidized water-insoluble dye molecule are in equilibrium, where the dye fixation takes place.

The general principle of indirect electrochemical reduction of a dye is illustrated in Fig. 7.2.

A simplified kinetic scheme of different essential steps of the dyeing process is illustrated in Fig. 7.3. For a reversible reaction, the equilibrium constant of each reaction step is calculated as the ratio of a pair of velocity constants k. The reduction and reoxidation velocity constants of a suspended dye are denoted by k_1 and k_2, respectively. k_1 must be larger than k_2 to achieve stable vatting of the dye. An affinity for reduced dye to cellulosic fibers is indicated by the ratio of k_3/k_4. A high k_3/k_4 ratio indicates that the dye has a good affinity and a high level of fixation. The quick exhaustion of the dyebath leads to issues in the dyehouse. The ratio k_8/k_7 indicates

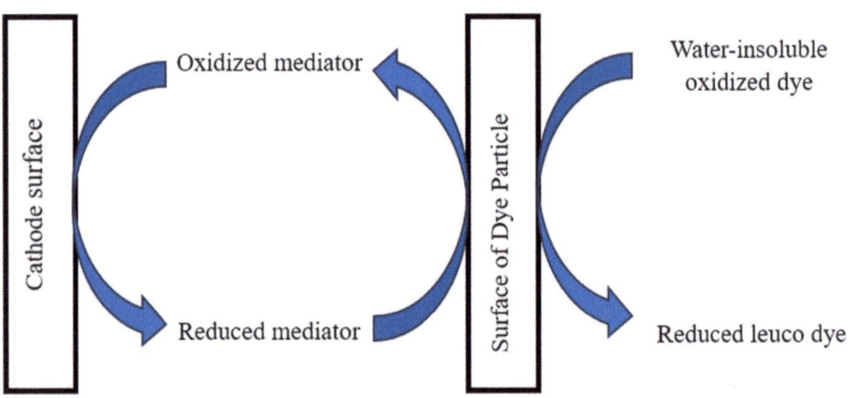

Fig. 7.2 General mechanism of indirect electrochemical reduction of a dye (Xu et al. 2014). Adapted with permission from Xu et al (2014) 'Indirect electrochemical reduction of indigo on carbon felt: Process optimization and reaction mechanism', Industrial and Engineering Chemistry Research, 53 (26), pp. 10,637–10,643 Copyright © American Chemical Society

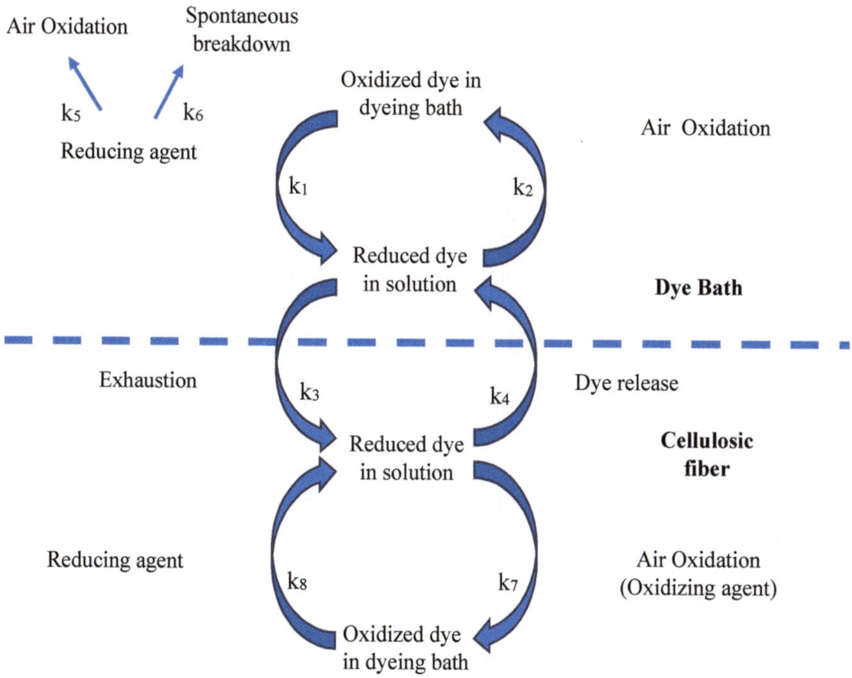

Fig. 7.3 Steps involved in the reaction scheme of a vat dye reduction process (Bechtold et al. 1994). Adapted with permission from Bechtold et al. (1994) 'The reduction of vat dyes by indirect electrolysis', Journal of the Society of Dyers and Colourists, 110(1), pp. 14–19 Copyright © 1994 Society of Dyers and Colourists

the stability of the dye adsorbed on the fiber in its reduced condition, and a high k_8/k_7 ratio indicates the higher stability of the reduced dye and the equilibrium achieved during the exhaustion of the dye bath. In cases where the ratio k_1/k_2 or k_8/k_7 is small, all oxidizing agents interfere with the dyeing process, resulting in uneven dyeing outcomes. To guarantee dyeing reproducibility, the dye reduction rates (k_1 and k_8) are crucial. During electrochemical reduction, the mediator complex reduces the dye at a rate of multiple orders of magnitude more rapidly than sodium dithionite, which results in a more stable reduced dye and more reproducible dyeings. The rate constants of the breakdown of the reducing agent through air oxidation and spontaneously are k_5 and k_6, respectively. If the value of k_5, is high, the oxygen in the dye liquor will be removed fast, preventing the oxidation of more vatted dye. Conversely, more reducing agents are used when oxygen diffuses into the dye bath. In alkaline solutions, a small k_6 value indicates the presence of a stable mediator system. Although Sodium dithionite is thermodynamically unstable during dyeing, it is used as a reducing agent because its breakdown is kinetically inhibited. Because of this instability, sodium dithionite stock solution is not kept in storage for an extended period (Bechtold et al. 1994).

7.4.1.1 Mediators

Mediators are reversible soluble redox systems that reduce the dye being oxidized in the process, are reproduced at the cathode, and are available again for the reduction of dye (Schnitzer et al. 1996). The mediators reduce the dye after reaching a redox potential value in the same or even more negative value than that of the leuco dye in an alkaline medium. (Bechtold et al. 1991, 1997c). The redox potential of vat, indigo, and sulfur dye is given in Table 7.3.

The mediators provide a continuously regenerable reduction potential, which is continuously monitored by measuring the dye bath redox potential and controlled by adjusting the cell current. Unlike the traditional dithionite reduction method, this avoids adding additional reducing chemicals to the dye bath. Hence, no successor products are produced from the reducing agents. The atmospheric oxygen remaining in the electrochemical cell is also reduced by this mediator, and after this process, the mediator is regenerated by the cathodically reduction process. Both reversible inorganic and organic redox systems are used as mediators. Most of the inorganic substances used as mediators are metal salt complexes (Bechtold 1993; Bechtold et al. 1997c; Roessler and Jin 2003; Bechtold et al. 2004; Ahmed and El-Shishtawy 2010). The highly preferred metal is iron. For the reduction of organic dyes, a group of Fe(II) complexes with various ligands was studied as mediators in many researches. The purpose was to identify redox systems that had a significant negative redox potential in homogeneous conditions, i.e., with no precipitation.

The redox potential of Fe(II/III)-compounds is frequently pH-dependent, shifting to more negative values as pH rises (Bechtold and Turcanu 2004). When selecting a complex metal salt as a mediator, the following experimental factors must be taken into account (Kulandainathan et al. 2007b):

- The typical redox potential of the mediator compound needs to be a little greater than that of the dye.
- The reduction potential must be considerably lower than the working potential of the electrode in the solution. Therefore, there should not be any noticeable overvoltage when the mediator is transformed from its oxidized to a reduced state at the electrode (i.e., with no evolution of hydrogen gas).
- Concerning the evolution of hydrogen gas, the working electrode must have the maximum overvoltage. Therefore, it is possible to obtain the highest cathodic peak potential.

Table 7.3 The redox potential of vat, indigo, and sulfur dye in the alkaline medium versus Ag/AgCl, 3 molL^{-1} KCl

Dye	Redox Potential (mV)	References
Vat	−600 to −1000	Bechtold et al. (1993), Roessler (2003)
Indigo	−600	Bechtold et al. (1993), Božič and Kokol (2008)
Sulfur	−400 to −500	Božič and Kokol (2008)

The mediators also have to fulfill the following requirements to be useful under dyebath conditions (Bechtold et al. 1994):

- Stable reduction conditions in the dye bath.
- Reproducible dyeing outcomes.
- Identical dyeing results when compared to traditional dyeing results.
- Use of minimum chemical concentration.
- Easy wastewater treatment.
- Chemical and wastewater recycling.

Triisopropanolamine, Triethanolamine, aliphatic hydroxy compounds, and bicine (N,N, bis(2-hydroxyethyl)glycine) were utilized as ligands in many researches. These molecules may possess many hydroxyl groups as well as functional groups such as carboxyl, keto, or aldehyde. Some examples of these substances are di- and polyalcohols, di- and polyhydroxyaldehydes, di- and polyhydroxyketones, di- and polyhydroxymono and dicarboxylic acids, di- and polysaccharides, and hydroxytricarboxylic acids. Sugar-based compounds were also preferred, particularly their acids and salts, such as gluconic and heptagluconic acid (Bechtold et al. 2004). The ligands must generally meet the following conditions to form coordination compounds and function as mediators (Bechtold et al. 1993):

(a) The main criterion is the stability of the Fe(III) and Fe(II) coordination compounds with ligands in an alkaline medium.
(b) The complex should undergo minimal dissociation when the reversible charge transfer occurs between Fe(III) and Fe(II).
(c) In an alkaline solution, they must be able to generate enough negative reduction potential that is at least −1000 mV or better for all vat dyestuffs, −600 mV or better versus Ag/AgCl/ 3 mol L^{-1} KCl, for indigo commonly utilized in the textile sector.
(d) An accelerated rate of electron transport to the Fe(III) complex from the cathode.
(e) A substantial amount of catalytic cycles, which could enable the dyeing liquor to be reused several times.
(f) Fast electrons transfer from the reduced mediator to the oxidized insoluble dye.

Table 7.4 lists values of the cathodic peak potential of some ligands with iron (which will not precipitate). Amino compounds and sugar acids were suitable ligands for Fe(II)- and Fe(III)-ions in an alkaline solution because they did not precipitate from the solution (Scholnick and Pfeffer, 1980). As for the Fe-(II) complex, only good stability in alkaline solution has been demonstrated by hydroxyethylethylenediaminetriacetic acids (HEDTA), bicine, diethanolamine, and triethanolamine (TEA) (Bechtold et al. 1993). The reduced form of these complexes was appropriate to reduce vat dyestuffs, and among them, the Fe(III)-TEA-based electrochemical approach is possibly the most widely studied green electrochemical technique being investigated as a conventional vat dyeing alternative (Bechtold 1993; Bechtold et al. 1996; Schnitzer et al. 1996; Bechtold et al. 1997a; WMA et al. 2014). This is because in an alkaline medium, this compound containing TEA as a ligand showed its highest stability and did not precipitate from the solution. This method was patented in the

Table 7.4 The cathodic peak potential (mV vs. Ag/AgCl/3 $molL^{-1}$ of KCl) and the potential use of several ligands for the formation of a stable Fe (II) complex in an alkaline medium (Bechtold et al. 1993; Kulandainathan et al. 2007b)

Ligand	Cathodic peak potential	The stability of the Fe (II)-ligand compounds
Nitilotriacetate	Precipitated	Not able to bind
Diethylenetriamine pentaacetate	Precipitated	Not able to bind
Ethylenediaminetetraacetic acid	Precipitated	Not able to bind
Hydroxyethylethylenediaminetriacetic acids	−650	High
Glycine	Precipitated	Not able to bind
Bicine	−850	High
Diethanolamine	−1000	Low
Triethanolamine	−1050	High
Triethylenetetramine	Precipitated	Not able to bind

early 1990s by Bechtold and collaborators (Senthil Kumar et al. 2009). It can also be used to reduce all existing sulfur, indigoid, and azo dyes without the need for additional reductants owing to the potential of Fe(II)-TEA equal to −1050 mV versus Ag/AgCl/ 3 $molL^{-1}$ KCl in an alkaline solution (Bechtold 1993). An important characteristic of the reduced Fe(III)-amine compound is the high dye reduction rate, which is significantly higher than that of $Na_2S_2O_4$ (Roessler 2003). Employing the metal ion in both low- and higher-valent forms is possible. For instance, Fe (II) salts may be used in conjunction with Fe(III) salts, which are first electrochemically reduced to Fe(II) salts (Bechtold et al. 2004). Another property of Fe-Amino compounds (e.g., G. Fe(III)-triethanolamine) is that they can provide a comparatively high current density and adequate negative redox potential at pH levels above 12, which are necessary for reducing all vat dyes currently in use appropriately; however, the stability of Fe(III)- and Fe(II)-TEA complexes is limited at pH levels below 12 (Mohr and Bechtold 2001) or 13 (Bechtold et al. 1997c). The Rinsing operation done after the vat dyeing process causes a decrease in pH which is lower than 12 and the dyeing of cellulose with indigo requires a pH of 11.5–11.7 both of which cause the precipitation of Fe(II)- and Fe(III)-TEA complexes as Fe-hydroxides from the dilute dye bath. This should be avoided as it is undesirable. To solve this problem Fe-sugar acid complexes (such as Fe(III)-D-gluconate) were used as they are the complexes with greater stability at pH values between 9.5 and 12.5 (Bechtold et al. 1997c). However, the problems of these complexes compared to Fe-amino complexes are that, they have a smaller negative redox potential and a lower attainable cathodic current density. To address this problem, a study looked into a combination of amino and sugar-based ligands to take advantage of these two redox couples. The advantages of these complexes were that they provided combined benefits of both ligands including enhanced stability at lower pH from the use of D-gluconate ligand and higher cathodic current density from the use of TEA, which allowed for more cell current in a technical electrolyzer. For these reasons, the complex combinations have significant importance in the reduction of vat dye in the indirect electrochemical method. (Mohr and Bechtold 2001).

After that, to find less expensive and environmentally friendly ligand alternatives, a novel system of mixed ligands of ferric-oxalate-gluconate was developed as a probable replacement for TEA. Fe(III) with oxalate and gluconate ligands is reasonably stable between the pH values of 8 and 13. Unexpectedly, the ligand system consisting of oxalic acid and gluconate combination demonstrated a remarkable conversion efficiency together with an adequate current efficiency. Fe(III)-oxalate-gluconate showed a quasi-reversible response in cyclic voltammetry, which proved that it could be used as a mediator. Under substantially identical conditions, the Fe(III)-oxalate-gluconate compound functions quite similarly to the Fe(III)-TEA-gluconate compound. It was also possible to reuse the electrolyte. The same mediator system could potentially be used to dye fabric and yarn successfully. The colored cotton fabrics exhibited satisfactory dyeability and showed strong colorfastness against light, washing, and chlorine bleaching (Senthil Kumar et al. 2009; Yu et al. 2013).

Cyclic voltammetry test is used to assess the redox potential, current density, and reversibility of a redox system. Cyclic voltammetry is an electrochemical method that measures the current response of the working electrode to a redox-active solution by sweeping over a linear potential. Starting from an initial value known as the primary potential, the potential is changed proportionally with time to the final potential and then returned to the primary potential. The current response is recorded as a function of the applied potential (Elgrishi et al. 2018). Across the scan range of applied potential for any reducible analytes, the forward scan produces a peak of current/ reduction peak. The current will rise as the potential gets closer to the analyte redox potential; however, it will drop as the analyte concentration decreases near the electrode surface. When the applied potential is returned to the primary value, the product produced during the first reduction phase will be reoxidized, resulting in a reverse-polarity current from the forward scan. Usually, the shape of this oxidation peak resembles that of the reduction peak (Roessler 2003; Westbroek 2005).

The peak current of a reversible system can be found from Eq. 7.1:

$$I_p = 0.4463\mathrm{nFA}\ (\mathrm{Da})^{1/2}\mathrm{C} \tag{7.1}$$

where I_p = Peak current (A)

n = Quantity of electrons transferred in the relevant reaction.

F = Faraday constant (Qmol^{-1}).

A = Electrode surface area (cm^2).

D = Coefficient of diffusion ($\mathrm{cm}^2\mathrm{s}^{-1}$).

C = Concentration of electroactive solution to be investigated (molL^{-1}).

And a can be calculated using Eq. (7.2):

$$\mathrm{a} = \mathrm{nFV/RT} \tag{7.2}$$

where V = Rate of scan (Vs^{-1})

R = Universal gas constant ($Jmol^{-1}\ K^{-1}$).
T = Temperature (K).
In cyclic voltammetry, the peak current of a reversible system in a forward scan expressed by the Sevcik–Randles equation is obtained by combining Eqs. (7.1) and (7.2) at T = 298.0 K:

$$I_p = 2.69 \times 10^5 n^{3/2} A D^{1/2} V^{1/2} C \tag{7.3}$$

The reversibility of the redox system under study can be determined by analyzing the curve obtained from cyclic voltammetry using the following criteria:

- A well-defined return peak indicates the emergence of the opposite reaction.
- The ratio of the anodic and cathodic peak currents should be 1. So, $I_{p,\,a} = I_{p,\,c}$, where $I_{p,\,c}$ and $I_{p,\,a}$ are the cathodic and anodic peak currents, respectively.
- The point at which the measured current achieves its highest value I_p is known as the peak potential, or E_p. Equation 7.4 can be used to link the peak potential to the half-wave potential $E_{1/2}$, which is the potential at a current equal to $I_p/2$:

$$E_p = E_{1/2} - 1.11RT/nF \tag{7.4}$$

- Half-wave potential for a reversible system can be expressed as Eq. (7.5) where $E_{1/2,a}$ and $E_{1/2,c}$ indicates the anodic and cathodic half wave potential respectively.

$$E_{1/2} = E_{1/2,\,a} = E_{1/2,\,c} \tag{7.5}$$

- From Eq. (7.6), the separation between the anodic and cathodic peaks of a reversible system can be determined:

$$|\Delta E_p| = |E_{p,a} - E_{p,c}| = 0.059/n \tag{7.6}$$

where the $E_{p,c}$ and $E_{p,a}$ represents the cathodic and anodic peak potential respectively. E is in Volt.

A good linear relationship between $I_{p,c}$, and the square root of the scan rate depicts a pure diffusion-controlled electrode reaction mechanism.

In Fig. 7.4, voltammograms of a solution containing 0.01 $molL^{-1}$ Fe^{3+}, 0.06 $molL^{-1}$ TEA, and 0.175 $mol\,L^{-1}$ NaOH at different scan rates are given, and the basic data describing these voltammograms are presented in Table 7.5. The voltammogram presents the expected features of a reversible redox couple. $(I_p)_a/(I_p)_c$, the ratio of the anodic to the cathodic peak current was almost 1, which is the condition of reversibility of a redox couple. $E_{p,c}$, the cathodic peak potential is not dependent on the scan rate, whereas $(I_p)_c$, the cathodic peak current, and the square root of the scan rate are in a linear relationship. From the separation of the cathodic half peak potential $(E_{p/2})_c$ and the cathodic peak potential $E_{p,}$ the number of transferred electrons was calculated as one. The catalytic activity of a mediator can be understood from the value of the enhancing factors. The ratio between the peak current of the mediator when the dyestuff is present and the peak current of the mediator by itself is known as the enhancement factor. The computed enhancement factors can be used to understand how the kind of dye and mediator employed impact the reduction rate. The higher the enhancement factor of a mediator complex with a dye, the higher the suitability of the mediator complex for that dye in indirect electrochemical reduction (Bechtold et al. 1991; Turcanu et al. 2011). Under the experimental conditions, the Fe(II/III)-TEA redox system forms a reversible redox couple that is well suited to be used as a mediator to reduce the insoluble organic molecules (Mohr and Bechtold 2001).

Instead of using two separate mononuclear complexes, a mixture of calcium-D-gluconate (Ca-DGL) and ferric-D-gluconate (Fe(III)-DGL), forming a binuclear complex can be used. The electrochemical behavior of the Fe-complex is altered for the incorporation of Ca^{2+} ions. So, the electrochemical behavior of the Fe(II)/(III) redox pair is markedly altered by the uptake of Ca^{2+}–ion into the complexes. The

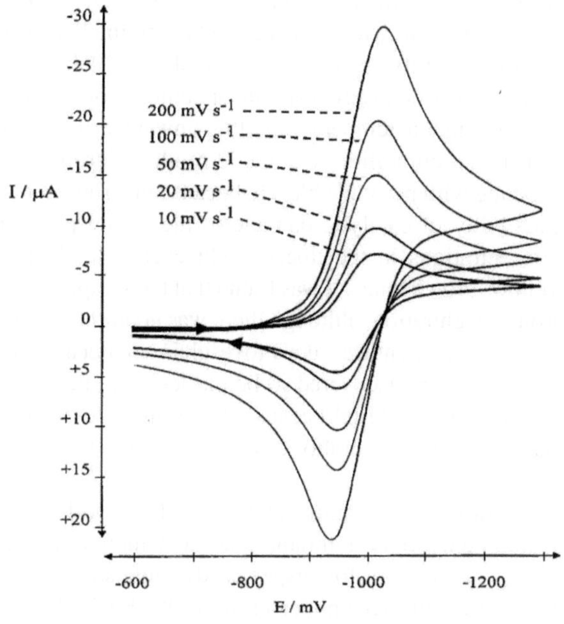

Fig. 7.4 Cyclic voltammograms of a solution containing 0.01 $molL^{-1}$ Fe^{3+}, 0.06 $molL^{-1}$ TEA, and 0.175 $molL^{-1}$ NaOH at different scan rates (Mohr and Bechtold 2001). Reprinted with permission from Mohr, S. and Bechtold, T. (2001) 'Electrochemical behaviour of iron-complexes in presence of competitive ligands: A strategy for optimization of current density', Journal of Applied Electrochemistry, 31(3), pp. 363–368 Copyright © 2001, Kluwer Academic Publisher, Springer Nature

Table 7.5 The cathodic peak potential and current were determined using a solution containing 0.01 $molL^{-1}$ Fe^{3+}, 0.06 $molL^{-1}$ TEA, and 0.175 $molL^{-1}$ NaOH at 10 and 100 mVs^{-1} scan rates (Mohr and Bechtold 2001)

Scan rate in mVs^{-1}	$(E_p)_c$ in mV	$(E_{p/2})_c$ in mV	$(I_p)_c$ in μA	$(I_p)_a/(I_p)_c$	n
10	−1015	−955	12.0	0.94	0.91
100	−1015	−955	33.1	0.98	0.93

incorporation of Ca^{2+}–ion shifts the formal redox potential of the Fe(II)/(III)-DGL compounds by about −150 mV in the direction of more negative values (Bechtold et al. 2002; Bechtold and Turcanu 2004). It was found that for the indirect cathodic reduction of indigo, the binuclear compounds with a 1:1:1.1 stoichiometry were most suitable as mediators, which allowed for effective usage of the Ca^{2+}, Fe^{3+}, and sodium D-gluconate compounds (Abdelileh et al. 2020). In the pH range of 11.5–12.5 for these complexes to attain -700 mV redox potential, comparatively a lower amount of Fe(II)-complex is needed. Lessening the ratio of Fe^{3+} to Ca^{2+} in solution, for example, to 1:0.8, raises the concentration of Fe^{2+} necessary to attain −700 mV redox potential, which in addition lowers the concentration of reduced molecules in the oxidation stage with air. This would considerably raise the required electrolyzer size (Abdelileh et al. 2020; Yi et al. 2020). However, the advantages of the mononuclear Fe^{3+}-DGL over the binuclear Ca^{2+}-Fe^{3+}-DGL have been verified by galvanostatic batch reduction studies using a lab multi-cathode cell. But, whereas full dyestuff reduction was possible with the Fe^{3+}-DGL mediator, the Ca^{2+}-Fe^{3+}-DGL mediator was unable to reduce the dispersed dye effectively in the catholyte under the same experimental conditions (Bechtold and Turcanu 2004).

The problem with these Fe-containing mediators is that cellulose materials have a special ability to form alkali-stable iron-cellulose complexes. Because of this, cellulose fibers absorb a significant amount of Fe-ions from the iron complex mediator. At higher temperatures, this unwanted deposition of Fe-ions increases. This affects the durability of a textile product during use, especially during household washing with peroxide bleach, because the iron deposited on the fabric increases the possibility of catalytic damage to the textile product. Extremely high iron oxide concentrations in the fibers might even visually produce a color change toward yellow–brown tones. It was found that the samples dyed in black hues had the highest iron concentration, although there was no noticeable color shift in these samples. An iron oxide content was likely to result in a noticeable color shift in lighter hues. Oxalic acid has been discovered to be particularly successful in removing iron ions from the cellulose material because it can dissolve complex iron, resulting in a colored material with a tolerably low iron content (Bechtold and Turcanu 2009; Yi et al. 2020).

In addition to Fe-amino and Fe-saccharic acid compounds, several anthraquinones including hydroxy-anthraquinones and anthraquinone-sulfonic acid compounds are used as mediators for dispersed dyestuff such as vat, indigo, and sulfur dyes in many literature (Schnitzer et al. 1996; Bechtold et al. 1999, 2010; Turcanu and Bechtold, 2011; Turcanu et al. 2011). Anthraquinone mediators are of interest over iron complexes when the precipitate formation is threatened by the by-products found in commercial dye samples, such as iron sulfide, for instance, the use of sulfur dyes with iron complexes. Anthraquinone compounds should have the following properties to be used as a reducing agent (Bechtold et al. 2010):

(a) The anthraquinone (ANQ) mediators must be colorless. Because, based on the substituents employed, anthraquinone (ANQ) molecules can also act as dyes, and when utilized as mediators for sulfur or vat dye reduction, they exhibit a

significant amount of adsorption on the cellulosic substrate, which may change the hue of the dye.
(b) The material should be dissolved in the alkaline dye bath, which can be attained by the incorporation of carboxylate or sulfonate substitute groups or by the presence of phenolic hydroxyl groups.
(c) In alkaline baths of sulfur and vats dye, ANQs must have sufficient stability.
(d) The oxidized form of ANQ must be capable of repeatedly undergoing cathodic electron transfer to its reduced form without decomposing.

Cathodic peak potentials of anthraquinoid systems can reach -849 mV versus Ag/AgCl/ 3 $molL^{-1}$ KCl. It is discovered that 1 g/l of anthraquinone compounds with an alkali level of 4 g/l NaOH are suitable to accomplish dye reduction. Table 7.6 represents the cathodic peak potential of several ANQ compounds. When calculating the required amount of mediator, the O_2 supplied from the air needs to be considered. Utilizing a closed apparatus, the amount of catalyst necessary can be decreased. The comparatively slow conversion rate at the surface of electrodes is one drawback of these organic redox systems (Kulandainathan, Patil, et al., 2007). The mechanism of ANQ mediators is illustrated in Eqs. (7.7)–(7.10) (Bechtold et al. 1999, 2010):

$$ANQ + e^{-} \leftrightarrow ANQ^{\bullet -} \tag{7.7}$$

$$ANQ^{\bullet -} + e^{-} \leftrightarrow ANQ^{2-} \tag{7.8}$$

$$2ANQ^{\bullet -} \leftrightarrow ANQ + ANQ^{2-} \tag{7.9}$$

$$ANQ^{2-} + Dye \leftrightarrow ANQ + Dye^{2-} \tag{7.10}$$

Equations 7.7 and 7.8 describe the consecutive quick transfer of two electrons, one at a time that results in the reduction of an ANQ in highly alkaline conditions. In aqueous solutions, radical anion generation is not observed. So it is a two-electron transfer process that produces the ANQ dianion. Equation 7.8 depicts how the radical anion receives the second electron, while Eq. (7.9) depicts how the radical anion disproportionates and becomes ANQ^{2-} anion. The reduction of the dispersed dye occurs by Eq. (7.10). It is possible to regenerate the oxidized form of the ANQ if the redox pair ANQ/ANQ^{2-} can provide an adequate negative reduction potential in solution.

The mediator is typically used in a quantity for dye reduction that is not greater than that specified by the dye reduction stoichiometry. Therefore, to be reduced, an oxidized dye molecule needs two electrons. According to this, one mole of oxidized dye requires two moles of a mediator, which is a redox-active metal ion providing one electron to the oxidized form of dye to be converted into the leuco form. It has been recognized that this mediator quantity can be decreased by electrochemical regeneration of the mediator (in the case of vat dyeing, approximately 0.1–1 mol L^{-1} of reduced mediator is required for each mole of dye). The higher the insufficiency of

Table 7.6 Cathodic peak potential of several anthraquinone mediator compounds using 4 g/l sodium hydroxide (Kulandainathan et al, 2007b)

Anthraquinone system	The potential of the cathodic peak (in mV vs. Ag/AgCl/ 3 $molL^{-1}$ KCl)
1,4-Dihydroxyanthraquinone	−760
1,2-Dihydroxyanthraquinone	−880
Anthraquinone-1,5-sulfonic acid	−750
1,2,5,8-Tetrahydroxyanthraquinone	−885
Anthraquinone-2-sulfonic acid	−640
1-Amino-2-carboxyanthraquinone	−749
Anthraquinone-2,6-sulfonic acid	−561
1,2-Dihyroxyanthraquinone-3-sulfonic acid	−869
1-Aminoanthraquinone-2-sulfonic acid	−720
1,8-Dihydroxyanthraquinone	−770

the mediator, the greater the demands the electrolytic cell has to fulfill. Considerable amounts of mediator systems are likely to be utilized in certain situations to effectively remove oxygen from the air. The amount of oxygen to be removed varies depending on the dyeing procedure and machinery used. For example, an additional 1–10 mol of reduced mediator is needed for dyeing of medium depths of shade with vat dyes in the exhaust method, and an additional 2–10 mol of reduced mediator is required for continuous indigo dyeing (Bechtold et al. 2004). As a consequence, even after many years, this mediator technique remains in development. As a result, new approaches are still needed to increase the eco-efficiency of applying vat dyes (Roessler and Jin 2003; Ahmed and El-Shishtawy 2010).

The concentrations of mediators can be minimized by employing an electrolyzer containing a large operational cathode area. A divided multicathode electrolyzer with three-dimensional flow-through electrodes is an example of such a type of electrolyzer that can offer a large cathode surface area (Bechtold and Turcanu 2002).

7.4.1.2 Indirect Electrochemical Dyeing with Vat Dye

Electrochemical Dyeing Apparatus

A modified multi-cathode electrolysis cell was connected to a commercial package dyeing apparatus that could hold up to 10 kg of goods. Figure 7.5 displays the construction of the dyeing apparatus.

The capacity of both the storage tanks used for fresh solution and the used dyebath was 400 l each. The total volume of the electrochemical dye bath was 200 l (the apparatus and the cell volume were 100 l of each). This dyeing apparatus could be used for alkaline scouring, bleaching, and also for oxidation of vat dyed substance

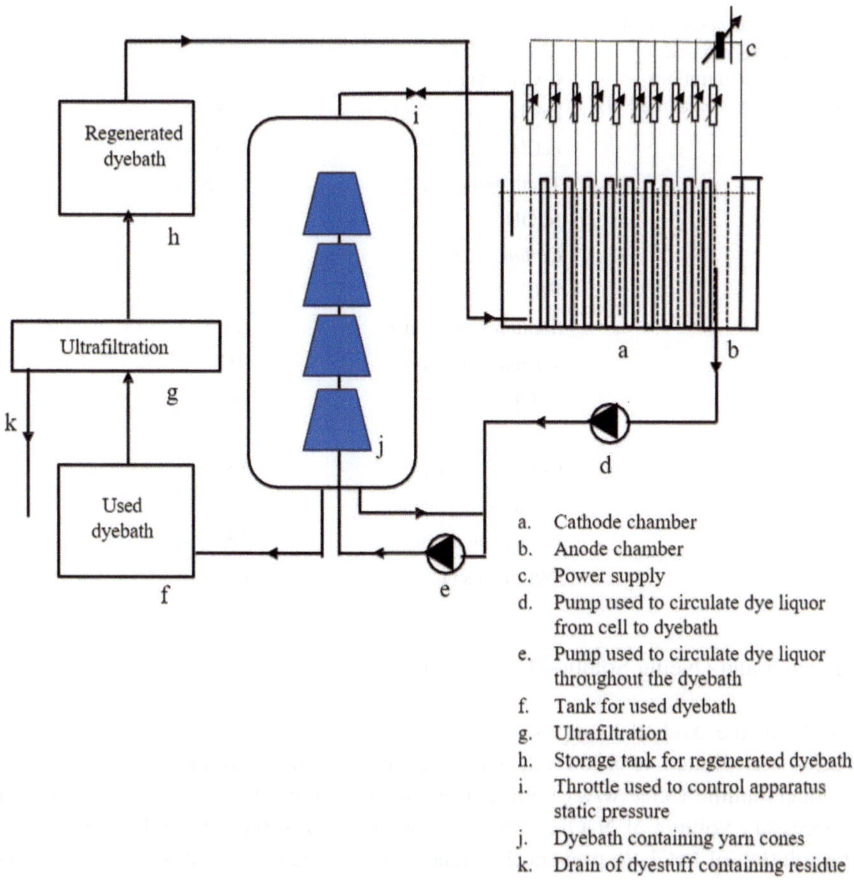

Fig. 7.5 Construction of electrochemical yarn package dyeing machine (Bechtold and Turcanu, 2009). Adapted with permission from Bechtold and Turcanu (2009) 'Electrochemical reduction in vat dyeing: greener chemistry replaces traditional processes', Journal of Cleaner Production, 17(18), pp. 1669–1679 Copyright © 2009 Elsevier Ltd

after the dyeing, soaping, and rinsing process. The liquor circulation rate between the dybath and cell was 150–200 l min^{-1}. Thus, it took less than one minute to exchange dyebath liquor between the apparatus and the cell. A 2 bar static pressure was developed in the dyebath due to the placement of a throttle in the main pipe connecting the dyebath to the cell. To ensure consistent dyeing throughout the yarn package, the internal flow of the dyeing liquor was regularly reversed. The cathode stack was built using ten individual cathode units, linked to a common anode and a separately adjustable power source (Bechtold et al. 1997b).

Table 7.7 Composition of catholyte/dyebath and mediator concentrations (Bechtold and Turcanu 2009)

Chemicals	Concentration in gl^{-1}
Concentrates	
$FeCl_3$ sodium-D-gluconate	12.5
$FeCl_3$	20.3
Triethanolamine	127.5
NaOH	172
Catholyte/Dyebath	
$FeCl_3$	1.62
Sodium-D-gluconate	1.0
Triethanolamine	10.2
NaOH	14
Auxiliaries	
Complexing agent	0.5
Sequestering agent	1
Dispersing agent	0.5–1.5
Wetting agent	1

Mediator and Dyeing Solution Composition

To obtain the dyebath composition, initially a volume of 80 ml 10% solution of $FeCl_3$ was diluted to 1000 ml. Then another solution was prepared by mixing triethanolamine (85% w/w) with a solution of sodium-D-gluconate in water, and a necessary volume of NaOH (50% w/w) which was further gradually poured into the solution of $FeCl_3$ to prevent the blend from overheating because this process is exothermic. Then, gradually, the remaining specified volume of NaOH was incorporated, and parallelly, deionized water was added to fill the volume to the necessary level. Table 7.7 shows the dyeing solution composition (Bechtold and Turcanu 2009).

Dyeing Process

The cell and the dyeing chamber were filled with mediator solution after the prepared yarn cones were inserted into the dyeing chamber. The dye bath and cell were connected by a pump as shown in Fig. 7.5d. This pump raised the dye bath static pressure to 1–2 bar, which was necessary for the internal circulation pump as shown in Fig. 7.5e to operate properly. The dye liquor was circulated across the yarn cones in the dyeing unit by the circulation pump. The dyeing was executed according to the established standard procedures for package dyeing. The dyebath was brought to the necessary temperature and, to create reducing circumstances, the dyebath's cathodic reduction was initiated simultaneously. With the use of a Pt-redox electrode (potentiometer Metrohm 654 pH-meter, Herisau, Switzerland, vs. Ag/AgCl, 3 M

KCl reference electrode), the redox potential in the dyebath was determined. Three phases must be completed during electrochemical dyeing, in which the cell must play different functions as mentioned below (Bechtold and Turcanu 2002):

- Phase 1: Reduction of dissolved O_2 and development of a minimum concentration of reduced mediator where the redox potential of the dyebath had been reached −910 to −920 mV.
- Phase 2: Indirect cathodic reduction of the dispersed dye by maintaining the dye bath redox potential the same as the leuco potential of the dye at the maximum cell current. Here, dyebath exhaustion started.
- Phase 3: The cell was then used to regulate the dye bath redox potential within a specified range by adjusting the charge flow, which was adjusted by cell current. In this phase, complete reduction of dyestuff, dyeing equilibrium, and leveling had been achieved.

So during phases 1 and 2, the cell acted as a reduction reactor, and in phase 3 it acted as a control device. Relevant process parameters of dyeings are presented in Fig. 7.6.

Initially, cathodic reduction was used in this procedure to remove the dissolved oxygen at a current strength of about 40A. The reduction potential of the catholyte dropped to -900 mV, indicating the complete elimination of dissolved oxygen. After that, the dispersed dye material was introduced to the dye bath and left there for around 40 minutes as unreduced pigment. The current of the cell was kept approximately at zero throughout this time to prevent an extensive dye reduction. Following the pigmentation stage, the dyestuff was cathodically reduced at maximum cell current. It was necessary to keep the cell current at its highest for the whole dyeing process while using dark dyes. It was not necessary to use the entire reduction capacity for lighter dyeings, as in Fig. 7.7, where the current of the cell may be lowered at the last stage of the coloring process. The bath was replaced with the fresh mediator solution after the dyeing process to prevent any interaction with the atmospheric oxygen. The storage tank illustrated in Fig. 7.5f stored the used dye bath. To prevent the possibility of reoxidation in the dark color dyeing, the mediator solution used for rinsing must be cathodically pre-reduced in the cell. After rinsing, this mediator solution was likewise stored in the exhausted dye bath storage tank. Subsequently, the dyeing process was completed by following the soaping, oxidation, rinsing, and acid rinsing processes required for vat dyeing.

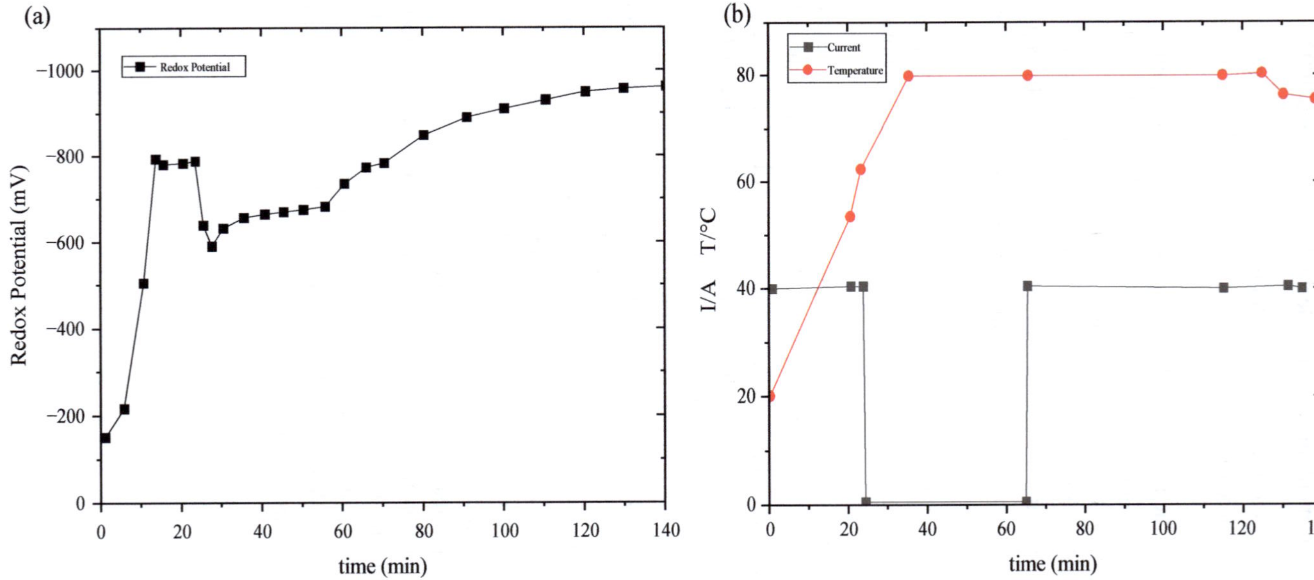

Fig. 7.6 Electrochemical dyeing of cotton yarn with indanthrene (Orange RRTS) dye in pigmentation method where pigmentation is done at time 35–55 minutes with a dye concentration of 1.8% at a liquor ratio of 1:55 **a** Redox potential as a function of time **b** current and temperature as a function of time (Bechtold and Turcanu 2009). Adapted with permission from Bechtold and Turcanu (2009) 'Electrochemical reduction in vat dyeing: greener chemistry replaces traditional processes', Journal of Cleaner Production, 17(18), pp. 1669–1679 Copyright © 2009 Elsevier Ltd.

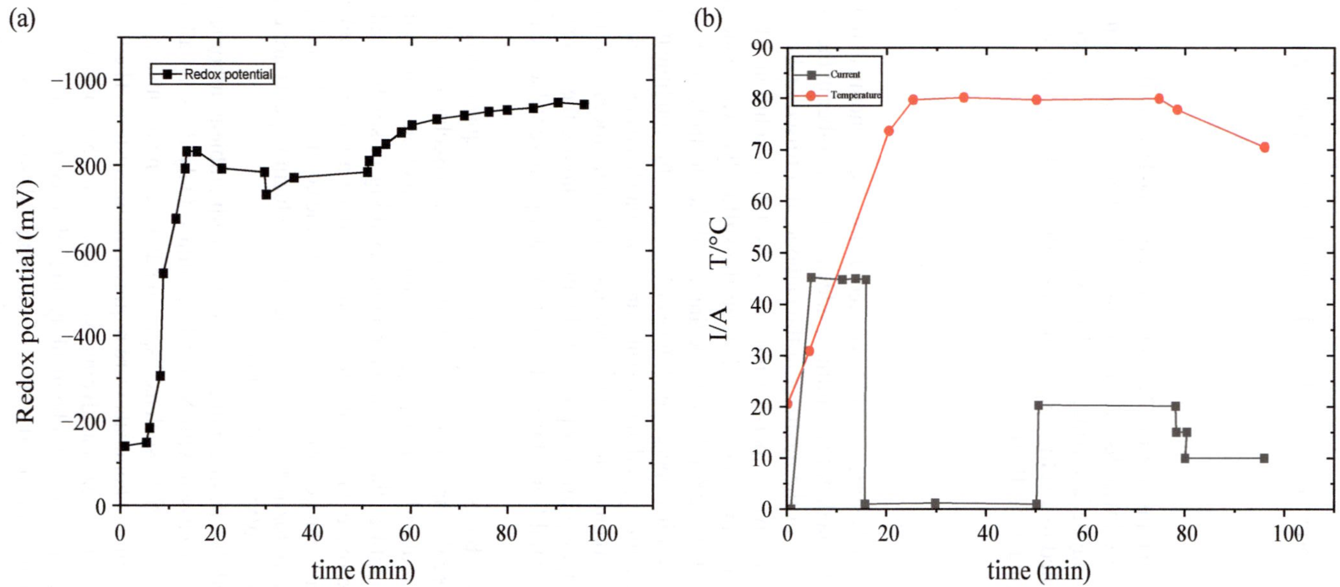

Fig. 7.7 Electrochemical dyeing of cotton yarn with Indanthrene (Blue BC) dye in pigmentation method where pigmentation is done at time 30–50 minutes with a dye concentration of 0.5% at a liquor ratio of 1:55 **a** Redox potential as a function of time **b** current and temperature as a function of time (Bechtold and Turcanu 2009). Adapted with permission from Bechtold and Turcanu (2009) 'Electrochemical reduction in vat dyeing: greener chemistry replaces traditional processes', Journal of Cleaner Production, 17(18), pp. 1669–1679 Copyright © 2009 Elsevier Ltd.

7.4.1.3 Indirect Electrochemical Dyeing with Sulfur Dye

The dyeing of cotton fabric with Sulfur Black B1 involves the following process:

Mediator Preparation

Indirect electrochemical reduction involved the use of Fe(III)-TEA mediator. At first, $Fe_2(SO_4)_3$ 7–8 H_2O 4.7 g/l was dissolved into distilled water to create the mediator solution. Next, under magnetic stirring, 20 g/l TEA and 34 g/l of NaOH were added. The solution was diluted to a sufficient volume after the iron hydroxide precipitate was dissolved (Bechtold et al. 1996).

Dyeing Procedure

Solid dyes on the cathode's surface can be reduced using electrochemical reduction (WMA et al. 2014). The mediator was initially added to the chamber in the oxidized state. After the reduction of the mediator at the cathode, it diffused and exchanged electrons with the sulfur dye, dispersing into microcrystals in solution. The mediator's oxidized state was then regenerated when the sulfur dye was reduced. The potential was measured at −200 mV at the beginning of the electrolysis process and increased to −700 mV during operation. Afterward, as the reduction time increased, the particles dissolved entirely, and the dye substance that had been electrochemically reduced kept its redox potential unchanged until the reduction process was finished. At that moment, the dyeing process proceeded without changing the current intensity to prevent the potential from shifting towards greater positive values (caused by oxygen intake).

When Sulfur Black B1 is in oxidized form, it exists in a quasi-polymer state where repeating molecules are joined by a disulfide bond (Bechtold et al. 2000). These disulfides are reduced to thiolate groups during reduction. After reduction, the sulfur molecules split at their disulfide bonds, resulting in smaller molecules. At this stage, the dye reacts faster (like the mediator) once a sufficient amount of reduced sulfur dye has been generated there. However, the color depth can decrease at −550 to −600 mV redox potential when chemical-reducing agents are used for Sulfur Black B1 dyeing (WMA et al. 2014).

To prevent reduced dye from coming into contact with ambient airborne O_2, all solutions need to first degassed with N_2 for 15 minutes. Only dyes, an alkali, and trace amounts of other materials, such as dispersing and wetting agents, were added to color the fabric. No more chemicals were required in this system. The fabric was put inside the dyeing chamber, and dyeing was carried out. Samples were taken out of the dye bath after the electrochemical dyeing, allowed to oxidize, then cold rinsed, boiled, and then cold rinsed again before being allowed to air dry (WMA et al. 2014). When the dyeing cycle was over, air oxidation was done to precipitate the unused dye, which was filtered off, making the electrochemical dyeing process simpler (Kulandainathan et al. 2007b).

The sulfur dye that was electrochemically reduced had a unique reduction potential and a modified chemical structure. Overall, the dyestuff exhibits essentially unchanged dyeing behavior. Compared to conventionally colored fabric, electrochemically dyed fabric appeared to have better color depth and a slightly lower color strength (K/S) value when analyzed for CIELab coordinates. The results for the fastness attributes were more positive. Concerning washing, rubbing, and light fastness, the fastness properties seem to be the same when using a traditional dyeing procedure.

7.4.1.4 The Recycling and Reusability of the Used Dyebath

Today, with growing concerns about dyehouse effluents, the prospect of recovering the reduction capacity of a used dyebath is appealing. The leftover dye needs to be eliminated from the dye solution if the mediator solution is to be reused. Since sulfur, vat, and indigo dyes are water-insoluble and may form a suspension in their oxidized form, they can be separated through a simpler filtration method from the oxidized dye solution. It has been found that to separate the dye, pressurized air was injected into the dye liquor to initiate oxidation in the tank. The redox potential was continuously measured to monitor the oxidation of the dye bath. For vat dyeing, the air supply ceased when the dye bath reduction potential reached a range of −300 mV to −200 mV from the reduced potential value. Oxidation caused the unused dye to precipitate (Bechtold and Turcanu 2009). Then, a pressure filtration process was used to separate the suspended dye molecules with a polytetrafluoroethylene filter having a pore dimension of 0.45 or 1.2 μm. In a study to check the reusability of the dye bath, the absorbances of the oxidized dye baths both before and after filtering were investigated and displayed in Table 7.8. A high level of dye removal was seen from the comparison of the absorbances of the filtered dye bath with a newly made one. Five dyeing experiments were conducted using the filtered solutions under identical experimental settings, including subsequent dyebath recycling stages. The dyeing performance was described by the CIELAB coordinates given in Table 7.9, and it showed a good consistency of shade of the dyed materials. It indicates that it is possible to reactivate the reducing agent electrochemically multiple times without a noticeable loss of electrochemical activity. Because of handling and processing, about 10–15% loss of the mediator solution occurred. It has to be refilled with a fresh solution in case of reusing the dyebath for further dyeing. The recycling process was run through up to nine times (Bechtold et al. 1994).

7.4.1.5 Factors Affecting Indirect Electrochemical Dyeing

Effect of the current intensity: The time required for the reduction of the dyestuff depends on the current intensity of the electrochemical cell. The requirement of time

Table 7.8 The absorbance of indanthren brilliant green FFB dyebath was measured at 620 nm both before and after pressure filtering via a Polytetrafluoroethylene membrane filter (Bechtold et al. 1994)

Absorbance before filtering	Absorbance after filtration
0.204	0.026
0.198	0.064
0.124	0.009
0.083	0.012

Table 7.9 CIELAB coordinates of indanthren blue GC dyed material from a series of dyebath recycling experiments conducted at 20 °C (Bechtold et al., 1994)

Recycling step	Potential (mV)	L*	a*	b*
1	−900	44.71	2.92	−30.43
2	-900	45.50	1.61	−27.98
3	−900/−920	45.88	1.39	−28.47
4	−870/−900	45.01	2.14	−28.72
5	−800/−875	45.55	3.05	−30.53

decreases with the increase of the current intensity, but high current intensity causes an increase in temperature due to the Joule effect (Said et al. 2008).

Influence of material-to-liquor ratio: The color depth of the dyed samples improves with shorter material-to-liquor ratios (Kulandainathan et al. 2007a).

Flow rate: The reduced species generated in the cathode must be transported into the dyeing unit by a constant flow of process fluid between the cell and the dyeing unit. As the flow rate rises, so does the rate of reduction. Equations 7.11–7.14 can be used to estimate the minimal catholyte exchange required between the electrolyzer and cell for getting the optimum reduction rate (Bechtold and Turcanu 2009).

$$m = \frac{\mathrm{I.d}}{n.\mathrm{F}} \tag{7.11}$$

$$\Delta C.s = \frac{m}{d} \tag{7.12}$$

$$\frac{\Delta p.C.s}{100} = \frac{m}{d} = \frac{I}{n.F} \tag{7.13}$$

$$s = \frac{100\ \mathrm{I}}{n.F.\Delta p.C} \tag{7.14}$$

where

ΔC = difference of concentration of the oxidized mediator between the dyeing unit and the cell, $molL^{-1}$;

C = initial concentration of mediator, $molL^{-1}$;

s = rate of electrolyte flow, Ls^{-1};

Δp = difference of concentration of the oxidized mediator between the dyeing unit and the cell in % of mediator concentration;

m = the oxidized mediator generated in moles;

I = current of the cell, A;

n = quantity of electrons transported during the electrochemical process;

d = time of electrochemical reaction, s;

F = Faraday's constant, 96,485 $Asmol^{-1}$.

Electrode materials: Very few electrode materials, such as copper mesh, silver mesh, stainless steel, and nickel, have been investigated for reducing dyes in the indirect electrochemical process. The current density and current efficiency (CE) obtained with these materials are not particularly enough as a consequence of their limited specific surface area. The current density and current efficiency can be improved by using a large specific surface area of the cathode. Multicathode membrane electrolyzers can be used owing to their large surface area. Besides the large surface area, the use of three-dimensional flow-through electrodes increases the cell current (Bechtold et al. 1997b; Bechtold et al. 1998). Nickel has a greater CE, electrochemical activity, and corrosion resistance than stainless steel during the electrochemical process. The catalytic feature of the nickel electrode increases the reduction rate of the dye. While deposits and corrosion affect the CE of stainless steel, the nickel electrode continues to retain a high rate of reduction over time (Yi et al. 2020).

7.4.1.6 Advantages of Indirect Electrochemical Dyeing

Indirect electrochemical dyeing has the following advantages (Bechtold 1993; Bechtold et al. 1994, 1996; Marte et al. 2003):

(a) This method allows for complete dyebath recycling.
(b) Electrochemical techniques can be employed to regulate the reduction parameters of the dye bath.
(c) The mediator and the rinse water can be reused.
(d) Reduce water consumption.
(e) The dyeing results are equivalent to those of traditional dyeing employing reductants such as $Na_2S_2O_4$.
(f) It is possible to control appropriate dyebath conditions at low temperatures.
(g) Salt-free dyeing, with improved fabric or thread quality, is possible.
(h) Easily reproducible, repeatable dyeing results.

7.4.1.7 Disadvantages of Indirect Electrochemical Dyeing

Indirect electrochemical dyeing has the following disadvantages (Schnitzer et al. 1996; Božič and Kokol 2008; Crettenand 2012; Yang et al. 2021):

(a) As with iron complex mediators, cellulose materials show a clear propensity to bind iron from the complexes through the creation of alkali-stable iron-cellulose complexes, which can change the shade of the dyed materials.
(b) Filtration significantly raises the expense of the vatting process and causes several technical problems, for instance, a significant pressure drop throughout the filtration process and a chance of reactor blockage.
(c) According to the quantities utilized and the ecological toxicity of the mediators, a severe environmental issue exists.
(d) The continuous need for extra mediator feeds to maintain a constant redox potential in the continuous dyeing process is another drawback of this approach.
(e) Compared to the conventional reduction procedure, it requires a lot of time to finish the dyestuff reduction.
(f) Sometimes significant amounts of conducting salt (about 65–30 g/l of Na_2SO_4 and 1.5 g/l of NaOH) are required to guarantee sufficient electrical conductivity and, thus, reduce resistance losses. Thus, the dyehouse wastewater has extremely high sulfate levels, which is not desirable.

7.4.1.8 Challenges to Establish the Indirect Electrochemical Dyeing Process

The increased production of the indirect electrochemical reduction process requires consideration of the following issues (Bechtold et al. 1996):

(i) It is necessary to build the electrolytic cell and the dyeing unit as two distinct units to keep using the dyeing equipment in the dyehouses.
(ii) To enable a constant replenishment of the reducing capacity in the dyebath, the electrolytic reactor needs to function as a flow cell.
(iii) It is necessary to build a cell with the highest specific cathode surface area.
(iv) The use of cheap electrode materials devoid of hazardous heavy metals is required.
(v) Three-dimensional electrodes that occupy a small cell volume and have a higher surface area need to be created.
(vi) To reduce separator expenses, a cell with a minimal separator area is required. It also prevents the reoxidation of dye due to bleeding of the separator and oxygen diffusion.

7.4.2 *Direct Electrochemical Dyeing*

A newly developed electrochemical vatting technique called "direct electrochemical dyeing" has been developed for sulfur, indigo, and vat dyes as an alternative to the requirement of the permanent presence of redox mediator, which is necessary to promote electrochemical reduction. Sodium hydroxide is the only chemical employed in this process. In autocatalysis, electrons are employed instead of chemical reducing agents to achieve both environmental and financial objectives. It is based on a comproportionation reaction where the dye and the leuco dye undergo a comproportionation reaction to produce a radical dye anion that is subsequently reduced electrochemically.

However, a small amount of soluble reducing agent is needed to produce a small amount of leuco dye, which is required to start the reduction reaction. Once the leuco dye is formed, it acts as an electron shuttle between the electrode and the dye particle, and the reduction of dyes continues without any interruption (Roessler et al. 2002; Roessler and Jin 2003; Crettenand, 012). Figure 7.8 shows a demonstration of the autocatalysis electrochemical reduction of indigo dye. From the figure, it is seen that a dye and a leuco dye undergo a comproportionation reaction and generate two radical dye anions, one of which is subsequently electrochemically reduced to leuco dye and the other continues the comproportionation reaction. The reduction rate of the intermediate radical anion is diffusion-controlled, and it is the rate-limiting stage of the reaction, which is largely dependent on the dye radical concentration in the solution and the thickness of the diffusion layer near the electrode surface (Xu et al. 2014; Yang et al. 2021).

However, it is very difficult to reduce the dye in an aqueous solution on a planar electrode because of minimal interaction between the dye and the electrode surface. To solve this problem, many researchers have conducted many experiments by designing new reactors, changing electrode materials, etc. Some of the processes for improving the kinetics of the reduction reaction in direct electrochemical dyeing are given below:

7.4.2.1 The Use of Precoat-Layer-Cell for Direct Electrochemical Reduction of Dye

The precoat layer cell is illustrated in Fig. 7.9, composed of a cathode, which is an electrically conductive porous woven filter fabric-like structure coated with a cathodically polarized and electrically conductive layer. The cathode can be made of any material that has electrical conductivity, such as carbon, stainless steel, nickel, graphite, plain steel, nickel alloys, tantalum, titanium, platinumized tantalum, and titanium alloy. Porous supporting material provides a very large open area to overcome the pressure drops in the execution of the process. The reactor operates based on the filtration principle, in which the reduction of the dye takes place when it comes into contact with the cathode in an alkaline medium (Roessler 2003). A metal,

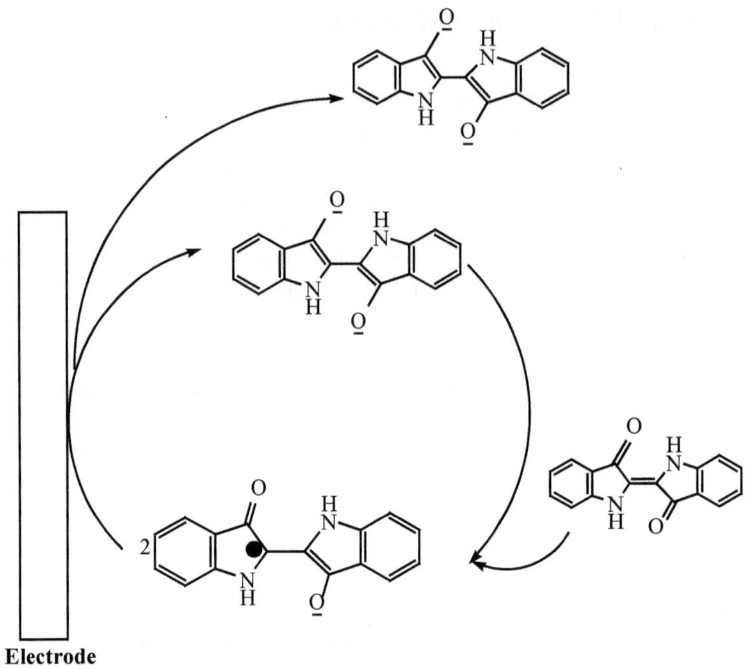

Fig. 7.8 Mechanism of direct electrochemical reduction of indigo dye (Božič and Kokol 2008). Adapted with permission from Božič and Kokol (2008) 'Ecological alternatives to the reduction and oxidation processes in dyeing with vat and sulfur dyes', Dyes and Pigments, 76(2), pp. 299–309 Copyright © 2006 Elsevier Ltd.

conductive metal oxides, carbon-based substances (for example, activated carbon, graphites, or carbon blacks), or a combination of two, finely divided activated metals (such as classic hydrogenation metals, especially those from the I, II, and VIII transition groups of the periodic table of elements) or nanoclusters are preferred components for the cathodically polarized coating layer. To enhance the addition between the cathodically polarized layer and the support or to enlarge the surface area of the cathode, the cathodically polarized layer may contain electrically conductive assistance. It is possible to conduct the reaction at standard pressure and temperature ranges of 0–100 °C, while 40–80 °C is the preferred range. The catalytically active layer can be regenerated to continue the reduction of the dye with a change in simple flow direction after the reduction has finished or after it is exhausted. The major disadvantage of this method appears to be the high-pressure drop that develops throughout the process of filtration and the continuous risk of reactor blockage. Also, the performance of the reactor is inadequate (Merk et al. 2003).

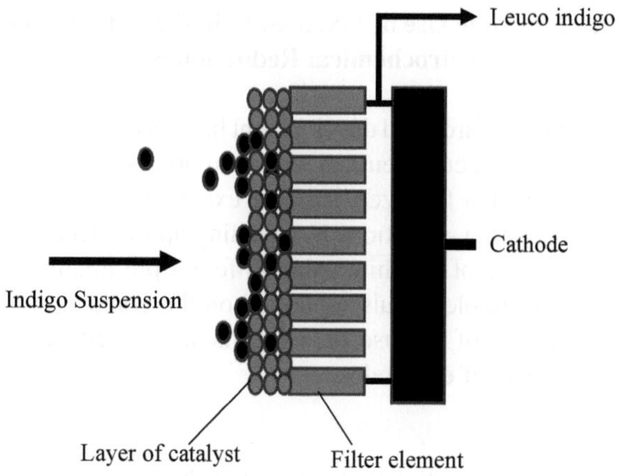

Fig. 7.9 Construction of the precoat-layer-cell (Roessler and Jin 2003). Adapted with permission from Roessler and Jin (2003) 'State of the art technologies and new electrochemical methods for the reduction of vat dyes', Dyes and Pigments, 59(3), pp. 223–235. Copyright © 2003 Elsevier Ltd.

7.4.2.2 Electrochemical Reduction of Dye in Laboratory-Scale Flow Cells

In the past, vat, and indigo dyes, was directly electrochemically reduced using traditional H-cell technology. The problems of this cell are the limited current and low concentration of radicals, which results in a low rate of reaction. The side reaction of hydrogen gas formation also provides a low current efficiency. To solve these issues, the direct electrochemical reduction of indigo has been investigated using a divided flow cell. By using this process, the concentration of reduced indigo was increased to 1 gl^{-1}. However, reactor effectiveness was still inadequate for large-scale industrial applications. The greatest constraint was the low concentration of radicals. Thus, more investigation was devoted to refining this novel electrochemical method and investigating the impact of crucial elements like surfactants and dispersion agents on the concentration of radical species. For this objective, the reduction of indigo radicals in the presence of surfactants was investigated using Ni electrodes. The anionic and nonionic surfactants were found to destabilize the indigo radicals and slow the electro-reduction process by forming less ordered surface aggregates that stopped the radical from reaching the electrode. However, cationic surfactants like hexadecyltrimethylammonium chloride favorably affected the electrode kinetics because they could stabilize the radical. Nevertheless, leuco dye precipitated with cationic surfactants was observed to cause several problems (Rusling, 1991). Setamol WS is generally used in electrochemical techniques for the reduction of vat dyes, and its usage as a dispersing agent is consistent with this since it has no negative effects on the electrochemical reduction process (Bechtold et al. 1997c; Information 2011). Unfortunately, it has been determined that the performance of the reactor is still insufficient for industrial usage. Consequently, it remains necessary to discover unique substitutes that enhance the sustainable approach of using vat dyes.

7.4.2.3 The Use of Fixed and Fluidized Bed of Graphite for Direct Electrochemical Reduction of Dye

Further research and development have been emphasized to accelerate the kinetics of the direct electrochemical reduction process, and for this purpose, graphite granules filled fixed or fluidized bed were examined. A lot of work has been done to optimize the operating conditions by speeding up the electron transfer reaction by modifying the surface of graphite using different pre-treatment techniques. Additionally, the effect of noble metals supported on the granules was examined. The following are the results of the use of noble metals, modification of electrodes, and oxidative treatments of electrodes.

Effect of Oxidative Pretreatment of the Electrode

Concentrated nitric acid, acidified potassium permanganate, acidified saturated sodium peroxodisulfate ($Na_2S_2O_8$), and acidified H_2O_2 solution were used to oxidize the graphite. Every other oxidizing agent, except H_2O_2, impacted the reduction of dye negatively. When graphite was immersed in H_2O_2, the rate of reduction was nearly doubled, reaching 40 nmol min^{-1} g^{-1} graphite. Most of the chemical reactivity of the carbon surface is associated with hydroquinone and quinone groups, whether they are directly bonded to the surface of carbon black or are components of more complicated structures.

The generation of quinone-like oxygen-containing functional groups on the graphite surface is likely correlated with increasing catalytic activity (Roessler and Crettenand 2004). Consequently, the production of specific quinone-like functional groups on the graphite electrode can speed up the reduction process.

Effect of Modification of the Electrode Surface

The attachment of quinoid molecules covalently onto the graphite electrode has provided another interesting method to investigate the increased electrocatalytic capabilities of graphite. On the carbon electrode, mediator-like substances were so immobilized. These species were immobilized with a functional hydroxyl or amine group by producing carboxylic groups on the graphite surface, and the process is illustrated in Fig. 7.10. The quinone was covalently bound to the electrode through the –OH group of the electrode surface as shown in Fig. 7.11 (Roessler 2003; Božič and Kokol 2008). The modification normally happens at the border of the graphite electrode, and physical adsorption favorably takes place at the basal planes.

Two different species of quinone and anthraquinone, one covalently bound to the graphite surface and the other physically adsorbed, may therefore be present in the functionalized material (Brett and Oliveira Brett 1984). Both the type of species connected to the electrodes and their surface concentration influenced their catalytic activity. When compared to the original reference material, Table 7.10 shows that

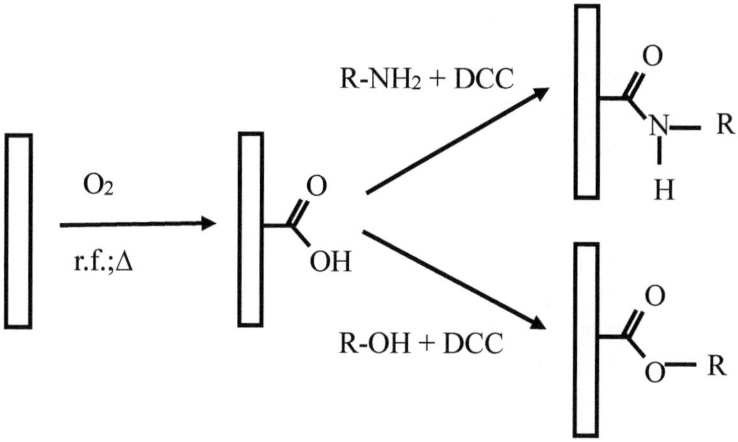

Fig. 7.10 Modification of carbon electrode through the covalent bond formation with the carboxylic surface group, where R = anthraquinone compound, Δ = thermal energy, r.f = radio frequency plasma (Roessler and Jin 2003). Adapted with permission from Roessler and Jin (2003) 'State of the art technologies and new electrochemical methods for the reduction of vat dyes', Dyes and Pigments, 59(3), pp. 223–235. Copyright © 2003 Elsevier Ltd.

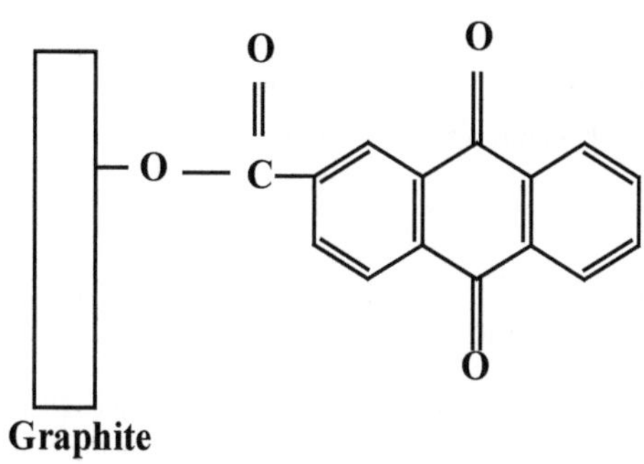

Fig. 7.11 Modification of the electrode with covalently bound redox-active quinone (Roessler and Jin 2003). Adapted with permission from Božič and Kokol (2003) 'Ecological alternatives to the reduction and oxidation processes in dyeing with vat and sulfur dyes', Dyes and Pigments, 76(2), pp. 299–309 Copyright © 2006 Elsevier Ltd

1,8-dihydroxyanthraquinone and 5-amino acenaphtheneanthraquinone are two of the most active catalysts, and the activity of all other immobilized species is lower. The catalytic properties of the electrode were increased further by first treating the surface with acid chloride groups rather than carboxylic groups. The increase in the catalytic activity was also done by treatment with an anchor molecule such as β - sulfato-ethylsulfonyl reactive groups or dichlortriazine that reacted with the hydroxy or amino groups at the surface of graphite.

Table 7.11 demonstrates that the modification using C.I. reactive blue 19, containing a 2-hydroxyethyl sulfone group at the 3'-position of aniline ring gives a

Table 7.10 Impact of some immobilized quinonoid and anthraquinoid compounds on the reduction rate r of indigo (40 g (enViro) graphite, pH 14, 50 °C, 2.0 cm s^{-1} fluid flow, and 1.50 mA cm^{-2} current density) (Roessler 2003; Roessler et al. 2003)

Mediator immobilized on the electrode	r, Rate of reduction /nmol min^{-1} meq^{-1}
enViro cell graphite (Reference electrode)	347
Naphthoquinone	57
Benzoquinone	24
2-Aminoanthraquinone	151
1-Aminoanthraquinone	230
5-Aminoacenaphtheneanthraquinone	450
Anthraquinone-2-carboxylic acid	244
1,8-Dihydroxyanthraquinone	389

Table 7.11 Effect of some immobilized compounds on the reduction rate r of indigo (40 g (enViro) graphite, pH 14, 50 °C, 2.0 cm s^{-1} fluid flow and 1.50 mA cm^{-2} current density) (Roessler 2003)

Mediator Immobilized on the electrode	r, Rate of reduction /μmol min^{-1} g^{-1} cat
enViro graphite cell (reference)[a]	0.021
5-Amino-acenaphtheneanthraquinone[b]	0.004
1,8- Dihydroxyanthraquinone[c]	0.002
1,8- Dihydroxyanthraquinone[c]	0.009
5-Amino-acenaphtheneanthraquinone[c]	0.016
C.I. Reactive Blue 4[d]	0.031
C.I. Reactive Blue 19[d]	0.019

[a]enViro cell graphite following a 4-h reaction at 100 °C with an H_2SO_4-HNO_3 mixture and then neutralization done using NaOH, and in specific circumstances, furthermore, following reduction using $NaBH_4$
[b]alteration of the electrode surface by the dicyclohexylcarbodnmide (DCC) and carboxylic groups
[c]alteration by the carboxylic groups acetylation at the electrode using $SOCl_2$
[d]values of the 2nd experimental run

reaction rate almost same to the reference material and C.I. Reactive Blue 4 containing a bifunctional dichlortriazine group, yields reaction rates that is higher than the reference material.

However, after a few experiments, the catalytic activity dropped to undetectable levels because covalent connections were unstable in alkaline environments. Redox-active molecules that were detached functioned as soluble mediator molecules and most likely positively affected the reduction rate (Roessler 2003; Roessler et al. 2003).

Table 7.12 Effects of different noble metals on the production rate of leuco indigo using graphite as a supporting material (Roessler 2003)

Electrode composition	Immobilized mediator rate of reduction in µmol min^{-1} meq^{-1}
enViro cell graphite as the reference electrode	0.025
Reference + 0.5 wt % Pd	0.048
Reference + 0.5 wt % Pt	0.033
PMC carbon + 0.5 wt % Pt	0.011

Effect of Noble Metal

In some experiments, the particles of noble metals were immobilized on the graphite granules and used for the reduction of indigo dye. This resulted in an enhanced electrocatalytic hydrogenation (ECH) of the dye. For the reduction of indigo dye, electrographite granules (enViro Cell) as supporting material for Pd (Palladium) and Pt (Platinum) or carbon granules of crushed coconut shell (PMC carbon) with 0.5 wt% Pt were used as catalyst materials. Table 7.12 illustrates how much more active the first material was against the indigo reduction, particularly when Pd was used as a catalyst. So, according to ECH, the type of noble metal distributed across the carbon determined how efficiently the reduction process occurred. The Pd gives better reduction efficiency than Pt because the rate of hydrogen desorption and hydrogenolysis are almost similar for Pd, which results in fair to good electrogenolysis effectiveness. On the other hand, for Pt, hydrogen desorption occurs much more rapidly than hydrogenolysis, which makes ECH less effective (Roessler 2003; Roessler et al. 2003).

Instead of using multiple cathodes, the electrodeposited high-surface area cathodes have been employed in direct electrochemical dyeing. The electrochemical reactivity of these electrodes depends on the Fe^{3+}/Fe^{2+} redox sites bound to the electrode surface, and the generated surface-bound Fe^{2+} acts as a reducing agent for the vat dye reduction. By using organic additives (such as animal glue and gelatin) and inorganic adsorbents (such as thiocyanate), the adhesion and reusability of the electrodeposited cathodes may be enhanced substantially. However, the wash fastness and dye intensity of the fabrics dyed using this method were found to be excellent (Anbu Kulandainathan et al. 2008).

To obtain a specifically required color, mixtures of many dyes are frequently applied. In the case of vat dyeing, it is required to reduce two dyes in a mixture simultaneously. As a result, research has been done on the potential for the reduction of dye mixes with comparable potential characteristics and nuances. C.I. Vat Blue 14 and C.I. Vat Green 1 in a 1:1 blend have been examined. It was found that both dyes undergo simultaneous reduction, with a lag in reduction starting time between the two colors. Most likely, there was competition between the color molecules for the graphite granules' adsorption. This causes somewhat decreased reduction rates in comparison to the results of single-dye reduction (Roessler and Crettenand 2004).

7.4.2.4 The Use of Carbon Felt for the Direct Electrochemical Reduction of Dye

However, the initial addition of leuco dye to start the reaction is an issue in all of the aforementioned procedures. Furthermore, the procedure of modifying electrodes is difficult and costly. A well-known three-dimensional carbon felt that provides the same chemical reactivity as carbon, together with a significant amount of specific surface area, is investigated as an alternative. It is also widely used in electrochemical systems for its cost-effectiveness. The direct electrochemical reduction of vat dye was investigated with this carbon felt electrode without the addition of any leuco dye in the electrochemical cell initially, and the dyeing results were evaluated. CI Vat Yellow 1, an easily reducible dye, was selected for this procedure. However, due to its porous structure, which enhanced dye adsorption on the active electrode surface, carbon felt was identified to be the perfect electrode for this procedure. The findings demonstrated that the dye reduction may occur naturally on the carbon felt electrode without the addition of any leuco dye to initiate the reduction process. The reduction mechanism of the dye was examined using both spectrophotometric analysis and potentiostatic electrolysis. Two reduction products were detected; one was the reduction product of the –C = N– group, and the other was the reduction product of the ketonic group. These reduction reactions were both one-step, two-electron processes. After experimenting with different potentials, it was discovered that the dyeing outcomes at–900 mV were similar to those obtained using the conventional dyeing technique. Dyeing at a more negative potential caused over-reduction of dyes, which resulted in a greenish tone on the fabric. To overcome this problem, the voltage must be regulated effectively. So, the rate of direct electrochemical reduction can be effectively increased by using a carbon-felt cathode. Further research is necessary to determine whether the use of carbon felt electrodes to reduce other vat dyes except CI Vat Yellow 1 in the direct electrochemical method is feasible or not (Yang et al. 2021).

7.4.2.5 Factors Affecting Direct Electrochemical Dyeing

Current density: According to Table 7.13, as current density rises, reaction rates rise as well, but experimental current efficiency falls off quickly. In electrochemical engineering, current efficiency is a crucial parameter that indicates the percentage of electrical charge required to achieve the intended reaction. A constant current efficiency of roughly 80% is observed when there is a minimal current density of 0.001 mA/cm^2. In all other cases, the side reaction of hydrogen evolution results in a much lower efficiency that decreases exponentially over time. The explanation behind this phenomenon is that without introducing more dye pigment, eventually, the radical concentration gets so low that the current density, which is galvanostatic, crosses the current density that limits it. As the electrode potential becomes more negative, a side reaction, like the release of hydrogen, occurs from the reduction of water (Roessler et al. 2001, 2002, 2003).

Table 7.13 The effect of current densities on the galvanostatic electrolysis experiments (50 °C, Ni-net-electrode (200 cm^2), the concentration of the radical at $t = 0$: 2×10^{-5} M) (Roessler et al. 2001)

Current density, j [$mA.cm^{-2}$]	Conversion, X [%] after 40 min	Current efficiency, Φ [%]
0.001	65	80
0.0025	70	55
0.005	79	40
0.01	85	20
0.1	90	5

Properties of dye: The lipophilic characteristics of the dye molecule affect the reduction rate of dye as well. As lipophilicity increases, the solubility and reduction rate of the dye sharply decrease. This is due to the generation of larger pigment aggregates as a result of higher hydrophobicity and a decreasing surface area of the dye. At the microscopic level, both dissolution and migration of the dye molecule to the electrode, followed by electron exchange across the solid/solid boundary layer, are possible processes for the reduction of vat dye, which are hindered by the agglomeration of pigments. So the reduction of dye can be improved by increasing its solubility.

Temperature: The reduction reaction occurs significantly faster at a higher temperature. The apparent activation energy of this electrode process can be determined by analyzing the relationship between the logarithm of current density versus 1/(Temperature) curve at a constant cathodic potential (Roessler et al. 2001).

pH: There is a noticeable rise in both the electrode kinetics as well the reaction rate as the pH of the catholyte rises. For instance, it has been shown that, for indigo dye, the equilibrium among indigo, leuco indigo, and radical moves toward the radical side with increased pH, which is mostly due to the higher concentration of radicals. Reduced indigo is not very water-soluble in its acid or nonionic forms, therefore, the stabilization action may be owing to the existence of the more readily soluble ionized forms of the indigo dye (Roessler et al. 2002, 2003).

Impact of the catholyte flow rate: Mass transfer of the radical dye species to the cathode is improved at greater velocities, which could enhance the reduction rate significantly(Crettenand, 2012).

Influence of dye concentration: As the dye concentration rises, the rate of reduction also rises. This is because of the greater chance of collision between the dye particles and the cathode, which results in better contact between them (Roessler et al. 2003).

Influence of ultrasound and particle size: The dye solution can be exposed to ultrasound before entering the reactor to improve the dispersion characteristics of the dye. The radical reduction rate is found to be somewhat increased in the presence of ultrasound. Ultrasound effectively breaks up larger dye particles or clusters of solid dye granules (>100 μm) into tiny fragments between 1 and 10 μm. No smaller particles are created when ultrasonic treatment is prolonged; it only modifies the particle surface. The decrease in particle diameter increases the active reaction surface, which increases the production rate of radical species. However, there is also a chance that an increased amount of radicals will adsorb at the greater water/organic interface present in the dyeing system, which lowers the available concentration of the radicals. Nevertheless, the radical concentration has never been seen to drop following the use of ultrasound in the UV–visible spectrum. Rather, a little increase was seen, perhaps suggesting that ultrasound can also produce radicals (Roessler et al. 2002).

Influence of organic solvents: The vatting process can be sped up by using organic solvents, which raises the rate of reaction. However, the long-lasting stabilizing effects of the radicals were only observed at concentrations of more than 20 vol% of the organic solvents (dimethylformamide, methanol, ethanol, isopropanol). However, it is difficult to use such a high amount of organic solvent in the dye bath (Roessler et al. 2002).

Electrode material: The electrode material selection is important for optimizing the process for industrial settings. Titanium is not useful for this technique because it produces extremely low current densities. When compared to the other materials, lead exhibits strong catalytic characteristics for the reduction of radicals; nevertheless, lead dissolves and produces greater background currents during electrolysis. Regarding reaction rate, copper proved to be the most appropriate material tested thus far (Roessler et al. 2001).

Electrode surface area: It is very challenging to reduce vat, indigo, and sulfur dye in direct electrochemical methods because they are contained in an aqueous suspension and have weak interaction with the solid cathode. The result is an extremely low reduction rate and reaction kinetics. The increase in the surface area of the cathode increases its adsorption capabilities, thus speeding up the reduction rate of the dye (Roessler et al. 2003; Yang et al. 2021).

7.4.2.6 Advantages of Direct Electrochemical Dyeing

(a) Requires neither the continuous presence of a redox mediator nor a soluble reducing agent.
(b) Relatively more environment-friendly than the indirect electrochemical process.

7.4.2.7 Disadvantages of Direct Electrochemical Dyeing

Direct electrochemical reduction has the following disadvantages (Xu et al. 2014):

(a) The slow diffusion rate of the dye radicals to the electrode surface and the low concentration of the dye radicals in the system will limit the rate of this electrochemical process.

7.4.3 Electrolytic Hydrogenation

The process of electrocatalytic reduction of indigo was established in the early years of this century. The chemical reaction between adsorbed hydrogen and vat dyes generates leuco vat dye at the electrode surface (such as platinum black or Raney Ni). However, there is still a need for improvement with this approach because of poor current efficiency and a limiting factor that results from the adsorbed hydrogen. It generates molecular hydrogen instead of a hydrogenation reaction in a competitive reaction (Xu et al. 2014). The procedure involves a series of reduction steps, which are mostly distinct from those of the steps discussed earlier for both indirect and direct electrochemical reduction methods. These involve the transmission of electrons to the leuco radical anion or mediator from the cathode whereas, in electrochemical hydrogenation, adsorbed hydrogen is produced from the reduction of water electrochemically, which then combines chemically with an organic substrate on the surface of an electrically conductive metal powder catalyst having a low hydrogen overpotential (e.g., Raney nickel or platinum black) as shown in Eqs. (7.15)–(7.16). In addition to acting as a catalyst for hydrogenation, the catalytic material also functions as an electrode to produce hydrogen. The competition between the hydrogenation and hydrogen gas evolution reaction, as shown in Eqs. (7.17)–(7.18), determines the performance of electrocatalytic hydrogenation.

$$2\,H_2O + 2e^- \rightarrow 2\,H_{ads} + 2\,OH^- \tag{7.15}$$

$$\text{Indigo} + 2\,H_{ads} \rightarrow \text{Leuco indigo} \tag{7.16}$$

$$H_2O + H_{ads} + e^- \rightarrow H_2 + OH^- \tag{7.17}$$

$$2\,H_{ads} \rightarrow H_2 \tag{7.18}$$

Compared to catalytic hydrogenation, electrocatalytic hydrogenation offers several advantages and has been successfully applied to numerous substrates throughout the years. Because of the complete avoidance of the kinetic resistance resulting from the breakdown of the hydrogen molecule, high pressures and temperatures can be avoided. Furthermore, the compression, transportation, and storage of hydrogen are avoided by the electrochemical synthesis of hydrogen directly at the catalyst surface (Roessler and Jin 2003; Blackburn ety al. 2009). Using Raney-nickel electrodes in a divided flow cell, the commercial effectiveness of the creation of water-soluble leuco indigo in an ecologically sound manner has been investigated. The process involved electrocatalytic hydrogenation of indigo in aqueous solutions. The ideal conditions for the system were attempted to be established, and the reduced indigo concentration was scaled up to 10 g/l. The reduction rate was accelerated by raising pH, and a maximum conversion was achieved by maximizing temperature and current density. Nevertheless, Raney nickel material was selected to use as an electrode substance for its outstanding affordability, alkali stability, and availability. The efficiency of the Raney nickel electrodes could be increased by doping them with Pt particles. The main obstacle up to this point has been the inadequate interaction between the Raney nickel and the indigo particles. Consequently, packed and fluidized bed reactors were investigated for further improvement (Roessler et al. 2003). It is known from earlier research that noble metal black electrodes (such as platinum black) can increase the efficiency of the current of electrocatalytic hydrogenation by as much as 70%. At atmospheric pressure, platinum is a well-known hydrogénation catalyst. Nevertheless, compared to Raney nickel, the stability of platinum-black electrodes is considerably lower. Consequently, as a substitute, the impact of varying the concentration of noble metal particles added to the Raney nickel electrodes was examined, and found that the stability of both of the electrodes was identical, but the catalytic activity of the new noble metal added electrodes was noticeably higher. The utilization of high catalyst concentrations in the plating electrolyte increased the amount of platinum in the electrode, and that improved both the conversion and the current efficiency, although notable improvements were only likely to be seen more of at least 0.02 mg per centimeter of platinum. However, the efficiency of the reactor is still inadequate, and the electrode lifespan is too short for the ECH to be a practical process on an industrial scale. However, the outcomes would provide a strong foundation for increasing process efficiency through the use of other electrode materials or reactor designs. Since the nature of the electrode material significantly impacted the kinetics of the electrohydrogenation process, the composite metallic powders were used in that process in the following way: A porous electrode material retaining the powder particles in its pores when it was cathodically polarized, essentially became inactive for the evolution of adsorbed and molecular hydrogen. These kinds of electrodes were made by combining powder particles that had composite coatings of catalyst metals on their surfaces. Because of having a well-developed three-dimensional network of large pores and being an electrochemically stable, outstanding electronic conductor, reticulated vitreous carbon (RVC) has great potential as the electrode material. The catalytic powder particles might access a huge number of big pores due to the RVC's high porosity. Forced convection allowed the electrolyte to enter the RVC, trapping the suspended catalytic particles within the pore in the electrolyte. Apart from assessing distinct electrode materials, the inquiries concentrated on an alternative reactor design. Packed and fluidized bed electrodes provided a significant specific surface area of the electrode, resulting in an increased interaction between the electrode and the dye particles. Thus, a variety of materials were employed as electrodes in a packed and fluidized bed electrode reactor, and a laboratory circuit with a flow channel cell and a particle bed electrode was used to study the ECH of indigo (Roessler 2003).

7.5 Electrochemical Salt-Free Dyeing of Cotton Fabric with Reactive Dye

The anionic reactive dye is widely used to dye textiles made of cotton and builds up an anionic charge in the aqueous dyeing solution. However, cotton fabric also builds up negative charges in an aqueous solution. This causes them to become repellent, which makes it challenging to dye textiles made of cotton with reactive dye. For this reason, a significant quantity of salt is added to the dyeing solution during the conventional dyeing procedure. Here, the negative charge of the fabric is mitigated by the cation part of the salt, which enables the reactive dye to migrate to the fabric surface and the subsequent dyeing. Nevertheless, the high-content salt addition leads to land salinization and raises the quantity of mineral salt present in the dyeing effluent, hence raising the post-treatment costs of wastewater. These shortcomings have inspired researchers worldwide to conduct a vast number of studies on the dyeing of cotton fabric with reactive dye with no or little salt. The secret to dyeing without salt is to remove the electrostatic repulsion that exists between the fabric and the dye and enable the dye to create an effective covalent bond to the fiber. One method of low salt/salt-free dyeing is the electrochemical dyeing technology, which replaces the salt by modifying the surface of fabrics and fibers using an electric field.

The principle behind the electrochemical salt-free or low salt coloration of cotton textiles with anionic dye is the elimination of the repulsive force between the fiber and dye anion by producing positive charges in the anode area to which the fabric is linked under the electric field system. In the anode region near the fabric surface, this raises the concentration of dye anions. This process causes complete contact of the fiber with the dye in the anode region without the addition of salt. Figures 7.12 and 7.13 illustrate the conventional and electrochemical dyeing mechanisms of cotton fabric with reactive dye, respectively.

7.5.1 *Dyeing Procedure*

The dying process took place in an electrolytic cell. The electrolytic dyeing system consisted of an anode, a cathode, and a DC-regulated power source, as shown in Fig. 7.13. The figure illustrates the electrochemical dyeing of cotton fabric with reactive dye. The fabric to be colored was coupled with the anode where the voltage was applied. The concentration of dye was 4% o.w.f, the applied voltage was (1–10 V), the concentration of anhydrous sodium carbonate was (10 g/L), and the material-to-liquor ratio was 1:100 (Fan et al. 2017). Because of the electric field effect, the negatively charged dye anions from the cathode region were drawn to the cotton fabric in the positive anode region, which resulted in an increased amount of dye anions near the fabric surface. This completed the dyeing process by allowing the dye to come into full contact with the fabric. Furthermore, in the dyeing solution, the applied electric field might enhance the reactive sites of the dyes and decrease those

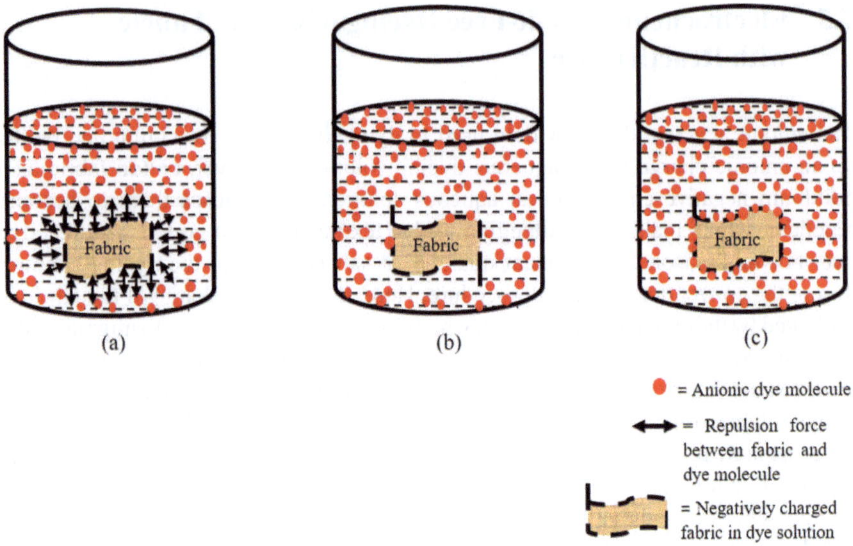

Fig. 7.12 Conventional dyeing mechanism of cotton fabric with reactive dye (**a**) Fabric in dyeing bath (**b**) dyeing of fabric without salt (**c**) dyeing of fabric with salt (Fan et al. 2017). Adapted from Fan et al. (2017) 'A research on electrochemical salt-free dyeing with KN reactive dyes: process optimization studies', *Journal of the Textile Institute*, 108(4), pp. 562–568

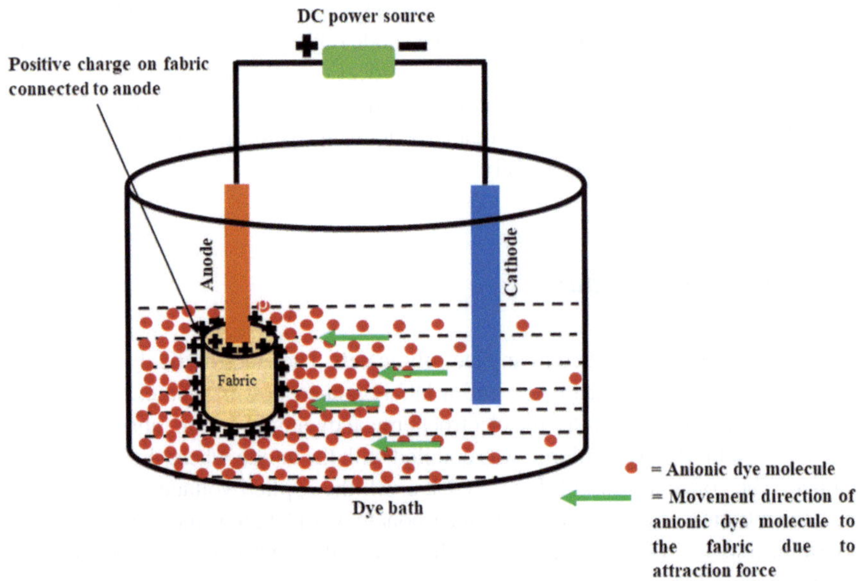

Fig. 7.13 Electrochemical dyeing apparatus for dyeing cotton fabric with reactive dye without salt (Fan et al. 2017). Adapted from Fan et al. (2017) 'A research on electrochemical salt-free dyeing with KN reactive dyes: process optimization studies', *Journal of the Textile Institute*, 108(4), pp. 562–568

of the fibers. Additionally, this improved the dye migration from the dye solution to the fabric surface, promoted the coefficient of diffusion of dye, increased dye adsorption, and improved dye exhaustion. Gradually, the dye and fiber reached an adsorption equilibrium where the adsorption and desorption rates were equal. Then, anhydrous Na_2CO_3 was incorporated into the dye bath, which raised the pH of the dyeing liquor, causing the hydroxyl groups of cellulose to ionize more readily into cellulosate anions. This disturbs the adsorption equilibria between the fibers and the dyes and lowers the rate of loss of dye from the fiber, enabling certain dyes to be retained on the fibers. However, sodium carbonate, as an extremely powerful electrolyte, substantially increases the electric field strength under the electrochemical action. Therefore, there's cause for optimism that electrochemical dyeing has led to a sustainable technique in the new era of textile dyeing technology (Xiong et al. 2021).

In a study on the salt-free electrochemical coloring of cotton fabric, four KN-type reactive dyes with different structures were employed to dye cotton fabric, and they were C.I. Reactive Blue 19, C.I. Reactive Violet 5, C.I. Reactive Blue 21, and C.I. Reactive Red 111. It has been found that as more sulfonic groups are present in a reactive dye, the less effective the dyeing method is because the molecular mass percentage of the sulfonic acid group in the dye is directly correlated with the repulsion of the dyes from the fiber, and require more applied voltage to be drawn to the anode electrode in the electrochemical dyeing process. By analyzing the structures of the above mentioned dyes it has been found that a significant amount of the total molecular weight of the dye is comprised of sulfonic acid molecules, which are represented as 32.88, 42.0, and 75.41% in the dyes C.I. Reactive Blue 19, C.I. Reactive Violet 5, and C.I. Reactive Red 111, respectively. Because of this, compared to C.I. Reactive Red 111, the other dyes exhibited better dyeing efficiency and required less applied voltage.

The KN-type reactive dyes possessed poor directness to the fabric without salt, and the repulsive force between the anionic fiber and the anionic dye molecule became stronger as the dyeing time increased. Hence, with the increase of dyeing time, the rise in color strength (K/S) value of salt-free electrochemically dyed fabric decreased. The (K/S) values of the above four dyes tended to increase steadily for 50 min, after which there was a small drop. This suggested that all four dyes had the same optimum fixing period of 50 min. The optimum dyeing temperatures at which the highest color strength (K/S) value was reached were 30 °C for all of the three dyes, C.I. Reactive Violet 5, C.I. Reactive Blue 19, C.I. Reactive Blue 21, and 35 °C for C.I. Reactive Red 111. The results of rubbing fastness, cotton spray, and wool spray fastness of salt-free electrochemical dyeing were about identical to conventional dyeing. Relative to conventional technology, electrochemical salt-free dyeing produced lower COD values (Fan et al. 2017).

A separate investigation used Reactive Blue 4 to dye cotton fabric without any separate heating from the outside because the applied potential caused both the electrodes and the dye solution to be heated. Except for applying heat, the electrochemical dyeing followed the same recipe conditions as the traditional dyeing method and dyed for 1 h. After connecting the fabric to the anode, a DC power supply unit was used to provide a 2 V potential. Graphite was utilized as an electrode material. Certain dyes showed less coloration when the applied voltage was increased; this could be because the dye structure may have deformed. Compared to the conventional dyeing procedure, the electrochemical method enhanced the dye uptake. This is due to the electrostatic interaction between the fabric and the dye (Kalapriya et al. 2017).

7.5.2 Advantages of Electrochemical Salt-Free Dyeing

- Easy to control,
- Requires no salt,
- Do not require external heating,
- Mild reaction conditions,
- Low emissions and energy consumption,
- Excellent dyeing effect and simple operation,
- Comparatively environmentally friendly process.

7.5.3 Disadvantages of Electrochemical Salt-Free Dyeing

- The fastness of dyed fabric needs to be improved,
- Loss of breaking strength of fabric occurs because of fiber damage,
- Need to modify the dye structure,
- Suitable dyeing auxiliaries should be developed that will be compatible with the modified dye,
- Uneven dyeing,
- Complicated operation,
- High cost.

7.6 Electrochemical Dyeing of Wool Fabric with Pigments

An ecologically conscious coloring approach, pigment coloring works on nearly every kind of fabric, including ones like glass and metal that are undyeable with traditional dyes. Pigments give color to textiles over the full-color spectrum with their brilliant shine. During pigment coloring, a dispersing agent is generally used to isolate the pigments, resulting in anionic species or negative charges in the dyeing solution. Aqueous solutions of widely used fibers like cotton and wool are anionic owing to the existence of –OH groups in their polymeric structure. During the coloring process, these anionic species of the fiber strongly repel the pigments, which leads to unsatisfactory dyeing results. To solve this problem, typically, cationic pretreatment has been employed, which converts the anionic fiber into cationic. As a result, the anionic pigments and cationic fibers are attracted to one another electrostatically, which makes the dyeing process easier and more effective.

While cationic pretreatment can improve pigment coloring, the high dyeing time, energy-intensive, and chemical-intensive nature of this process makes it relatively ineffective and detrimental to the environment. As electrochemical coloring modifies the fiber surface in a similar way to cationic pretreatment, it can be employed instead of a cationic pretreatment process. To be more precise, the DC source in the dye bath creates an electric field that allows the wool fabric to exchange protons and electrons, creating a cationic surface, as demonstrated in Fig. 7.14b–c. The surface of wool materials is charged through the application of this electrical field to the dyeing bath for a brief duration. The dispersing agent encompasses the pigments, causing them to show negative charges in the bath. These colors are attracted to the positively charged fiber surface through ionic interactions as illustrated in Fig. 7.14c–d.

In a study, a wool fabric was dyed in an electrochemical method with a solution containing 1 L of purified water, 5 g of sodium sulfate, 40 g of binder, and 20 g

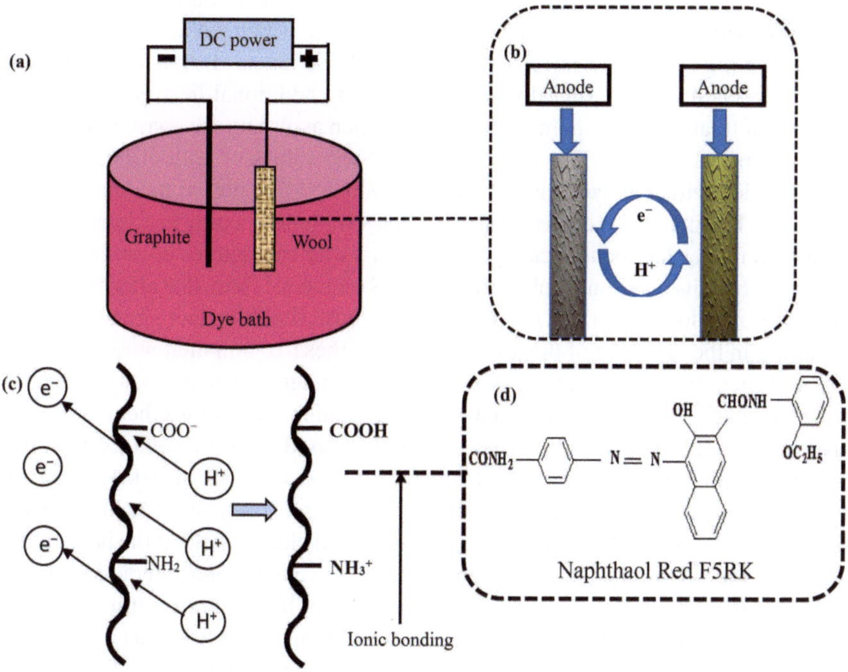

Fig. 7.14 Electrochemical dyeing mechanism of wool with pigments; **a** dyeing process, **b** electrons and hydrogen exchange on the surface of wool, **c** interaction between pigments and wool fiber, and **d** Naphthol Red F5RK chemical structure (Cao et al., 2018). Adapted with permission from Cao et al. (2018) 'Electrochemical Pigment Coloring of Wool Fabrics', Fibers and Polymers, 19(1), pp. 65–70 Copyright © 2018, The Korean Fiber Society and Springer Science Business Media B.V., part of Springer Nature

of pigment. An electric field was employed together with the traditional dyeing procedure to carry out the electrochemical dyeing of the wool fabric without the need for a cationic pretreatment. As seen in Fig. 7.14a, the coloring bath was supplied with electricity by a DC source (20 V), and dyeing was done at 40 °C. The cathode was linked to a graphite placed 5 cm away from the anode, and the wool fabric to be colored was linked to the anode. The DC source was turned on once the wool fabric was linked to the anode. This resulted in the generation of a current of 1.04 A, which confirmed that the treatment was proceeding. After ten minutes, the power was turned off, and padding was done with a pickup of 80%. After that, the dyed fabric was dried at 70 °C and cured for three minutes at 140 °C.

The surface of the electrochemically dyed wool fabric had a significant amount of aggregated pigments. The pigments were uniformly distributed in the areas between its fibers and throughout the wool scale. Compared to both cationic pretreatment and regular coloring, the surface of the wool fabric showed a high pigment deposition and superior dyeing performance due to the efficient pigment migration to the wool fabric at the time of electrochemical pigment coloring. The bath temperature impacted the color efficiency of electrochemical pigment coloration negatively. A temperature rise caused a slight loss in the color strength (K/S) values because higher temperatures could speed up the aggregation of pigments in the bath, which yielded colored fabrics with fewer pigments (and lower color strength K/S values). It thus makes it clear that electrochemical coloring does not require any additional heating energy. The duration of treatment was an important factor when assessing the color efficiency of electrochemical coloring. The color strength (K/S) values of both electrochemical and cationic pretreated dyed fabrics are displayed as a function of treatment time in Fig. 7.15b. From this, it is evident that 10 minutes was enough time to achieve the best coloration in the electrochemical dyeing process, whereas cationic coloring required 20 min to reach its maximum color strength (K/S) value. It is seen that after 10 minutes the value of color strength (K/S) decreases because the pigments aggregate while migrating in the direction of the textiles and after these 10 min their attraction to the textiles is neutralized by the increased gravity of the aggregated pigments inhibiting the pigments from attaching to the fabric surface and instead causes them to become suspended in solution. This is the reason for decreased color strength (K/S) values beyond 10 min. The color strength (K/S) value of the dyed fabric also depends on the applied voltage, and it has been found that there was a linear relationship between them. The increased voltage created a more powerful electric field in the bath, which increased the possibility of charge production, movement, and interaction on the surface of the wool fabric. Consequently, the pigment had a greater attraction to the fabric, which produced better dyeing quality with higher color strength (K/S) values. However, increased voltage also necessitates high energy utilization. Therefore, a reasonable voltage of 20 V was a sufficient solution in that case.

Another variable that influences color performance is the concentration of the electrolyte Na_2SO_4. As the concentration of electrolyte rises, the value of color strength (K/S) rises about linearly. Immediately, the concentration approaches 5 g/l, the value of color strength (K/S) exhibited its maximum, and subsequent improvement had no noticeable impact on the color strength (K/S) value. The deposition of pigment on

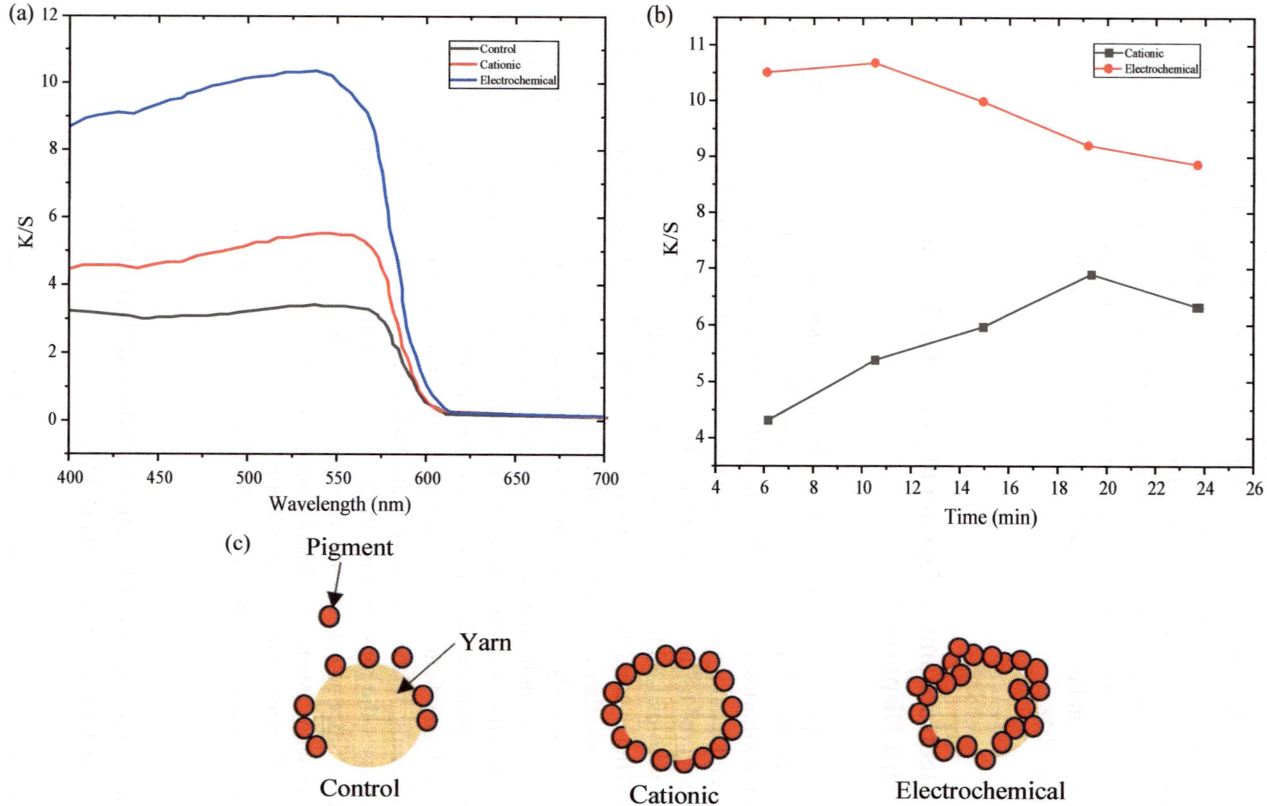

Fig. 7.15 **a** Color strength value (K/S) of fabrics dyed conventionally (control), cationic pretreated, and electrochemically, **b** color strength values (K/S) of dyed fabrics with respect to time, and **c** pigment arrangement on the fabric and the colored fabrics (Cao et al. 2018). Adapted with permission from Cao et al. (2018) 'Electrochemical Pigment Coloring of Wool Fabrics', Fibers and Polymers, 19(1), pp. 65–70. Copyright © 2018, The Korean Fiber Society and Springer Science Business Media B.V., part of Springer Nature

normal, cationic pretreatment, and electrochemically dyed fabrics is demonstrated in Fig. 7.15c. Since most of the pigment is deposited on electrochemically dyed fabric, as seen in Fig. 7.15c, it is the most vivid and glossy colored fabric. All coloring techniques yielded identical dry rubbing fastness grades 4. Both the electrochemical and cationic dyeing processes provided grade 4 wet rubbing fastness, which was better than grades 3–4 for the conventional coloring method. The attraction between the oppositely charged treated fiber surface and pigments improved the wet rubbing fastness of vat textiles. So this technique saves time and energy without polluting the environment because it is quick, easy, and does not require any extra chemicals (Cao et al. 2018).

7.7 Electrochemical Dyeing of Polyester Fabrics with Disperse Dye

The low-temperature electrochemical dyeing of polyester was investigated with Disperse Orange 3, Disperse Orange 13, and Disperse Red 11. In this dyeing, the dye and auxiliaries were added to the cell according to the recipe of conventional dyeing, and the polyester fabric was immersed in that cell. Two electrodes made of graphite served as the cathode and anode, respectively, and dyeing was done with a liquor ratio of 1:30 for an hour by applying a voltage of 2, 3, and 4 V with a DC source. After dyeing, the fabrics were removed from the cell and thoroughly washed. Compared to other dyes, Disperse Orange 13 provided improved dyeability. It has been found that dyeability diminishes as voltage increases because of the distortion of the dye molecular structure. In the cyclic voltammetry investigation, it has been observed that disperse dyes with a wide separation of the anodic peaks from the cathodic peaks produced more promising results. It was discovered that electrochemical dyeing beats traditional dyeing in terms of dyeability and dye absorption. This is because of the ionization of dye molecules by the applied electric potential, which causes the ions to move in the direction of the fabric through electrostatic attraction. Since no external heat is employed, the electrochemical method of dyeing polyester with disperse dye has the additional advantage of energy saving, and the dyeing is performed at a lower potential with improved absorption. This makes the process more environmentally friendly (Kalapriya et al. 2017, 2019).

7.8 Electrochemical Dyeing of Silk with Acid Dye and Cotton with Direct Dye

Silk and cotton fabric were dyed in the electrochemical method by placing the cloth on the graphite anode and applying a voltage of 2 V with a DC source without external heating with Acid Yellow 17 and Direct Red 7, respectively. Both the direct dye and acid dye yielded a greater color strength (K/S) value compared to conventional dyeing. Higher color strength (K/S) values in this embedded system for both Direct Red 7 on cotton and Acid Yellow 17 on silk may be the result of the preferred anodic electronic transfer, and therefore, a decreased energy requirement for anodic electronic transfer was calculated for the dye from the voltammetric study of the dye. The applied potential ionizes the dyes efficiently, making them easier to trap onto textile surfaces, which is responsible for the better coloration when compared to the traditional dying procedure. When compared to other dyes in terms of washing fastness and dye uptake, an acid dye generated poor washing fastness but superior dyeability (Kalapriya et al. 2017).

7.9 The Environmentally Friendly Concept and Savings of Electrochemical Dyeing

The estimated savings for a 100 kg yarn package dying machine in the indirect electrochemical dyeing are displayed in Table 7.14. In this calculation, only process chemicals NaOH, $Na_2S_2O_4$, and loss of mediator of 15% of dyebath volume at the time of recycling were considered (Bechtold et al. no date). For the actual dyebath step, the chemical waste was decreased by around 80%, and for the entire process, it was reduced by 20%. It saves 30% water, and 70% energy, uses about 0% hazardous chemicals, and emits 90% less CO_2, especially in the direct electrochemical reduction of indigo dye.

7.10 Prospects of Electrochemical Dyeing

Electrochemical dyeing has great prospects in denim dyeing. Over a decade, DyStar estimated that 67% of denim was primarily dyed with indigo, 26% with sulfur black, and 6% with other sulfur colors. The ongoing development of fashion trends has also utilized reactive, direct, and vat dyes, and pigments, but these each accounted for less than 1% of the total amount of dye applied to denim. Industrialists can grab this huge market of denim dyed with indigo and sulfur dye using a sustainable electrochemical dyeing process.

The 40% indigo dye, a patented pre-reduced dye, is produced by the catalytic hydrogenation method. it was developed by BASF in 1993, and now it is marketed by DyStar. Despite being pre-reduced, it still requires a small amount of sodium dithionite to prevent it from oxidizing in the dye bath. To address this issue, DyStar and RedElec, a Swiss company founded in 2007, decided to work together to develop indigo electrochemical dyeing, a sustainable solution that will get rid of dangerous materials from wastewater in 2012. Throughout the development process, the group

Table 7.14 Savings and chemical usage, IN-process, 100 kg cotton, material to liquor ratio 1:10. Here, Conc. means concentration (Bechtold et al. no date)

	Electrochemical reduction		Reduction with $Na_2S_2O_4$					
Depth of color	0.3–4%		0.3%		2%		4%	
	Conc	Weight	Conc	Weight	Conc	Weight	Conc	Weight
Dyeing								
Mediator	13.60 g/L	2.05 kg						
Sodium dithionite			3.0 g/L	3.0 kg	5.0 g/L	5.0 kg	8.0 g/L	8.0 kg
Sodium hydroxide	32.0 ml/L	2.10 kg	14.0 ml/L	6.16 kg	19.0 ml/L	8.36 kg	24.0 ml/L	10.56 kg
Reductive rinse:								
Sodium dithionite	–	–	–	–	2.0 g/L	2.0 kg	2.0 g/L	2.0 kg
Sodium hydroxide	–	–	–	–	2.0 ml/L	0.88 kg	2.0 ml/L	0.88 kg
Total amount	4.15 kg		9.16 kg		16.24 kg		21.44 kg	
Savings			55%		75%		80%	
Water-dyeing	150 L		1×10^3 L		2×10^3 L		2×10^3 L	

employed RedElec's electrochemical methods and DyStar Indigo® Vat 40% Solution. The technique uses a 3D electrode of carbon particles with quinine-like groups on their surface. These structures are changed by electrons from the power source to resemble the hydroquinone-reducing agent. RedElec claims that its electrode is less expensive than electrodes using precious metals and maintains its stability over several months of use. However, it is a matter of worry that the first implementation of this new technology will be high (Chavan, 2015).

Sedo Engineering SA was established in 2014 as a part of VANDEWIELE in Switzerland to develop sustainable technology for the denim industry and introduce Smart-Indigo™. In the denim industry, Smart-Indigo™ provides a liquid-indigo production process that uses electrical energy rather than chemicals and produces only oxygen as waste. It offers many benefits, including the following (SA 2023):

(a) 0% hazardous chemicals,
(b) 90% less CO_2 emission,
(c) 70% energy savings,
(d) 30% water savings,
(e) No costs associated with chemicals or reduced liquid indigo,
(f) Higher reproducibility through better color consistency,
(g) Fire hazards,
(h) Does not contain any dangerous chemicals, causing no risk to health,
(i) Dyed denim clothing is free of harmful residues.

Anywhere indigo dyeing is used, Smart-Indigo can be used:

- Cotton yarns and their blends,
- Silk,
- Wool,
- Modal,
- Viscose,
- Tencel,
- Recycled yarn.

In addition to helping the environment, the technology produces constant quality and color shade because of its greater production rate and completely automated Industry 4.0 design. Additionally, the Indigo™ unit is directly connected to the dye bath. There are already 17 Smart-Indigo™ machines in operation worldwide, 5 of which are located in China and seven of which are located in Pakistan. One ton of "leuco" indigo is produced daily on average by a Smart-Indigo™ unit. The amazing pay-back period of a Smart-Indigo™ machine is estimated to be approximately 2.54 years. For the past 1.5 years, the first machine has been successfully in full production at Italdenim, Inveruno, close to Milan, Italy. Using Smart-Indigo™, the denim industry, one of the major polluters on earth, will be able to protect the environment to unprecedented levels.

7.11 Conclusion

An eco-friendly substitute for traditional dyeing is electrochemical dyeing. This chapter has explored the variety of electrochemical ways that textile producers can apply to lessen the harmful effects of textile dying with various dye classes on the environment. A few of these have already found commercial adoption, and upon further research in the field, there will be more in the future.

References

Abdelileh M et al (2020) 'Calcium-iron-D-gluconate complexes for the indirect cathodic reduction of indigo in denim dyeing: a greener alternative to non-regenerable chemicals. J Clean Prod 266. https://doi.org/10.1016/j.jclepro.2020.121753

Ahmed NSE, El-Shishtawy RM (2010) 'The use of new technologies in coloration of textile fibers. J Mater Sci 45(5):1143–1153. https://doi.org/10.1007/s10853-009-4111-6

Al-Mousawi SM, El-Apasery MA, Elnagdi MH (2013) Microwave assisted dyeing of polyester fabrics with disperse dyes. Molecules 18(9):11033–11043. https://doi.org/10.3390/molecules 180911033

Ammayappan L, Jose S, Arputha Raj A (2016) Sustainable production processes in textile dyeing. In: Environmental footprints and eco-design of products and processes. https://doi.org/10.1007/978-981-10-0111-6_8

Anbu Kulandainathan M et al (2008) Preparation of iron-deposited graphite surface for application as cathode material during electrochemical vat-dyeing process. Mater Chem Phys 112(2):478–484. https://doi.org/10.1016/j.matchemphys.2008.06.010

Bechtold T et al (1991) The redox-catalysed reduction of dispersed organic compounds. J Electroanal Chem Interfacial Electrochem 306(1–2):169–183. https://doi.org/10.1016/0022-0728(91)85229-i

Bechtold T et al (1993) Alkali-stable iron complexes as mediators for the electrochemical reduction of dispersed organic dyestuffs. J Chem Soc Faraday Trans 89(14):2451–2456. https://doi.org/10.1039/FT9938902451

Bechtold T (1993) Process for the reduction of dyes. Google Patents

Bechtold T et al (1994) The reduction of vat dyes by indirect electrolysis. J Soc Dyers Colour 110(1):14–19. https://doi.org/10.1111/j.1478-4408.1994.tb01586.x

Bechtold T et al (1996) Indirect electrochemical reduction of dispersed indigo dyestuff. J Electrochem Soc 143(8):2411–2416. https://doi.org/10.1149/1.1837023

Bechtold T, Burtscher E, Turcanu A et al (1997a) Dyeing behavior of Indigo reduced by indirect electrolysis. Textile Res J 67(9):635–642. https://doi.org/10.1177/004051759706700902

Bechtold T, Burtscher E, Kühnel G et al (1997) Electrochemical reduction processes in indigo dyeing. J Soc Dyers Colour 113(4):135–144. https://doi.org/10.1111/j.1478-4408.1997.tb01886.x

Bechtold T, Burtscher E, Turcanu A et al (1997b) Multi-cathode cell with flow-through electrodes for the production of iron(II)-triethanolamine complexes. J Appl Electrochem 27(9):1021–1028. https://doi.org/10.1023/A:1018426307725

Bechtold T et al (1998) Optimization of multi-cathode membrane electrolysers for the indirect electrochemical reduction of indigo. Chem Eng Technol 21(11):877–880. https://doi.org/10.1002/(SICI)1521-4125(199811)21:11<877::AID-CEAT877>3.0.CO;2-F

Bechtold T et al (2004) Mediator systems based on mixed metal complexes, used for reducing dyes. Google Patents

Bechtold T, Berktold F, Turcanu A (2000) The redox behaviour of CI Sulphur Black 1-a'basis for improved understanding of sulphur dyeing. Color Technol 116(7–8):215–221. https://doi.org/10.1111/j.1478-4408.2000.tb00041.x

Bechtold T, Burtscher E, Turcanu A (1999) Anthraquinones as mediators for the indirect cathodic reduction of dispersed organic dyestuffs. J Electroanal Chem 465(1):80–87. https://doi.org/10.1016/S0022-0728(99)00057-1

Bechtold T, Burtscher E, Turcanu A (2002) Ca2+-Fe3+-D-gluconate-complexes in alkaline solution. Complex stabilities and electrochemical properties. J Chem Soc Dalton Trans (13):2683–2688. https://doi.org/10.1039/b202086f

Bechtold T, Fitz-Binder C, Turcanu A (2010) Electrochemical characteristics and dyeing properties of selected 9,10-anthraquinones as mediators for the indirect cathodic reduction of dyes. Dyes Pigment 87(3):194–203. https://doi.org/10.1016/j.dyepig.2010.03.026

Bechtold T, Turcanu A (2002) Electrochemical vat dyeing combination of an electrolyzer with a dyeing apparatus. J Electrochem Soc 149(1):D7. https://doi.org/10.1149/1.1424288

Bechtold T, Turcanu A (2004) Fe3+--gluconate and Ca2+-Fe3+--gluconate complexes as mediators for indirect cathodic reduction of vat dyes ? Cyclic voltammetry and batch electrolysis experiments. J Appl Electrochem 34(12):1221–1227. https://doi.org/10.1007/s10800-004-1707-z

Bechtold T, Turcanu A (2009) Electrochemical reduction in vat dyeing: greener chemistry replaces traditional processes. J Clean Prod 17(18):1669–1679. https://doi.org/10.1016/j.jclepro.2009.08.004

Bechtold T, Turcanu A, Schrott W (no date) Electrochemical processes in textile dyeing-State of the research and development activities

Blackburn RS, Bechtold T, John P (2009) The development of indigo reduction methods and pre-reduced indigo products. Color Technol 125(4):193–207. https://doi.org/10.1111/j.1478-4408.2009.00197.x

Blackburn RS, Harvey A (2004) Green chemistry methods in sulfur dyeing: application of various reducing D-sugars and analysis of the importance of optimum redox potential. Environ Sci Technol 38(14):4034–4039. https://doi.org/10.1021/es0498484

Bokhari TH et al (2023) Eco-friendly sustainable dyeing of cotton fabric using reactive violet 05 and direct violet 09 dyes. Coatings 13(4). https://doi.org/10.3390/coatings13040677

Božič M, Kokol V (2008) Ecological alternatives to the reduction and oxidation processes in dyeing with vat and sulphur dyes. Dyes Pigment 76(2):299–309. https://doi.org/10.1016/j.dyepig.2006.05.041

Brett CMA, Oliveira Brett AMCF (1984) Surface modified electrodes-reasons and advantages. In: NATO ASI series, series E: applied sciences, vol 85, pp 656–664. https://doi.org/10.1007/978-94-009-6216-3_42

Cao G et al (2018) Electrochemical pigment coloring of wool fabrics. Fibers Polym 19(1):65–70. https://doi.org/10.1007/s12221-018-7502-3

Chavan RB.(2015) Indigo dye and reduction techniques, Denim: Manufacture, Finishing and Applications. Elsevier Ltd. https://doi.org/10.1016/B978-0-85709-843-6.00003-2

Choudhury AKR (2018) Eco-friendly dyes and dyeing. AdvMatTechEnv 2(1):145–176

Crettenand D (2012) Electrochemical reactor. Google Patents

Elgrishi N et al (2018) 'A practical beginner's guide to cyclic voltammetry. J Chem Educ 95(2):197–206. https://doi.org/10.1021/acs.jchemed.7b00361

Fan Z, Cai X, Li Q (2017) A research on electrochemical salt-free dyeing with KN reactive dyes: process optimization studies. J Textile Inst 108(4):562–568. https://doi.org/10.1080/00405000.2016.1174457

Ge K et al (2018) Effects of liposome assisted dyeing on PET fabric properties. J Text Sci Eng 08(01):8–11. https://doi.org/10.4172/2165-8064.1000333

Hou Y et al. (2022) The influence of electrode for the electrochemical dyeing of cotton fabric. Adv Transdiscip Eng 24:724–729. https://doi.org/10.3233/ATDE220505.

Information T (2011) Setamol disperse ws (June), pp 1–5

Jiang H (2011) Indirect electrochemical reduction of indanthrene brilliant green FFB. Chin J Chem Eng 19(2):199–204. https://doi.org/10.1016/S1004-9541(11)60154-7

Kalapriya K, Gurumallesh Prabu H, Nithya S (2017) Comparative study on the dyeability of cotton, silk and polyester fabrics by conventional and electrochemical methods. Rasayan J Chem 10(4):1330–1333. https://doi.org/10.7324/RJC.2017.1041923

Kalapriya K, Prabu HG, Ramakrishnan G (2019) Electrochemical dyeing of polyester fabric using disperse dyes. Rasayan J Chem 12(1):319–323. https://doi.org/10.31788/RJC.2019.1215004

Kartal GE et al (2020) The effect of liposome on dyeing mohair/wool blends. Color Technol 136(2):167–176. https://doi.org/10.1111/cote.12461

Khatri A et al (2015) A review on developments in dyeing cotton fabrics with reactive dyes for reducing effluent pollution. J Clean Prod 87(1):50–57. https://doi.org/10.1016/j.jclepro.2014.09.017

Khatri A, Jhatial RA (2015) Advances in reactive dyeing of cellulose textiles

Kulandainathan MA, Muthukumaran A et al (2007a) Potentiostatic studies on indirect electrochemical reduction of vat dyes. Dyes Pigment 73(1):47–54. https://doi.org/10.1016/j.dyepig.2005.10.007

Kulandainathan MA, Patil K et al (2007b) Review of the process development aspects of electrochemical dyeing: its impact and commercial applications. Color Technol 123(3):143–151. https://doi.org/10.1111/j.1478-4408.2007.00082.x

Marti M et al (2004) Physicochemical aspects of the Liposome-Wool interaction in wool dyeing. Langmuir 20(8):3068–3073. https://doi.org/10.1021/la030385+

Marte W, Dossenbach O, Meyer U (2003) Method and apparatus for reducing vat and sulfur dyes. Google Patents

Merk C et al (2003) Method for electrochemically reducing reducible dyes. Google Patents

Mohr S, Bechtold T (2001) Electrochemical behaviour of iron-complexes in presence of competitive ligands: a strategy for optimization of current density. J Appl Electrochem 31(3):363–368. https://doi.org/10.1023/A:1017536204530

Motaghi Z (2012) The comparison between a natural reducing agent and sodium dithionite in vat, indigo and sulphur dyeing on cotton fabric. Adv Mater Res 441(Jan 2012):207–211. https://doi.org/10.4028/www.scientific.net/AMR.441.207

Patra SK et al (2018) Vat dyeing at room temperature. Cellulose 25(9):5349–5359. https://doi.org/10.1007/s10570-018-1901-5

Periyasamy AP (2024) Recent advances in the remediation of textile-dye-containing wastewater: prioritizing human health and sustainable wastewater treatment. Sustainability (Switzerland) 16(2). https://doi.org/10.3390/su16020495

Popescu V et al (2019) Sustainable and cleaner microwave-assisted dyeing process for obtaining eco-friendly and fluorescent acrylic knitted fabrics. J Clean Prod 232:451–461. https://doi.org/10.1016/j.jclepro.2019.05.281

Publishers M (2024) Global textile dyes market 2024 by manufacturer, regions, type and application, forecast to 2030. https://pdf.marketpublishers.com/globalinfo/global-textile-dyes-market-by-manufacturers-forecast-to-2023.pdf. Accessed 17 Jun 2024

Roessler A et al (2001) Direct electrochemical reduction of indigo. Chimia 55(10):879–882. https://doi.org/10.2533/chimia.2001.879

Roessler A et al (2002) Direct electrochemical reduction of indigo: process optimization and scale-up in a flow cell. J Appl Electrochem 32(6):647–651. https://doi.org/10.1023/A:1020198116170

Roessler A et al (2003) Electrochemical reduction of indigo in fixed and fluidized beds of graphite granules. J Appl Electrochem 33(10), pp. 901–908. Available at: https://doi.org/10.1023/A:1025876114390.

Roessler A (2003) New electrochemical methods for the reduction of vat dyes, Swiss Federal Institute of Technology (ETH)

Roessler A, Crettenand D (2004) Direct electrochemical reduction of vat dyes in a fixed bed of graphite granules. Dyes Pigment 63(1):29–37. https://doi.org/10.1016/j.dyepig.2004.01.005

Roessler A, Dossenbach O, Rys P (2003) Electrocatalytic hydrogenation of indigo. J Electrochem Soc 150(1):D1. https://doi.org/10.1149/1.1521757

Roessler A, Jin X (2003) State of the art technologies and new electrochemical methods for the reduction of vat dyes. Dyes Pigment 59(3):223–235. https://doi.org/10.1016/S0143-7208(03)00108-6

Ru J, Qian X, Wang Y (2018) Low-salt or salt-free dyeing of cotton fibers with reactive dyes using liposomes as dyeing/level-dyeing promotors. Sci Rep 8(1):1–10. https://doi.org/10.1038/s41598-018-31501-7

Rusling JF (1991) Controlling electrochemical catalysis with surfactant microstructures. Acc Chem Res 24(3):75–81

Sa SE (2023) SMART-INDIGO. https://www.smartindigo.com/wp-content/uploads/2023/05/Smart-Indigo-2023.pdf. Accessed 11 Jun 2024

Said AH et al (2008) New prospects in the electrochemical dyeing of indigo. J Appl Sci 8(13):2456–2461

Schnitzer G et al (1996) Dyeing cellulose-containing textile material with hydrogenated indigo. Google Patents

Scholnick F, Pfeffer PE (1980) Iron chelating capacity of gluconamides and lactobionamides. J Dairy Sci 63(3):471–473. https://doi.org/10.3168/jds.S0022-0302(80)82955-9

Senthil Kumar R et al (2009) Redox mediated electrochemical method for vat dyeing in ferric-oxalate-gluconate system: process optimization studies. J Appl Electrochem 39(12):2569–2577. https://doi.org/10.1007/s10800-009-9952-9

Shrwardi MHS, Das SC, Mamun MAAl (2023) A green approach of vat dyeing of cotton fabric with natural reducing agents. Heliyon 9(9):e19663. https://doi.org/10.1016/j.heliyon.2023.e19663

Srivastava JK (2023) Mass coloration of polyester fiber–emerging opportunities for green sustainable dyeing, vol 12, no 13, pp 1182–1191. https://doi.org/10.53555/ecb/2023.12.Si13.20810.53555/ecb/2023.12.Si13.208

Techology R (2024) Electrochemical cell. https://www.redelec.ch/en/technology-en-2/electrochemical-cell.html. Accessed: 5 Nov 2024

Turcanu A, Bechtold T (2011) Indirect cathodic reduction of dispersed indigo by 1,2-dihydroxy-9,10- anthraquinone-3-sulphonate (Alizarin Red S). J Solid State Electrochem 15(9):1875–1884. https://doi.org/10.1007/s10008-010-1204-8

Turcanu A, Fitz-Binder C, Bechtold T (2011) Indirect cathodic reduction of dispersed CI Vat Blue 1 (indigo) by dihydroxy-9,10-anthraquinones in cyclic voltammetry experiments. J Electroanal Chem 654(1–2):29–37. https://doi.org/10.1016/j.jelechem.2011.01.040

WMA, HS, SR (2014) Electrochemical reduction behavior of vat and sulphur dyes: a comparitive study. Int J Appl Res Text (ijartex) 2(2):1–10

Westbroek P (2005) In Westbroek P, Priniotakis G, Kiekens P (eds) 2-Electrochemical methods, analytical electrochemistry in textiles. Woodhead Publishing. Woodhead Publishing Series in Textiles. https://doi.org/10.1533/9781845690878.1.37

Xiong W et al (2021) Electrochemical dyeing of cotton fiber for environmental protection. Adv Transdiscip Eng 19:602–607. https://doi.org/10.3233/ATDE210218

Xu Y et al (2014) Indirect electrochemical reduction of indigo on carbon felt: process optimization and reaction mechanism. Ind Eng Chem Res 53(26):10637–10643. https://doi.org/10.1021/ie500603y

Yang Z et al. (2021) Direct electrochemical reduction and dyeing properties of CI Vat Yellow 1 using carbon felt electrode. Dyes Pigment 184(Feb 2020):108835. https://doi.org/10.1016/j.dyepig.2020.108835

Yi C et al (2020) Practical and environment-friendly indirect electrochemical reduction of indigo and dyeing. Sci Rep 10(1):1–8. https://doi.org/10.1038/s41598-020-61795-5

Yu D et al (2013) Study on electrochemistry property of ferric (III)-oxalate-gluconate complex and its dyeing application. Adv Mater Res 641–642(1):508–512. https://doi.org/10.4028/www.scientific.net/AMR.641-642.508

Chapter 8
Supercritical Fluid Dyeing Technology: Innovations and Applications

Astha, Neha, and Amit Madhu

Abstract Traditional textile dyeing techniques pose significant environmental challenges, primarily due to excessive water consumption and the discharge of untreated dye effluents into aquatic ecosystems. These effluents, often containing azo dyes, heavy metals, and other hazardous compounds, threaten water quality and biodiversity. As a result, there is a growing emphasis on sustainable dyeing technologies that minimize environmental impact. Supercritical fluid dyeing, particularly using supercritical carbon dioxide ($SCCO_2$), has emerged as a promising alternative to conventional water-based dyeing methods. $SCCO_2$ acts as an efficient solvent, easily penetrating textile fibers due to its low viscosity and high diffusivity. This technique eliminates wastewater generation, significantly reducing environmental pollution. The process allows precise control over temperature and pressure, ensuring uniform and vibrant coloration. The non-toxic nature of $SCCO_2$ and its recyclability make this approach an eco-friendly solution for textile coloration. Supercritical fluid dyeing applies to various textile materials, including natural, synthetic, and high-performance fibers, offering versatility across the industry. Despite its advantages, challenges remain in industrial adoption, such as high initial investment costs and the need for specialized equipment. Addressing these barriers is essential for widespread commercialization. This chapter overviews supercritical fluids and their role in textile coloration. It explores recent innovations in the field, discussing this technology's advantages, challenges, and future prospects in achieving sustainable textile manufacturing.

Keywords Supercritical fluid dyeing · Sustainable textile coloration · Supercritical carbon dioxide · Eco-friendly dyeing technology · Wastewater-free dyeing

---

Astha · Neha · A. Madhu (✉)
Textile Chemistry Department, The Technological Institute of Textile and Sciences, Bhiwani, India
e-mail: amit.madhu5@gmail.com

Astha
e-mail: asthabecool@gmail.com

Neha
e-mail: poonianeha503@gmail.com

S. Maiti et al. (eds.), *Sustainable Coloration Techniques in Textiles*,
https://doi.org/10.1007/978-981-96-4975-4_8

8.1 Introduction

A large amount of fresh water is needed to color textiles. Water acts as a solvent to dissolve various dyes, chemicals, and auxiliary materials when textiles are treated wet, from the beginning to the end (pretreatment to finishing). The wastewater is extensively contaminated by harmful and non-biodegradable compounds, including salts, finishing chemicals, residual colors, and other auxiliary chemicals. If these wastes are disposed of without the proper treatment, they can seriously harm the environment (Madhav et al. 2018; Lakshmanan and Raghavendran 2017). The actual quantity of water required for conventional aqueous dyeing depends on the nature of the fiber, dye, and the type of dyeing equipment and can range from 100 to 150 L for one kilogram of cloth (Miah et al. 2013). After dyeing, about 10% to 15% of the dye content remains in the wastewater, releasing an immense quantity of loose dye to the surroundings (Lu et al. 2010). Global concerns about water scarcity and rising ecological awareness are driving up the costs of water consumption and disposal, putting increasing pressure on industries, particularly the textile sector, which remains a significant water user despite efforts to implement sustainable practices. To address this challenge, the textile industry must adopt innovative, resource-efficient methods that reduce water usage, improve recycling, and minimize environmental impact. By cutting water consumption, manufacturers can lower operational costs related to water intake, treatment, and disposal while alleviating the strain on local water supplies. This shift offers cost savings and supports environmental preservation, ensuring that future generations can access this vital resource. Ultimately, reducing water usage in textile processing benefits the industry and the planet, aligning with global sustainability goals (Saravanan 2006; Moore and Ausley 2004; Su 2012; Lakshmanan and Raghavendran 2017; Ali and Gaballa 2014–2015; Desai 2008).

8.1.1 Water Saving: A Sustainable Approach

Supercritical carbon dioxide-aided dyeing offers a waterless solution, providing ecological and economic benefits (Agrawal 2015). The textile industry is focusing on eco-friendly chemical agents and green technologies to promote sustainable development (Miah et al. 2013; Saling et al. 2002; Rebitzer et al. 2004; Pennington et al. 2004; Stewart and Jolliet 2004; Razzak 2021). Figure 8.1 depicts the principle of sustainability.

Supercritical fluids, e.g., carbon dioxide, are materials that exceed their critical temperature and pressure in a closed system. This technique uses recyclable CO_2 as a medium, requiring efficient, low-energy, chemical, and waste output. Figure 8.2 demonstrates the idea of conventional and supercritical dyeing processes. $SCCO_2$ has "*lower mass transfer resistance, higher diffusion rates*", and waterless, making it a cost-effective and environmentally friendly process for wet textile processing (Abou Elmaaty et al. 2020; Özcan et al. 1998; Knittel and Schollmeyer 1995).

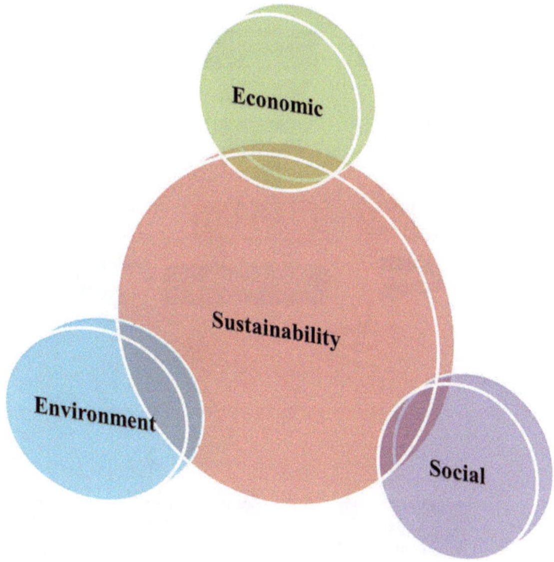

Fig. 8.1 Principle of "Sustainability"

The supercritical CO_2 dyeing concept was pioneered by "*Professor Schollmeyer* and the team at *Deutsches Textilforschungszentrum Nord-West*" (DTNW) in Krefeld, Germany (Banchero 2013; Schollmeyer et al. 1990) in the late 1980s. The literature of investigation findings and further patents regarding the $SCCO_2$ dyeing of diverse textiles has been published (Abou Elmaaty and Abd El 2018; Bach et al. 2002; Banchero 2013; Banchero 2020). This chapter covers the development of carbon dioxide dyeing technology, including its basic principles, the dyeing of synthetic and natural fibers, and the ongoing challenges facing this technique.

8.2 Fundamentals of Supercritical Fluid

8.2.1 Fluid

A supercritical fluid (SCF) is a state of matter that occurs when a substance is subjected to temperature and pressure conditions beyond its critical temperature (Tc) and critical pressure (Pc). (Hassabo et al. 2022; Abou Elmaaty and Abd El 2018; Montero et al. 2000; Hendrix 2001). At this supercritical state, the material exhibits unique properties that are intermediate between gases and liquids. Specifically, it possesses the diffusivity and low viscosity of a gas while maintaining the density and solvating power of a liquid. This combination of characteristics makes SCF highly versatile for various industrial applications.

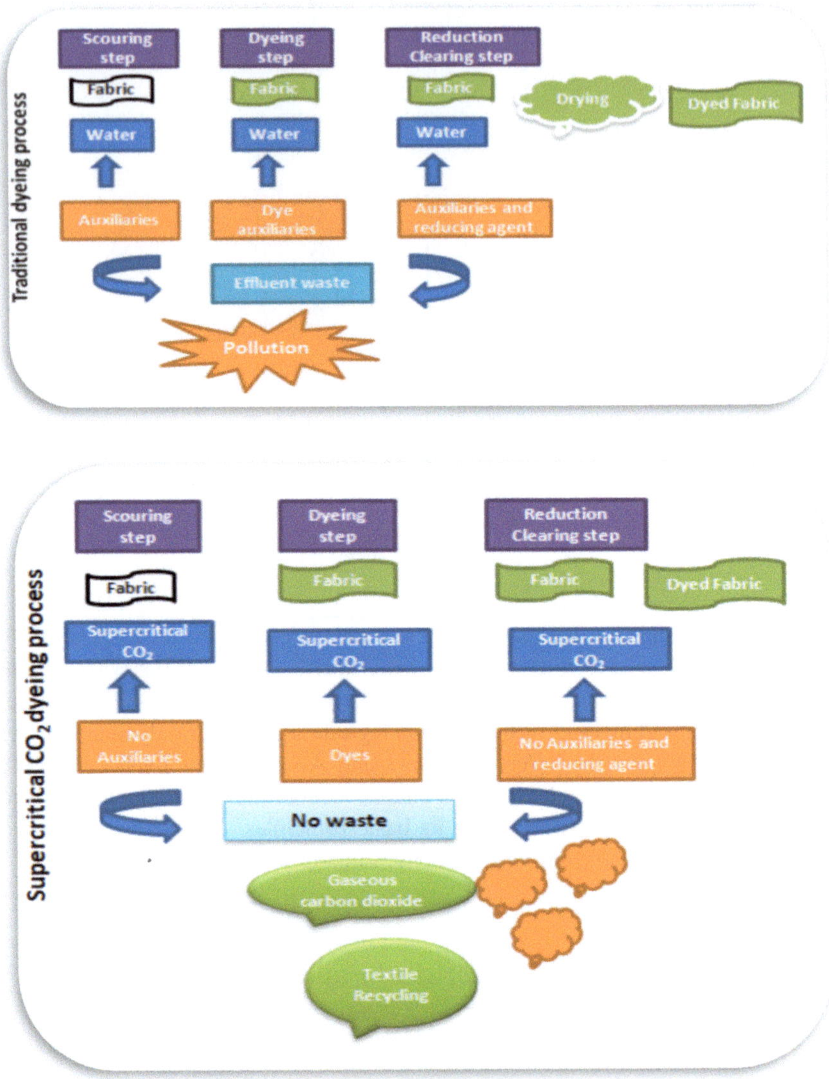

Fig. 8.2 Comparison between traditional and supercritical dyeing process

8.2.2 *Critical Point*

The critical point is defined as the condition at which a substance reaches its highest temperature and pressure, where the liquid and vapor phases exist in equilibrium and become indistinguishable, meaning the substance neither condenses nor evaporates (Abou Elmaaty and Abd El 2018; Montero et al. 2000; Hendrix 2001).

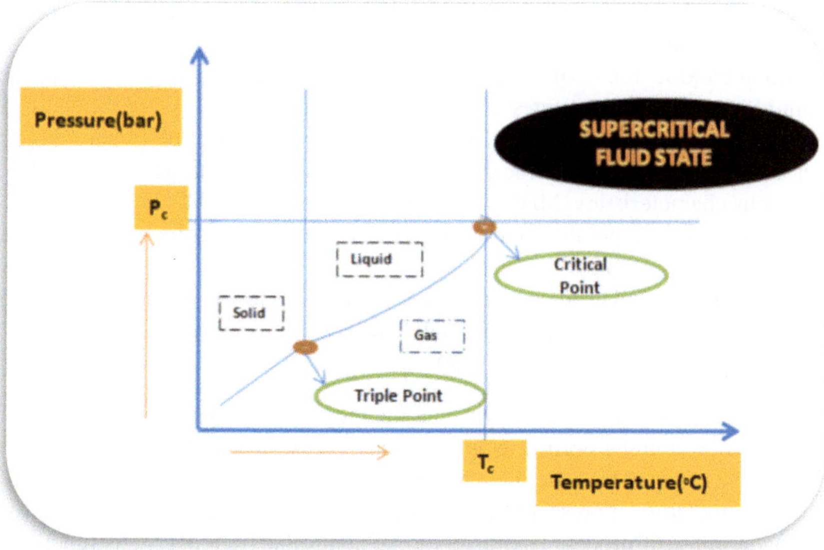

Fig. 8.3 Phase diagram of CO_2

8.2.3 *Phase Diagram*

The pressure–temperature phase diagram of pure carbon dioxide is commonly used to illustrate the supercritical fluid phenomenon. Figure 8.3 depicts the different phases of CO_2, including gas, liquid, solid, and supercritical fluid. A substance enters the supercritical state when the vapor and liquid phases merge into a single phase, existing in equilibrium at a temperature and pressure beyond its critical point. In this state, the substance exhibits both gas-like and liquid-like properties, making it highly effective for various industrial applications (Abou Elmaaty and Abd El 2018; Montero et al. 2000; Hendrix 2001; Kazarian 2000).

8.2.4 *Properties of Supercritical Fluids (SCFs)*

- As a substance approaches its critical point, its isothermal compressibility tends toward limitlessness. "The two-phase boundary between solid and liquid diminishes as the substance behaves as a single phase, representing the transformation into a supercritical fluid" (Jonin et al. 2009; Montero et al. 2000; Hendrix 2001). As a result, the substance's specific thickness or volume changes significantly. A material in the critical region exhibits the properties of a gas with "liquid-type density and higher solvent capacity" (Miah et al. 2013; Montero et al. 2000; Hendrix 2001). This occurs because the increase in density reduces the average intermolecular distance, thus increasing the number of intersections between the solute and solvent.

- Carbon dioxide transforms into a supercritical fluid that resembles an "expanded liquid or a strongly compressed gas" (Hassabo et al. 2022) when heated to its critical temperature (Tc) of 310 K and critical pressure (Pc) of 74 bar (Hassabo et al. 2022). Above the critical value, carbon dioxide exhibits both liquid and gaseous characteristics (Miah et al. 2013; Montero et al. 2000; Hendrix 2001).
- Depending on temperature and pressure, CO_2 can exist in solid, liquid, gaseous, and supercritical states.
- Carbon dioxide and water are the most valuable SCFs for green chemistry because they are non-flammable and renewable. Water has a Tc of 647 K (374.1 °C) and a Pc of 218 atm (Kerton and Marriott 2013).

Additionally, a system's dielectric constant increases with pressure, enhancing its dissolving power under normal temperature and pressure conditions. The density of SCFs can be adjusted from liquid to gas by simply manipulating the phase boundary through temperature or pressure without altering the molecular structure. This allows for a wide range of process parameters to be controlled, making SCFs function as hybrid solvents. As a result, research into the underlying characteristics of SCFs is particularly compelling (Jessop and Leitner 1999; Kemmere 2005; Eckert et al. 1996).

8.3 History of $SCCO_2$ Dyeing

The primary industrial application of supercritical carbon dioxide ($SCCO_2$) remains extraction, with the first patent filed in 1986 focusing on pharmacological compositions, impregnation of thermoplastic polymers with fragrances, and related processes (Sand 1986). In 1988, another patent introduced the concept of dyeing textile substrates in pure $SCCO_2$ by incorporating polar co-solvents to enhance dye solubility (Schollmeyer et al. 1990). Over time, more comprehensive patents have been developed, detailing improved dyeing techniques and dyes specifically formulated for $SCCO_2$, further advancing its application in sustainable textile processing.

8.4 Supercritical CO_2 Dyeing Instrument

Saus et al. explain the dyeing process in supercritical fluids, which involves wrapping a sample in an autoclave, filling it with powdered dye, maintaining critical temperature and pressure, and measuring dye uptake spectrophotometrically (Saus et al. 1993).

8.5 Variables Impacting $SCCO_2$ in Textiles

Brantley et al. examined how the dyeing medium influences key fiber characteristics, such as crystallinity, shrinkage, damage, and polymer orientation, which are crucial for the successful commercial implementation of $SCCO_2$ dyeing technology. These factors directly impact the uniformity and quality of dye absorption in textile substrates. Additionally, their study emphasized the importance of understanding the rate of CO_2 sorption, its solubility, and its effects on critical variables such as dye diffusivity and polymer free volume, which play a significant role in optimizing the $SCCO_2$ dyeing process (Brantley et al. 2000).

8.6 Dyeing of $SCCO_2$ in Man-Made and Natural Textiles

Synthetic fibers such as polyamide (PA), polypropylene (PP), poly-lactic acid (PLA), ultrahigh molecular weight polyethylene (UHMWPE), polyacrylonitrile, aramids, and cellulose acetate can be dyed using $SCCO_2$, representing a larger share of applications compared to natural fibers. However, research on other fiber types remains limited. Conventional dyeing processes contribute to wastewater contamination, whereas $SCCO_2$ dyeing offers a safer and more environmentally sustainable alternative (Abate and Tadesse 2021). A key factor influencing the success of $SCCO_2$ dyeing is the solubility of dyes in the supercritical medium. Various studies have investigated this aspect, examining how molecular interactions between dyes and $SCCO_2$ affect diffusion, absorption, and overall dyeing efficiency. The concept of dye solubility and its implications for $SCCO_2$ dyeing can be found in the literature (Abou Elmaaty et al. 2020; Bach et al. 2002; Banchero 2013, 2020).

8.6.1 Polyester (PET) Dyeing Under $SCCO_2$

Polyester dyeing under carbon dioxide (a non-polar medium) operates similarly to dyeing in an aqueous medium, although the diffusivities in aqueous solutions are observed to be three orders of magnitude lower than in dyed PET. Since CO_2 is a byproduct of processes like fermentation, combustion, and ammonia production, no additional CO_2 needs to be produced for the dyeing process (Vincent et al. 1998; Von Schnitzler and Eggers 1999; Das 2023). To facilitate the solubilization of solid dyes (low vapor pressure compounds), CO_2 must reach a density of around 600 kg/m3 and a temperature of approximately 120 °C. These conditions require the temperature and pressure to be much higher than the critical point of 31°C and 74 bar, which are necessary for effective dyeing (Vincent et al. 1998; Von Schnitzler and Eggers 1999; Das 2023). Chakraborty (2014) identified two primary functions of CO_2 dyeing: first, to dissolve and transport the dye onto the surface, and second, to transfer the dye to the substrate. The dyeing process involves dissolving the dye, transferring it, adsorbing it onto the surface, diffusing it into the fiber, and depressurizing the dye (Zheng et al. 2015). High diffusivity and low viscosity are key parameters for $SCCO_2$, making it a promising alternative to water-based dyeing methods. A slight shift in temperature near the CO_2 critical point can cause a significant decrease in density, reducing the solvating power of the dye molecule. This phenomenon is especially relevant near the critical point, where solvent density decreases, increasing the dye's sublimation pressure as temperature rises. The most important factor in this process is that the hydrophobic polymer is plasticized, which lowers the glass transition temperature (Tg) (Chiou et al. 1985) and swelling due to the supercritical state of carbon dioxide. This plasticization increases dye uptake and more uniform dyeing (Van der Kraan et al. 2007). Supercritical CO_2 also acts as a molecular lubricant, enhancing chain motion in the amorphous regions of the textile substrate, which further facilitates dye penetration and distribution (Vincent et al. 1998; Von Schnitzler and Eggers 1999; Das 2023). Abate et al. studied the "one-step dyeing and finishing process" using supercritical CO_2 dyeing with natural curcumin on polyethylene terephthalate (PET) fabric without needing mordants or other auxiliaries. This research demonstrated that the treated polyester fabric exhibited several unique properties, including ultraviolet protection, antibacterial characteristics, and antioxidant activity. Additionally, the fabric showed improved color fastness, highlighting the potential of $SCCO_2$ dyeing as a sustainable and functional method for enhancing textile performance (Abate et al. 2019).

8.6.2 Dyeing of Other Synthetic Fibers Under Supercritical Carbon Dioxide (scCO$_2$)

8.6.2.1 Nylon

Nylon can be dyed using $SCCO_2$, but it typically shows lower sorption and fastness than other fibers due to its higher crystallinity (Bach et al. 2002; Park et al. 2010; Liao et al. 2000a). However, adding water to the $SCCO_2$ medium can enhance the color intensity and dye fixation percentage, influencing the dye-substrate system (Van der Kraan et al. 2007). Researchers have also improved the fastness performance of nylon 6,6 by using both commercial disperse dyes and newly synthesized dyes, such as disperse reactive dyes, new azo-based disperse dyes, and new antibacterial hydrazonopropanenitrile disperse dyes (Abou Elmaaty et al. 2015). These dyes form covalent bonds between the polymer and the dye molecule, improving color retention and durability. The dyeing of nylon was achieved at 100–120 °C and 230–245 bar using a reactive vinyl sulphone azo dye, resulting in excellent color strength and fastness (Abou Elmaaty et al. 2015; Bach et al. 2002). The dyeing parameters for $SCCO_2$ dyeing of nylon 6,6 were optimized at 120 °C and 315 bar pressure (Lakshmanan and Raghavendran 2017; Liao et al. 2000a).

8.6.2.2 Polypropylene

Polypropylene dyeing is challenging due to its crystallinity and non-polar aliphatic structure (Miyazaki et al. 2012a, 2012b). The material's high hydrophobicity and aliphaticity make it difficult to dye effectively, so supercritical dyeing has been considered a suitable alternative (Kim et al. 2008). The dyeability of unmodified polypropylene was improved with long alkyl chain components of 1,4-bis (alkylamino) anthraquinone dyestuffs under a supercritical CO_2 ($SCCO_2$) medium (Yuan et al. 2002). Various commercial disperse dyestuffs were used, and dye uptake was better in carbon dioxide than in water (Miyazaki et al. 2012b; Liao et al. 2000b). Garay et al. (2010) studied co-solvent-based Isolan 2S-RB® acid dye, while Abou Elmaaty et al. (2018) used hydrozonopropanenitrile dyes for dyeing unmodified polypropylene substrates using aqueous and supercritical mediums. These findings suggest that supercritical dyeing could be a promising alternative for dyeing polypropylene.

8.6.2.3 Cellulose Acetate

In contrast to the conventional aqueous-based disperse dyeing process, it was found that the coloration of cellulose diacetate, synthetic cellulose-based, uses a variety of Corangar $SCCO_2$ solvent dyes. The outcomes demonstrated enhanced color fastness and improved color strength (Broadbent et al. 2023).

8.6.2.4 Polylactic Acid (PLA)

Disperse dyes have shown promising results in dyeing polylactic acid (PLA) fibers using supercritical fluid (SCF) technology, marking an advancement in sustainable textile processing. Banchero (2013) highlighted that PLA is an eco-friendly alternative to traditional fibers, but its poor resistance to conventional processing conditions presents challenges for its broader use. The use of non-aqueous media, such as supercritical CO_2 with disperse dyes, has proven advantageous for PLA dyeing, as it enables comparable dye uptake and fastness performance to that of conventional dyeing methods. This method improves the color quality of PLA fibers and offers additional benefits such as reducing PLA shrinkage and preserving mechanical properties. However, while the SCF treatment demonstrates positive effects, some studies report varying results regarding the retention of mechanical properties, particularly at different temperatures. The dyeing process under SCF conditions can cause a loss in mechanical properties at certain temperatures (Bach et al. 2006; Wen and Dai 2007; Yuan et al. 2002). This suggests that the temperature during the supercritical treatment plays a significant role in determining the final performance of PLA fibers after dyeing. Despite these challenges, SCF dyeing of PLA continues to show promise as a more sustainable and efficient alternative to traditional methods, particularly in applications where environmental concerns and sustainability are prioritized. Further research is needed to optimize the dyeing process for PLA to ensure its mechanical properties are not compromised while achieving high-quality coloration.

8.6.2.5 Polyacrylonitrile (PAN), Ultra High Molecular Weight Polyethylene (UHMWPE) and Aramids

Many researchers have explored the coloration of polyacrylonitrile (PAN) fibers using supercritical fluids (SCF), with studies from Zheng et al. (2015), Banchero (2020), Zheng et al. (2016), Jun et al. (2005), and Zheng et al. (2017b). The high crystallinity and dense structures of high-performance fibers, such as PAN, make them challenging to dye (Miyazaki et al. 2012a, b; Ma et al. 2019; Banchero 2020). Co-solvents and carriers have been employed to enhance dye uptake and improve the dyeing process (Banchero 2013; Abou Elmaaty et al. 2015; Zheng et al. 2017a; Zheng and Zheng 2014). Despite these advancements, the dyed fibers exhibit fair washing fastness but often have insufficient light fastness ratings, limiting their practical applications (Banchero 2013).

8.6.3 Dyeing of Natural Fibers Under Supercritical Carbon

Studies have shown that dyeing natural fibers, such as cotton, wool, silk, and ramie, in supercritical CO_2 ($SCCO_2$) presents significant challenges due to the difficulty of swelling polar fibers and the inability to diffuse dye molecules effectively. Polar dyes, such as vat, acid, direct, and basic dyes, are inherently insoluble in $SCCO_2$ (Bach et al. 2002). Several strategies have been developed to overcome these challenges to improve the dyeability of natural fibers in $SCCO_2$. These strategies include treating the fabric with swelling agents (Gebert et al., 1994; Guzel and Akgerman 2000; Beltrame et al. 1998), chemically modifying the fibers by introducing hydrophobic functional groups or polar modifiers (Liu et al. 2006; Schmidt et al. 2003), and using a water/CO_2 reverse micellar system to enhance the solubility of reactive and conventional polar dyes (Jun et al. 2004; Sagisaka et al. 2019). While water micro-emulsions in $SCCO_2$ reverse micellar systems have shown stability (Sagisaka et al. 2019), poor results and environmental concerns have made this approach less favorable (Bach et al. 2002; Long et al. 2012). Reactive disperse dyes, soluble in $SCCO_2$, hold promise for dyeing natural fibers. Despite the promising results, the commercialization of this process is still far from realization. Researchers, including Guzel and Akgerman (2000), have proposed using mordant dyes like C.I. Mordant Brown, C.I. Mordant Yellow 12, and C.I. Mordant Red 11 for dyeing metal-mordanted wool. Many studies have focused on modifying natural fibers to improve dye uptake, such as cotton treated with trichlorotriazine and mercerized ramie modified with benzoyl chloride to enhance dye affinity (Liu et al. 2006; Schmidt et al. 2003; Sawada and Ueda 2007). However, the chemical modification of natural fibers has been limited in the literature, highlighting the need for further research to optimize these processes.

8.7 Advantages and Disadvantages of CO_2 Dyeing

In contrast to conventional textile dyeing methods, the supercritical CO_2 (SCF) dyeing technique offers numerous advantages. Supercritical CO_2 is a sustainable, cost-effective solution for dyeing man-made fibers (Bach et al. 2002). It provides benefits such as low energy consumption, energy preservation, and the absence of air pollution. The low viscosity and high diffusivity of CO_2 enable faster dye diffusion into the voids between fibers, reducing processing time and improving performance. Additionally, it does not require dispersants, adulterants, or other chemicals, making it an environmentally friendly alternative (Agrawal 2015; Saus et al. 1993; Bach et al. 2002). The process eliminates the need for freshwater, wastewater, and auxiliary chemicals, making it especially suitable for regions with limited water supplies. Compared to traditional dyeing methods, $SCCO_2$ dyeing results in better dye uptake and uniformity, enhancing resource efficiency. It dissolves dyes on a mono-molecular level, eliminating the need for filtration systems (Saus et al. 1993). CO_2 is a cheap, non-toxic, and incombustible solvent, with 95% of it being recycled during the dyeing process. Overall, $SCCO_2$ dyeing is environmentally friendly and can reduce costs by 40–60% compared to conventional methods (Agrawal 2015; Abate and Tadesse 2021; Bach et al. 2002; Rani and Arya 2019). However, the method does have drawbacks, such as the high-pressure requirements for dye solubility, challenges in eliminating trimers, and limited knowledge about dye solubility in waterless systems (Agrawal 2015).

8.8 Applications of $SCCO_2$ Fluids

Supercritical fluid dyeing technology has diverse potential applications across various industries, including food, health, cosmetics, environmental cleaning, petroleum, and medicine (Fig. 8.4). One of the key application areas is in the food industry, where it can be used for processes like decaffeination, hops, lipids, and fragrance extraction, as Hendrix (2001) highlighted. In the textile industry, supercritical CO_2 ($SCCO_2$) serves as a non-polar solvent in dyeing, scouring, and UV stabilizer transfer, offering an eco-friendly alternative to traditional methods. Its excellent extraction efficiency makes it a viable solution for dry-cleaning applications (Agrawal 2015). Supercritical fluids (SCFs) have also gained traction in the pharmaceutical and medical fields. They are increasingly used in tissue engineering, image processing, and sterilization. For instance, Deutsches Textilforschungszentrum Nord-West (DTNW) is researching supercritical fluid dyeing technology to commercialize textile sterilization and disinfection for medical applications (Bach et al. 2002). SCF technology is also valuable in chromatographic applications and the dissolution of organic molecules, expanding its versatility beyond textiles to other industries such as leather dyeing (Lakshmanan and Raghavendran 2017; Prokein et al. 2021). Additionally, SCF technology is applied in coloring zipper tapes, offering

Fig. 8.4 Application areas of SCF

a cost-effective solution while boosting competitiveness in the market (Banchero 2020). These various applications demonstrate the broad potential of supercritical fluids in enhancing efficiency and sustainability across numerous sectors, positioning SCF dyeing technology as a key innovation with wide-reaching implications for both industrial processes and environmental impact reduction.

8.9 Future Prospects and Challenges of Dyeing Textiles Under a Carbon Dioxide System

Strict environmental standards and the growing demand for end-of-pipe treatments drive textile dyeing in supercritical CO_2 ($SCCO_2$). The need for environmentally friendly dyeing methods is increasing due to both financial and environmental incentives. Before implementing $SCCO_2$ dyeing, factors such as economic and environmental effects, product quality, and process stability must be carefully considered (Van der Kraan 2005). Supercritical carbon dioxide dyeing is commercially feasible for synthetic fibers like nylon, polyester, and elastomeric materials. However, challenges such as high pressures, large holding capacities, and associated costs still remain. $SCCO_2$ dyeing is currently limited to synthetic textiles and cannot break the polar groups (hydrogen bonds) in natural fibers (Agrawal 2015). Additional research is needed to modify the fibers or develop new fixation mechanisms that improve dyeing efficiency to effectively dye natural fibers. Despite these challenges, supercritical CO_2 offers significant advantages, including energy and water savings, positioning it as a promising alternative to traditional dyeing methods (Agrawal 2015). $SCCO_2$ dyeing has shown significant environmental benefits, including a 20% reduction in operating expenses compared to traditional aqueous dyeing methods, owing to shorter dyeing times and reduced energy consumption (Kant 2012). The operational cost of using $SCCO_2$ to dye polyester (PET) ranges from $0.76 to $1.37 per kilogram, resulting in less dye loss than water-based methods (Barletta et al. 2017). Accurate control of $SCCO_2$ dyeing conditions is essential for optimizing energy usage and ensuring the financial feasibility of the process. $SCCO_2$ dyeing units are recyclable, which helps reduce waste, lowering transportation costs, and simplifying product distribution (Das 2023). Research into supercritical dyeing (SFD) has advanced the understanding of dye uptake and solubility, meeting commercial requirements for PET. However, further research is needed to improve mechanical resistance and fastness performance for fabrics such as aramids, polypropylene (PP), and poly-lactic acid (PLA). Future efforts should focus on minimizing solvent effects and exploring dyeing blends (Banchero 2013).

8.9.1 Conclusions

This review covers the development of supercritical carbon dioxide dyeing technology, including its fundamental principles, synthetic and natural fibers dyeing, and the ongoing challenges facing this technique. Supercritical carbon dioxide dyeing technology has become a high-tech response to the escalating call for textile production techniques that are environmentally friendly, efficient, and sustainable. Significant advantages over conventional dyeing techniques include a huge drop in water use, eliminating unsafe chemicals, and lessening ecological pollution using supercritical carbon dioxide ($SCCO_2$) as a solvent. Recent developments in $SCCO_2$ have increased the technology's scalability and profitable feasibility related to improvements in process variables, dye structures, and equipment design related to PET dyeing. The constraints related to the large-scale setup of SFD include hefty initial expenditures and specialized machinery. Further, under the supercritical dyeing technology approach, studies are required for the "coloration of high-performance and other fibers" (PP and PLA) for improved color fastness and physical properties. Natural fibers such as cotton, wool, and silk face challenges with single-stage supercritical fluid dyeing, which is fruitful for the textile industry. Prospect studies should consider enhancing colorant absorption and cutting water consumption using ecologically aware processing techniques.

Declaration The authors declare that they have no known competing financial interests or personal relationships that could have appeared to influence the work reported in this chapter. Generative AI and AI-assisted technologies have been used in the writing process to improve the readability and language of the work only. After using this tool, the authors carefully reviewed and revised the content to ensure accuracy and quality. The authors take full responsibility for the final version of this publication.

References

Abate MT, Tadesse MG (2021) Airflow, foam, and supercritical carbon dioxide dyeing technologies. In: Innovative and emerging technologies for textile dyeing and finishing, Rather LJ, Haji A and Shabbir M (eds), Scrivener Publishing LLC, pp 137–164

Abate MT, Ferri A, Guan J, Chen G, Nierstrasz V (2019) Colouration and bio-activation of polyester fabric with curcumin in supercritical CO_2: Part I-Investigating colouration properties. J Supercrit Fluids 152:104548

Abou Elmaaty T, Abd El-Aziz E (2018) Supercritical carbon dioxide as a green media in textile dyeing: a review. Text Res J 88(10):1184–1212

Abou Elmaaty T, Abd El-Aziz E, Ma J, El-Taweel F, Okubayashi S (2015) Eco-friendly disperse dyeing and functional finishing of nylon 6 using supercritical carbon dioxide. Fibers 3(3):309–322

Abou Elmaaty T, El-Taweel F, Elsisi H, Okubayashi S (2018) Water free dyeing of polypropylene fabric under supercritical carbon dioxide and comparison with its aqueous analogue. J Supercrit Fluids 139:114–121

Abou Elmaaty T, Mousa A, Gaafar H, Hebeish A, Sorour H (2020) Development of textile dyeing using the green supercritical fluid technology: a review. Mater Int 2(3):373–390

Agrawal BJ (2015) Supercritical carbon-dioxide assisted dyeing of textiles: an environmental benign waterless dyeing process. Int J Innov Res Creat Technol 1(2):201–205

Ali RB, Gaballa AS (2014–2015) Studying the functional properties of Linen fabrics treated with microwave and natural green dyes. Int J ChemTech Res 7(1):119–130

Bach E, Cleve E, Schollmeyer E (2002) Past, present and future of supercritical fluid dyeing technology–an overview. Rev Prog Color Relat Top 32(1):88–102

Bach E, Knittel D, Schollmeyer E (2006) Dyeing poly (lactic acid) s in supercritical carbon dioxide. Color Technol 122(5):252–258

Banchero M (2013) Supercritical fluid dyeing of synthetic and natural textiles–a review. Color Technol 129(1):2–17

Banchero M (2020) Recent advances in supercritical fluid dyeing. Color Technol 136(4):317–335

Barletta, Kraan,VDM, Kronholm J et al. (2017) Climate implications of water free dyeing of biofibers (ClimaDYE). https://publications.lib.chalmers.se/publication/254430

Beltrame PL, Castelli A, Selli E, Mossa A, Testa G, Bonfatti AM, Seves A (1998) Dyeing of cotton in supercritical carbon dioxide. Dyes Pigm 39(4):335–340

Brantley NH, Kazarian SG, Eckert CA (2000) In situ FTIR measurement of carbon dioxide sorption into poly (ethylene terephthalate) at elevated pressures. J Appl Polym Sci 77(4):764–775

Broadbent PJ, Carr CM, Lewis DM, Rigout ML, Siewers EJ, Shojai Kaveh N (2023) Supercritical carbon dioxide (SC-CO2) dyeing of cellulose acetate: an opportunity for a "greener" circular textile economy. Color Technol 139(4):475–488

Chakraborty JN (2014) Dyeing in supercritical carbon dioxide. In: Fundamentals and practices in colouration of textiles, 2nd Ed New Delhi, India: Woodhead Publishing India Pvt. Ltd, pp 356–364

Chiou JS, Barlow JW, Paul DR (1985) Plasticization of glassy polymers by CO_2. J Appl Polym Sci 30(6):2633–2642

Das T (2023) Influence of supercritical carbon dioxide ($ScCO_2$) dyeing parameters on color fastness of polyester fabric dyed with disperse dye: a study towards minimizing the energy consumption of polyester (pet) fabric dyeing with azo disperse dye, pp 1–64

Desai A (2008) Plasma technology: a review. Indian Text J 118(4):25–35

Eckert CA, Knutson BL, Debenedetti PG (1996) Supercritical fluids as solvents for chemical and materials processing. Nature 383(6598):313–318

Garay I, Pocheville A, Hernando I, Fages E, Gironés S, Ferrándiz M (2010) Polyamide and polypropylene textile dyeing using supercritical carbon dioxide ($scCO_2$). In: 12th European Meeting on SCF, Gratz, Austria

Gebert B, Saus W, Knittel D, Buschmann HJ, Schollmeyer E (1994) Dyeing natural fibers with disperse dyes in supercritical carbon dioxide. Text Res J 64(7):371–374

Guzel B, Akgerman A (2000) Mordant dyeing of wool by supercritical processing. J Supercrit Fluids 18(3):247–252

Hassabo AG, Zayed M, Bakr M, El-Aziz A, Othman H (2022) Applications of supercritical carbon dioxide in textile finishing: a review. J Text, Color Polym Sci 19(2):179–187

Hendrix WA (2001) Progress in supercritical CO_2 Dyeing. J Ind Text 31(1):43–56

Jessop PG, Leitner W (1999) Supercritical fluids as media for chemical reactions. In: Chemical synthesis using supercritical fluids WILEY-VCH Verlag GmbH, pp 1–36

Jonin TM, Adjadj LP, Rizvi SS (2009) Supercritical extraction. Food Eng 3:209

Jun JH, Sawada K, Ueda M (2004) Application of perfluoropolyether reverse micelles in supercritical CO_2 to dyeing process. Dyes Pigm 61(1):17–22

Jun JH, Sawada K, Takagi T, Kim GB, Park CH, Ueda M (2005) Effects of pressure and temperature on dyeing acrylic fibres with basic dyes in supercritical carbon dioxide. Color Technol 121(1):25–28

Kant R (2012) Textile dyeing industry an environmental hazard. Nat Sci 4(1):22–26

Kazarian SG (2000) Polymer processing with supercritical fluids. Polym Sci, Ser C 42(1):78–101

Kemmere MF (2005) Supercritical carbon dioxide for sustainable polymer processes. Supercritical carbon dioxide: In: Polymer reaction engineering WILEY-VCH Verlag GmbH, pp 1–14

Kerton FM, Marriott R (2013) Alternative solvents for green chemistry. In: the royal society of chemistry, 2nd edn. By Francesca M. Kerton, pp 115–148

Kim T, Jung J, Son S, Yoon S, Kim M, Bae JS (2008) Synthesis and application of alkyl-substituted disazo yellow dyes for unmodified polypropylene fiber. Fibers Polym 9:538–543

Van der Kraan M, Cid MF, Woerlee GF, Veugelers WJT, Witkamp GJ (2007) Dyeing of natural and synthetic textiles in supercritical carbon dioxide with disperse reactive dyes. J Supercrit Fluids 40(3):470–476

Knittel D, Schollmeyer E (1995) Environmentally friendly dyeing of synthetic fibres and textile accessories. Int J Cloth Sci Technol 7(1):17–35

Lakshmanan SO, Raghavendran G (2017) Low water-consumption technologies for textile production. In: Sustainable fibres and textiles. Woodhead publishing, pp 243–265

Liao SK, Chang PS, Lin YC (2000b) Analysis on the dyeing of polypropylene fibers in supercritical carbon dioxide. J Polym Res 7:155–159

Liao SK, Ho YC, Chang PS (2000a). Dyeing of nylon 66 with a disperse-reactive dye using supercritical carbon dioxide as the transport medium. Color Tech 116(12):403–407

Liu ZT, Zhang L, Liu Z, Gao Z, Dong W, Xiong H, Peng Y, Tang S (2006) Supercritical CO2 dyeing of ramie fiber with disperse dye. Ind Eng Chem Res 45(26):8932–8938

Long JJ, Xiao GD, Xu HM, Wang L, Cui CL, Liu J, Yang MY, Wang K, Chen C, Ren YM, Luan T, Ding ZF (2012) Dyeing of cotton fabric with a reactive disperse dye in supercritical carbon dioxide. J Supercrit Fluids 69:13–20

Lu X, Liu L, Liu R, Chen J (2010) Textile wastewater reuse as an alternative water source for dyeing and finishing processes: a case study. Desalination 258(1–3):229–232

Ma J, Elmaaty TA, Okubayashi S (2019) Effect of supercritical carbon dioxide on dyeability and physical properties of ultra-high-molecular-weight polyethylene fiber. Autex Research J 19(3):1–8

Madhav S, Ahamad A, Singh P, Mishra PK (2018) A review of textile industry: wet processing, environmental impacts, and effluent treatment methods. Environ Qual Manage 27(3):31–41

Miah L, Ferdous N, Azad MM (2013) Textiles material dyeing with supercritical carbon dioxide (CO_2) without using water. Chem Mater Res 3(5):38–40
Miyazaki K, Tabata I, Hori T (2012a) Effects of molecular structure on dyeing performance and colour fastness of yellow dyestuffs applied to polypropylene fibres in supercritical carbon dioxide. Color Techn 128(1):51–59
Miyazaki K, Tabata I, Hori T (2012b) Relationship between colour fastness and colour strength of polypropylene fabrics dyed in supercritical carbon dioxide: effect of chemical structure in 1, 4-bis (alkylamino) anthraquinone dyestuffs on dyeing performance. Color Technol 128(1):60–67
Montero GA, Smith CB, Hendrix WA, Butcher DL (2000) Supercritical fluid technology in textile processing: an overview. Ind Eng Chem Res 39(12):4806–4812
Moore SB, Ausley LW (2004) Systems thinking and green chemistry in the textile industry: concepts, technologies and benefits. J Clean Prod 12(6):585–601
Özcan AS, Clifford AA, Bartle KD, Broadbent PJ, Lewis DM (1998) Dyeing of modified cotton fibres with disperse dyes from supercritical carbon dioxide. J Soc Dyers Colour 114(5–6):169–173
Park SC, Tuma D, Kim S, Lee YR, Shim JJ (2010) Sorption of CI Disperse Red 60 in polystyrene and PMMA films and polyester and Nylon 6 textiles in the presence of supercritical carbon dioxide. Korean J Chem Eng 27:299–309
Pennington DW, Potting J, Finnveden G, Lindeijer E, Jolliet O, Rydberg T, Rebitzer G (2004) Life cycle assessment Part 2: Current impact assessment practice. Environ Int 30(5):721–739
Prokein M, Dyes T, Renner M, Weidner E (2021) Waterless leather dyeing with dense carbon dioxide as solvent for dyes. J Supercrit Fluids 178:1–11
Rani L, Arya N (2019) Dyeing and finishing of fabrics with low or without water. Pharma Innov J 8(5):1159–1165
Razzak FSA (2021) Review on green solvent supercritical carbon dioxide and its chemical reactions. J Sci Res 65(8):17–22
Rebitzer G, Ekvall T, Frischknecht R, Hunkeler D, Norris G, Rydberg T, Schmidt WP, Suh S, Weidema BP, Pennington DW (2004) Life cycle assessment: Part 1: Framework, goal and scope definition, inventory analysis, and applications. Environ Int 30(5):701–720
Sagisaka M, Saito T, Yoshizawa A, Rogers SE, Guittard F, Hill C, Eastoe J, Blesic M (2019) Water-in-CO_2 micro emulsions stabilized by fluorinated Cation–anion surfactant pairs. Langmuir 35(9):3445–3454
Saling P, Kicherer A, Dittrich-Krämer B, Wittlinger R, Zombik W, Schmidt I, Schrott W, Schmidt S (2002) Eco-efficiency analysis by BASF: the method. Int J Life Cycle Assess 7:203–218
Sand ML (1986) Method for Impregnating Thermoplastic Polymer. Off Gaz US4598006A
Saravanan D (2006) Ultrasonic assisted textile processing: an update. Colourage 53(4):111–116
Saus W, Knittel D, Schollmeyer E (1993) Dyeing of textiles in supercritical carbon dioxide. Text Res J 63(3):135–142
Sawada K, Ueda M (2007) Chemical fixation of disperse dyes on protein fibers. Dyes Pigm 75(3):580–584
Schmidt A, Bach E, Schollmeyer E (2003) Supercritical fluid dyeing of cotton modified with 2, 4, 6-trichloro-1, 3, 5-triazine. Color Technol 119(1):31–36
Schollmeyer E, Knittel D, Buschmann HG, Schneider GM, Poulakis K (1990) Deutsches Textilforschzentrum, Germany Patent DE3906724A1, D 06 P 1 assigned to Deutsches Textilforschungszentrum
Von Schnitzler J, Eggers R (1999) Mass transfer in polymers in a supercritical CO_2-mass atmosphere. J Supercrit Fluids 16(1):81–92
Stewart M, Jolliet O (2004) User needs analysis and development of priorities for life cycle impact assessment. Int J Life Cycle Assess 9:153–160
Su Y (2012) Cotton pad dyeing microwave fixation technology research. J Dye Dye 49(2):23–26
Van der Kraan M (2005) Process and equipment development for textile dyeing in supercritical carbon dioxide 1–138

Vincent MF, Kazarian SG, West BL, Berkner JA, Bright FV, Liotta CL, Eckert CA (1998) Cosolvent effects of modified supercritical carbon dioxide on cross-linked poly (dimethylsiloxane). J Phys Chem B 102(12):2176–2186

Wen H, Dai JJ (2007) Dyeing of polylactide fibers in supercritical carbon dioxide. J Appl Polym Sci 105(4):1903–1907

Yuan X, Mak AF, Yao K (2002) Comparative observation of accelerated degradation of poly (l-lactic acid) fibres in phosphate buffered saline and a dilute alkaline solution. Polym Degrad Stab 75(1):45–53

Zheng HD, Zhang J, Yan J, Zheng LJ (2017a) Investigations on the effect of carriers on meta-aramid fabric dyeing properties in supercritical carbon dioxide. RSC Advances 7(6):3470–3479

Zheng H, Zheng L (2014) Dyeing of meta-aramid fibers with disperse dyes in supercritical carbon dioxide. Fibers Polym 15:1627–1634

Zheng H, Zhang J, Zheng L (2017b) Optimization of an ecofriendly dyeing process in an industrialized supercritical carbon dioxide unit for acrylic fibers. Text Res J 87(15):1818–1827

Zheng H, Zhang J, Yan J, Zheng L (2016) An industrial scale multiple supercritical carbon dioxide apparatus and its eco-friendly dyeing production. J CO_2 Util 16:272–281

Zheng L, Zheng H, Du B, Wei J, Gao S, Zhang J (2015) Dyeing procedures of polyester fiber in supercritical carbon dioxide using a special dyeing frame. J Eng Fibers Fabr 10(4):37–46

Chapter 9
Salt-Free Dyeing Techniques: Minimizing Environmental Impact

Ashraful Alam Aulia, Md. Reazuddin Repon, Latifa Noushin, Md. Shalit Hasan, Pranto Biswas, and Meherin Sultana

Abstract Salt-free dyeing aims to decrease the environmental impact during the dyeing process and targets moving towards an eco-friendly and sustainable dyeing process. It is an eligible, more impactful way of minimizing salt discharge into water and conserving water as much as possible. The materials used for salt-free dyeing are cotton or modified cotton fabrics that contain cellulosic groups, which aid in the fixation of salt. Cationization plays a key role in this method as it turns the negative parts of the cellulosic groups into positive parts, which are more strongly bonded with the negative parts of the reactive dyes. The more the positively and negatively charged molecules in the medium, the faster the fixation rate will increase. It can also be regarded as an alternative dye fixation method of reactive dyes. Cationization improves characteristics such as color fastness, strength, uniformity, absorption, and fixation. This paper aims to provide a critical overview of the cationic process of the salt-free dyeing process in different fabrics. The salt-free dyeing methods are thoroughly discussed, and various dyeing properties and enhancements for different fabrics are comprehensively analyzed.

Keywords Absorption · Cationization · Coloration · Fixation · Textile dyes

A. A. Aulia
Department of Dyes and Chemical Engineering, Bangladesh University of Textiles, Dhaka, Bangladesh

Md. R. Repon (✉)
Department of Bioproducts and Biosystems, School of Chemical Engineering, Aalto University, Vuorimiehentie 1, 02150 Espoo, Finland
e-mail: reazmbstu.te@gmail.com

L. Noushin · Md. S. Hasan
Department of Wet Processing Engineering, Bangladesh University of Textiles, Dhaka, Bangladesh

P. Biswas
Department of Textile Machinery Design and Maintenance, Bangladesh University of Textiles, Dhaka, Bangladesh

M. Sultana
Department of Textile Fashion and Design, Bangladesh University of Textiles, Dhaka, Bangladesh

S. Maiti et al. (eds.), *Sustainable Coloration Techniques in Textiles*,
https://doi.org/10.1007/978-981-96-4975-4_9

9.1 Introduction

Salts play a pivotal role in dyeing, particularly when dyeing fibers like cotton. This fiber possesses a negative charge, which naturally repels the negatively charged dye anions. This repulsion can hinder the dye's ability to bond effectively with the fibers, leading to poor dye uptake and inadequate color fastness (Ad 2021). To overcome this challenge, salts are introduced into the dye bath to counteract the repulsive forces between the fiber and dye molecules. Salt provides a high ionic strength in the dye bath. This high ionic strength helps to shield the negative charges on the dye anions and the fibers, reducing their mutual repulsion. By doing so, it facilitates better interaction between the dye molecules and the fiber, leading to improved dye uptake and deeper, more vibrant colors. Salt also accelerates the rate at which dye molecules are absorbed by the fibers, thereby enhancing the overall efficiency of the dyeing process (Sun et al. 2017). Salts are indispensable in dyeing as they neutralize the repulsive forces between dye anions and fiber charges. This neutralization allows for more efficient dye uptake, better color uniformity, and enhanced color fastness in the finished fabric (Wolela 2019; Periyasamy et al. 2024).

Salt-free dyeing has been extensively studied through the cationic modification of fibers, addressing the high salt and alkali requirements in conventional reactive dyeing methods (Nallathambi and Venkateshwarapuram Rengaswami 2017; Islam et al. 2023). The negative charges of reactive dyes and cellulose fibers repel each other, necessitating large amounts of salt to overcome this repulsion, which leads to high dye concentration in wastewater and environmental issues. To mitigate these problems, cotton fabrics are treated with cationic reagents like quaternary ammonium compounds and bio-based polymers. These treatments enhance dye uptake without excessive salts, reducing environmental impact and improving dye exhaustion and fixation (Repon et al. 2023; Hiwot et al. 2024).

This chapter outlines the evolution of fabric dyeing methods with various dyes, highlighting early innovations such as epoxy derivatives and cationic reagents and advancements including CHPTAC and CHPTMAC for salt-free dyeing. It discusses recent techniques such as betaine modification, chitosan hybrids, plasma treatments, non-aqueous media, and biodegradable polymers. These developments improve dye uptake and color depth while addressing environmental concerns by reducing electrolyte use, wastewater, and chemical consumption. The chapter also notes ongoing challenges, such as high costs and maintaining dye fastness, emphasizing a shift towards more sustainable and eco-friendly dyeing practices.

9.2 Chemicals Used in Textile Dyeing

Various chemicals and auxiliaries are used in textile processing for specific purposes. Basic chemicals like soda ash and hydrochloric acid aid in dyeing and pH adjustment, while washing agents remove residual chemicals. Detergents and scouring agents clean fabrics, and leveling agents ensure even dye application. Salts enhance dye fixation, and sequestering agents prevent interference from metal ions (Uddin 2021). Whitening agents improve brightness, and fixing agents enhance color fastness. Softeners make fabrics softer while bleaching agents remove color. Reducing agents aid in color removal, and stabilizers maintain dye stability. Enzymes break down organic matter for better processing, anticreasing agents prevent fabric creasing, and antifoaming agents reduce foam during processing (Said et al. 2020).

9.3 Impacts of Salt in Textile Dyeing

The dyeing process should be conducted in an alkaline environment and necessitates the inclusion of a significant amount of inorganic electrolyte, typically NaCl or Na2SO4. The electrolyte concentration ranges from 0.6 to 0.8 kg of salt per kg of fiber, depending on the dye's structure, form, and the dyeing method employed (Buscio et al. 2019).

Electrochemical oxidation of dyes can occur directly on the anode surface or indirectly through the use of electro-generated oxidants like chlorine or hypochlorite, as outlined in Eq. 9.1.

$$2Cl^- \rightarrow Cl_{2(aq)} + 2e^- Cl_{2(aq)} + H_2O \rightarrow HClO + Cl^- + H^+ HClO \leftrightarrow H^+ + ClO^- \tag{9.1}$$

The effectiveness of the electrochemical process is influenced by the type of anode material used. As the concentration of NaCl in the spent liquor increased from 10 to 80 g/L, the membrane's rejection rate for NaCl initially decreased; however, it later increased over time, which was unexpected (Depeursinge et al. 2010). Additionally, research indicated that recovering water through reverse osmosis and sodium chloride from the solution via nanofiltration yields significantly higher direct profits compared to recovering Na_2SO_4 (Vishnu et al. 2008).

9.4 Alternative Textile Dyeing to Salt

The textile wet process requires a significant amount of water and chemicals, leading to increased awareness of cleaner production methods to safeguard the environment from industrial effluents. The residual salt and alkali in these effluents can have detrimental effects on the ecosystem (Uddin 2021). To overcome this problem, there are many alternatives to salt and alkali.

A novel dyeing technique has been developed that employs a hydrophobic deep eutectic solvent made from thymol, which is derived from thyme leaves, and menthol extracted from Mentha haplocalyx bing. These two natural compounds are known for their medicinal properties and effectively address issues associated with the reactive dyeing of cotton. In recent years, organic solvent dyeing has gained popularity. To overcome the limitations of using a single non-nucleophilic organic solvent for swelling cotton, researchers have created polar-non-polar systems to reduce the hydrolysis of reactive dyes. Examples of these systems include N,N-dimethylacetamide combined with dimethyl carbonate, ethyl octanoate with dimethyl sulfoxide, and N,N-dimethylformamide paired with trichloroethylene. By employing ball-milling and reversed micellar techniques, the dye can be integrated with D5 to create an effective dyeing system that allows for cotton dyeing without the need for salt (Pizzicato et al. 2023).

Trisodium nitrilotriacetate has been suggested as a versatile agent in the dyeing of cotton fabrics with reactive dyes, serving both as an exhausting agent (organic salt) and a fixing agent. Overall, the findings indicate that TNA is a suitable and environmentally friendly choice for acting as both an exhausting and fixing agent for cellulosic fabrics (Khatri et al. 2013).

Liposome technology is utilized in various applications, including wool dyeing, long-lasting fragrance finishing, textile protection, preparation of parasiticides and antibacterial drugs, and the development of phase-transition materials. Cationic liposomes have shown promise as innovative enhancers for dyeing processes and achieving uniform coloration, as well as in creating microencapsulated dye wall materials for reactive dye applications. The use of liposomes provides several benefits, such as a notable reduction in dyeing temperature (approximately 10 °C lower than traditional wool dyeing), improved quality, smoothness, and mechanical properties of dyed fibers, along with decreased contamination levels in the dye bath (Pizzicato et al. 2023).

9.5 Salt-Free Dyeing

Since the 1930s, researchers have explored various cationic compounds to boost the dyeability of cotton fabrics with reactive dyes. Early work by Schlack introduced epoxy derivatives to enhance the attraction of cellulose for acid dyes. Subsequent studies modified cellulose using different cationic reagents, including epoxy diethylamine-3-propane, glycidyl trimethyl ammonium chloride (Glytac), and polyepichlorohydrin dimethylamine (PECH-amine) (Mousa et al. 2024). These treatments generally improved dyeability, allowing for salt-free dyeing under certain conditions. Cationic reagents fall into two categories: monomeric (e.g., epoxy compounds, chlorotriazine quaternary compounds) and polymeric (e.g., polyepichlorohydrin, polyamide epichlorohydrin, biopolymers like chitosan). While these reagents improve dye uptake and reduce salt use, they have limitations, including high costs, complexity, and non-biodegradability. Formaldehyde-releasing agents and the need for cross-linkers also pose challenges (Wolela 2019; Momotaz et al. 2024).

9.5.1 Cationizing Using Quaternary Ammonium Compounds

Cationization offers the advantage of removing the need for salt in traditional processes. It is estimated that this cationic method can decrease the salt concentration in reactive dyeing effluent by at least 70%. Four cationic reagents were tested for cationizing cotton fabrics: Poly diallyldimethylammonium chloride (PDDACl), Poly acrylamide-co-diallyldimethylammonium chloride (PAcD), Polyquaternium-2 (P42), and 3-chloro-2-hydroxypropyl trimethyl ammonium chloride (CHPTAC). PDDACl and PAcD, both with quaternary ammonium groups, are notable for their textile treatment potential. PDDACl is cost-effective and eco-friendly, while PAcD offers high charge density but limited textile use. P42, used in cosmetics, shows promise due to its antimicrobial properties and dyeing efficiency, though dye uniformity needs improvement. CHPTAC is a common commercial agent but is less effective at higher pH levels compared to PDDACl and P42. Overall, PDDACl and P42 are strong alternatives to CHPTAC for industrial textile applications (Correia et al. 2021).

2,3-Epoxypropyltrimethylammonium chloride (CHPTMAC) is generated on-site by reacting 3-chloro-2-hydroxypropyltrimethylammonium chloride with an alkaline solution. This compound reacts with cotton fibers to introduce cationic sites, which attract anionic reactive dyes, thereby reducing the need for large quantities of electrolytes during dyeing.

Cationization of cotton was confirmed by an increase in nitrogen content to about 0.06%, demonstrating successful modification. The treated cotton shows significantly improved dye uptake with reactive dyes even without the use of salt, and its wash and dry rubbing fastness are similar to untreated cotton, with often better light fastness. However, the addition of salt negatively affects dye exhaustion; for example, with **2% o.w.f. Cibacron Red HF**, dye uptake dropped from 60% without salt to 45% with 20 g/L sodium sulfate. The salt competes with dye anions for the cationic sites on the cotton, reducing dye absorption and making the treated cotton perform like untreated cotton (Roy Choudhury 2013; Gunasekar and Ponnusami 2015; Jabar et al. 2023).

In recent years, CHPTAC has emerged as a preferred cationic agent for cotton cationization due to its effectiveness in enhancing dye uptake and color strength without the need for salt. Historically, CHPTAC was applied using the cold pad-batch (CPB) process, which, while effective, is energy-intensive and time-consuming, involving a lengthy 18–24 h fixation period. To address these limitations and reduce environmental impact, researchers have developed an exhaust method for applying CHPTAC, which allows for salt-free dyeing. This optimized approach not only achieves nearly complete dye fixation but also significantly decreases dye waste and the associated salt sludge, which can leach into groundwater and contribute to environmental pollution. The exhaust method has proven to be a more viable and efficient alternative in the textile industry, offering reduced alkali consumption and improved fabric fastness, making it a promising solution for the challenges posed by traditional dyeing processes (Nallathambi and Venkateshwarapuram Rengaswami 2016).

The paper discusses the synthesis and application of a novel finishing agent, organosilicon-loaded hyperbranched poly(amidoamine) (Si@PAMAM), for cotton fabrics. This dual-function compound combines the properties of hyperbranched polymers and silica sol. Si@PAMAM was synthesized and characterized using ^1H NMR and FT-IR spectroscopy. Cotton fabrics modified with Si@PAMAM through a sol–gel process showed improved salt-free dyeing properties. The treated fabrics demonstrated better color depth, color fastness, leveling properties, and washing endurance compared to untreated samples. Thermogravimetric analysis revealed a significant increase in thermal stability. Scanning electron microscopy confirmed effective deposition of Si@PAMAM on fiber surfaces. The modification enhances dye uptake due to electrostatic attraction between the cationic sites on the fiber and the anionic reactive dyes, eliminating the need for traditional electrolytes in the dyeing process. The use of Si@PAMAM simplifies the grafting process and avoids the need for pre-treatment steps commonly required for other hyperbranched polymers (Zhang et al. 2021).

A two-bath pad-bake process was developed to cationize cotton using CHPTAC. Key parameters were 8% reagent concentration, a 1:1 molar ratio of NaOH to CHPTAC, and a baking temperature of 60 °C for 6 min. This method allows for the recycling of CHPTAC and continuous processing of cotton (Wang et al. 2009).

Tests showed that cationized cotton had higher dye utilization and color yield compared to untreated cotton, with good dyeing levelness and fastness properties. The process also reduced environmental impact by minimizing effluent discharge. Recycled CHPTAC solutions maintain effectiveness over multiple uses, with consistent dye fixation and color strength. XRD and SEM established that cationization appeared mainly on the cotton surface with minimal changes in fiber morphology. This method eliminates the need for salt in dyeing, enhances dye efficiency, and maintains fabric strength and color fastness, making it a viable option for environmentally friendly, salt-free reactive dyeing (Wang et al. 2009; Mikucioniene et al. 2024).

9.5.2 *Cationizing Using Bio-Based Polymers*

Bleached cotton fabric treated with a natural amino acid mixture derived from soybean seed waste influences its dyeing properties. By incorporating nucleophilic amine/aldimine groups or cationized quaternary ammonium groups into the cotton cellulose, the fabric's affinity for reactive dyes is improved, promoting an eco-friendly, salt-free dyeing process. The absorption of C.I. Reactive Red 141 at 550 nm indicates that during the acid bath dyeing process, the majority of the dye is absorbed within 15 to 30 min. In contrast, in an alkali bath, dye exhaustion begins 15 min after adding electrolytes, peaks at 45 min, and then gradually declines to about 10% at 90 min. The soya-treated fabric, which undergoes cationization from soybean extract, shows better dye exhaustion compared to the control fabric. Additionally, dye penetration is superior in the soya-treated fabric during alkali bath dyeing, as opposed to the lower dye penetration observed with the acid bath method (Samanta et al. 2016).

The cationization of the cotton was confirmed by a zeta potential. The study optimized various grafting conditions, including CS-PPI concentration, pH, temperature, and time, finding the best parameters to be 20% of concentration, pH 6, 50 °C, and 18 h. This modification significantly enhanced the K/S and dye uptake of the cotton fabric in a salt-free dyeing process compared to traditional methods. The treated cotton showed higher dye fixation and better color yield without the need for electrolytes or alkali. Additionally, the modified fabrics exhibited good wash, light, and rubbing fastness, as well as satisfactory mechanical properties. The CS-PPI hybrid proved to be an effective and eco-friendly alternative for reactive dyeing, improving both environmental sustainability and dye utilization efficiency. The improved performance is attributed to the introduction of amine groups on the cotton, which enhances the fabric's affinity for anionic dyes, making salt-free dyeing more efficient (Sadeghi-Kiakhani and Safapour 2015; Cai et al. 2024).

An eco-friendly method for cotton coloration with direct dyes using chitosan-polypropylene imine dendrimer (CS-PPI) has been introduced. By optimizing factors such as CS-PPI concentration, dye, dyeing time, temperature, and salt concentration, the method enhances dye uptake and reduces the need for high dye concentrations. The process eliminates the use of salt, significantly reduces dyeing time and temperature, and maintains color hue and fastness. This approach saves time, energy, chemicals, and costs while addressing dyeing effluent issues (Sadeghi-Kiakhani and Safapour 2018).

In another study, a new approach for salt-free cotton coloration using betaine has been explored, which is a biodegradable and biocompatible natural compound. Betaine modifies cotton fibers through cationization by reacting its carboxyl group with the hydroxyl groups in cotton, forming ester bonds. These bonds facilitate dye uptake but can be hydrolyzed under alkaline and high-temperature conditions, aligning with dye fixation requirements and potentially reducing color staining. A pad-dry-bake method was developed for applying betaine, and a steaming fixation process was designed to enhance dye fixation. Results showed that betaine-modified cotton achieved higher dye fixation compared to untreated fabrics, with improvements of 6.7% to 9.5% for various dyes. While the wash and wet rub fastness of cationic cotton were slightly lower, overall color fastness remained acceptable.

The study demonstrates that betaine-modified fabrics offer effective salt-free dyeing with enhanced dye uptake and environmental benefits, although further improvements in color fastness are needed (Ma et al. 2016).

9.5.3 Modification of Textile Using Other Compounds

A study, a method has been described where cotton fibers pretreated with sodium periodate to create aldehyde groups were then grafted with an amino-terminated hyperbranched polymer. This process introduced cationic groups that enhanced the interaction with reactive dyes. Key factors for optimal dyeing included polymer concentration, temperature, and reaction time, with higher temperatures accelerating the reaction and reducing processing time. Hyperbranched polymers, with their three-dimensional dendritic structures, offer unique properties and applications. Amino-terminated hyperbranched polymers, in particular, have abundant imino and amino groups that improve dyeability. By grafting these polymers onto oxidized cotton, high color strength can be achieved with reactive dyes, even without electrolytes. This method also ensures good fastness properties and reduces the need for salt in dyeing (Hebeish et al. 2019).

The study evaluated how pre-treating cotton with amino groups enhances its dyeing performance with cold, hot brand, and HE reactive dyes across different shades. The treatment significantly improved dye absorption and fixation without the need for salt, with pre-treated cotton showing better or comparable dye uptake compared to untreated cotton dyed with salt. Specifically, cold brand dyes exhibited markedly improved absorption without salt, and HE dyes achieved superior results across all concentrations. The pre-treatment method, which introduces free amino groups to the cotton, enhances dye affinity and reduces environmental impact by enabling salt-free dyeing, thus lowering dye effluent (Chattopadhyay et al. 2007).

This research investigated the use of PVAmHCl as a modification agent for modal fabrics, aiming to enhance dyeability with reactive dyes without using electrolytes. The pad-batch process with PVAmHCl improved the ionic interaction between the dyes and the fabric, leading to better dye uptake and color strength compared to conventional methods with salt. The pretreated fabrics demonstrated comparable or better fastness properties (light, wash, rubbing) than those dyed with salt, and showed improved crease recovery and flexural rigidity. PVAmHCl treatment also reduced water retention and dissolved solids, contributing to environmental benefits and cost savings. Fastness properties were generally excellent, with some variations in perspiration fastness, indicating effective dye fixation and reduced pollution (Latif et al. 2019).

A study explored the modification of bleached cotton muslin fabrics using glycine to enhance color exhaustion with reactive dyes under no salt conditions. Both oxidized and unoxidized cotton were treated with glycine and dyed with reactive dyes in an acid bath, revealing that unoxidized cotton exhibited superior dye uptake and color strength (K/S value) compared to oxidized cotton. Glycine treatment improved dye fixation and overall color depth, especially when combined with a fixing agent. The research demonstrated that unoxidized cotton modified with glycine achieved higher dye uptake and better dye fastness compared to both untreated cotton and oxidized cotton. Optimization of dyeing conditions was essential for maximizing color strength. Overall, glycine modification provides an effective method for salt-free dyeing with high exhaustion reactive dyes, offering an eco-friendly alternative to conventional dyeing processes (Samanta et al. 2015).

Another study focused on the eco-friendly modification of cotton fibers to enhance their dyeability with anionic dyes without salt. Chitosan, a biopolymer, was used to introduce cationic groups to natural fibers, and plasma treatment with oxygen was employed to create functional sites for chitosan attachment. The modified cotton exhibited improved dye uptake and color strength with both direct and acid dyes, compared to untreated cotton. The presence of chitosan significantly reduced the need for electrolytes, enabling effective dyeing at lower salt concentrations. This method not only enhanced dyeability and wash fastness but also demonstrated the environmental benefits of using plasma and chitosan, as it eliminates the need for traditional salt and alkali treatments in the dyeing process (Gunasekar and Ponnusami 2015).

To address issues in conventional coloration with reactive dyes, a study explored a salt-free dyeing method using liquid paraffin. The optimal conditions for dyeing Reactive Black KN-B included a pickup rate of 120%, 30 g/L sodium carbonate, and fixation at 75 °C for 30 min. This method achieved nearly 100% dye exhaustion and about 90% fixation without adding salt (An et al. 2020).

9.6 Environmental Impacts of Salt-Free Dyeing

The textile coloration process significantly impacts the environment, affecting water, chemicals, energy, and air quality. This process contributes 17% to 20% of global water pollution due to high concentrations of chemicals like salts, dyes, and surfactants, which increase DS, BOD, and COD. These pollutants can be hazardous, causing harm to humans and ecosystems, and potentially entering the food chain through bioaccumulation. Additionally, the process consumes substantial energy, primarily from fossil fuels, exacerbating greenhouse gas emissions and contributing to climate change. The textile industry is responsible for 10% of global greenhouse gas emissions, and it is crucial to adopt cleaner technologies and improve wastewater treatment to reduce environmental impact and energy consumption (Lara et al. 2022; Benli 2024).

Cotton dyeing with reactive dyes, while producing vibrant colors and high wash fastness, results in significant wastewater pollution due to the use of inorganic salts and alkalis. To address this, two main approaches are considered: adopting alternative dyeing technologies and salt-free dyeing process. Sustainable options include developing bifunctional and low-salt dyes, using ultra-low liquor ratio machines, and employing biodegradable polymers or organic compounds in the modification of fabric. Although these methods may be costlier, they can reduce overall pollution and may become more feasible with stricter environmental regulations and increased treatment costs (Khatri et al. 2015).

A study investigated a salt-free dyeing method using liquid paraffin to tackle existing problems. This approach significantly cuts water and chemical usage, easing wastewater treatment and supporting eco-friendly practices (An et al. 2020). Cotton coloration results in significant, hard-to-treat wastewater due to low dye uptake and high electrolyte use. This research developed modified nanoparticles for a salt-free and that is near zero-discharge. The study examined how PPN composition affects color properties and compared the new method's results with traditional salt-free dyeing (Periyasamy et al. 2011).

9.7 Conclusion

Research has made significant strides in improving the color strength of cotton fabrics with reactive dyes by exploring various cationic compounds and methods to reduce salt use. Early innovations included the use of epoxy derivatives and cationic reagents like glycidyl trimethyl ammonium chloride and PECH-amine, which improved the affinity of cotton for dyes. Over time, the focus shifted to more advanced techniques, such as the use of CHPTAC and CHPTMAC, which enhance dye uptake and facilitate salt-free dyeing processes. Recent advancements have introduced even more sophisticated methods, including betaine modification, chitosan hybrids, and plasma treatments. These methods improve dye uptake and color depth and address environmental concerns by reducing the need for electrolytes and significantly reducing wastewater and chemical usage. Further innovations include the use of liquid paraffin for dyeing and the development of biodegradable polymers, which enhance sustainability. Despite these advancements, challenges remain, including the high costs of these new technologies and the consistent fastness properties. Overall, these developments represent a major shift towards eco-friendlier and more efficient textile dyeing practices, reflecting a growing commitment to reducing environmental impact and improving sustainability in the textile industry.

9.8 Declaration

The authors declare that they have no known competing financial interests or personal relationships that could have appeared to influence the work reported in this chapter. Generative AI and AI-assisted technologies have been used in the writing process to improve the readability and language of the work only. After using this tool, the authors carefully reviewed and revised the content to ensure accuracy and quality. The authors take full responsibility for the final version of this publication.

References

Ad W (2021) Effect and role of salt in cellulosic fabric dyeing. Adv Res Text Eng 6:2–6

An Y, Ma J, Zhu Z et al (2020) Study on a water-saving and salt-free reactive dyeing of cotton fabrics in non-aqueous medium of liquid paraffin system. J Text Inst 111:1538–1545. https://doi.org/10.1080/00405000.2020.1711552

Benli H (2024) Bio-mordants: a review. Environ Sci Pollut Res 31:20714–20771. https://doi.org/10.1007/s11356-024-32174-8

Buscio V, López-Grimau V, Álvarez MD, Gutiérrez-Bouzán C (2019) Reducing the environmental impact of textile industry by reusing residual salts and water: ECUVal system. Chem Eng J 373:161–170. https://doi.org/10.1016/j.cej.2019.04.146

Cai Y, Xiong J, Li L et al (2024) Green whitening of polyester fabric using fluorescent brightener OB-1 in a mixture of water and decamethylcyclopentasiloxane media. Heliyon 10:1–12. https://doi.org/10.1016/j.heliyon.2024.e36926

Chattopadhyay DP, Chavan RB, Sharma JK (2007) Salt-free reactive dyeing of cotton. Int J Cloth Sci Technol 19:99–108. https://doi.org/10.1108/09556220710725702

Correia J, Oliveira FR, de Cássia Siqueira Curto Valle R, Valle JAB (2021) Preparation of cationic cotton through reaction with different polyelectrolytes. Cellulose 28:11679–11700. https://doi.org/10.1007/s10570-021-04260-4

Depeursinge A, Racoceanu D, Iavindrasana J et al (2010) Fusing visual and clinical information for lung tissue classification in HRCT data. Artif Intell Med 172:235–243. https://doi.org/10.1016/j.artmed.2010.04.006

Gunasekar V, Ponnusami V (2015) Eco-friendly textile dyeing processes. In: Lichtfouse E, Schwarzbauer J, Robert D (eds) Hydrogen production and remediation of carbon and pollutants. Springer, Cham, pp 255–287

Hebeish A, El-Naggar ME, Tawfik S et al (2019) Hyperbranched polymer–silver nanohybrid induce super antibacterial activity and high performance to cotton fabric. Cellulose 26:3543–3555. https://doi.org/10.1007/s10570-019-02319-x

Hiwot TG, Zewde S, Mulugeta E (2024) Salt free reactive dyeing of cotton fabric by cationization with keratin obtained from sheep wool. Adv Res Text Eng 9:1–9

Islam T, Repon MR, Islam T et al (2023) Impact of textile dyes on health and ecosystem: a review of structure, causes, and potential solutions. Environ Sci Pollut Res 30:9207–9242. https://doi.org/10.1007/s11356-022-24398-3

Jabar JM, Ogunsade AF, Odusote YA, Yılmaz M (2023) Utilization of Nigerian mango (Mangifera indica L) leaves dye extract for silk fabric coloration: Influence of extraction technique, mordant and mordanting type on the fabric color attributes. Ind Crops Prod 193:116235. https://doi.org/10.1016/j.indcrop.2022.116235

Khatri A, Padhye R, White M (2013) The use of trisodium nitrilo triacetate in the pad-steam dyeing of cotton with reactive dyes. Color Technol 129:76–81. https://doi.org/10.1111/j.1478-4408.2012.00410.x

Khatri A, Peerzada MH, Mohsin M, White M (2015) A review on developments in dyeing cotton fabrics with reactive dyes for reducing effluent pollution. J Clean Prod 87:50–57. https://doi.org/10.1016/j.jclepro.2014.09.017

Lara L, Cabral I, Cunha J (2022) Ecological approaches to textile dyeing: a review. Sustainability 14:1–17. https://doi.org/10.3390/su14148353

Latif W, Basit A, Rehman A et al (2019) Study of mechanical and comfort properties of modal with cotton and regenerated fibers blended woven fabrics. J Nat Fibers 16:836–845. https://doi.org/10.1080/15440478.2018.1441084

Ma W, Meng M, Yan S, Zhang S (2016) Salt-free reactive dyeing of betaine-modified cationic cotton fabrics with enhanced dye fixation. Chinese J Chem Eng 24:175–179. https://doi.org/10.1016/j.cjche.2015.07.008

Mikucioniene D, García DM, Repon R et al (2024) Understanding and addressing the water footprint in the textile sector: a review. Autex Res J 24:1–27. https://doi.org/10.1515/aut

Momotaz F, Repon MR, Prapti US et al (2024) Dyeing performance and antimicrobial activity of cellulose-based biomaterials. Cellulose. https://doi.org/10.1007/s10570-024-06270-4

Mousa AA, Almetwally AA, Nassar SH et al (2024) A novel green approach for reactive printing of cotton/cellulosic regenerated blended fabrics using trisodium nitrilotriacetate. Sci Rep 14:1–17. https://doi.org/10.1038/s41598-024-75633-5

Nallathambi A, Venkateshwarapuram Rengaswami GD (2017) Industrial scale salt-free reactive dyeing of cationized cotton fabric with different reactive dye chemistry. Carbohydr Polym 174:137–145. https://doi.org/10.1016/j.carbpol.2017.06.045

Nallathambi A, Venkateshwarapuram Rengaswami GD (2016) Salt-free reactive dyeing of cotton hosiery fabrics by exhaust application of cationic agent. Carbohydr Polym 152:1–11. https://doi.org/10.1016/j.carbpol.2016.06.087

Periyasamy AP, Dhurai B, Thangamani K (2011) Salt-free dyeing—a new method of dyeing on lyocell/cotton blended fabrics with reactive dyes. Autex Res J 11:14–17. https://doi.org/10.1515/aut-2011-110103

Periyasamy AP, Dutta S, Tehrani-Bagha AR (2024) Solvent-assisted salt-free reactive dyeing of cotton fabric. Cellulose 31:7101–7118. https://doi.org/10.1007/s10570-024-05928-3

Pizzicato B, Pacifico S, Cayuela D et al (2023) Advancements in sustainable natural dyes for textile applications: a review. Molecules 28:1–22. https://doi.org/10.3390/molecules28165954

Repon MR, Islam T, Islam T et al (2023) Cleaner pathway for developing bioactive textile materials using natural dyes: a review. Environ Sci Pollut Res 1–31. https://doi.org/10.1007/s11356-023-26131-0

Roy Choudhury AK (2013) Green chemistry and the textile industry. Text Prog 45:3–143. https://doi.org/10.1080/00405167.2013.807601

Sadeghi-Kiakhani M, Safapour S (2018) Salt-free dyeing of cotton fabric modified with prepared chitosan-poly (propylene) imine dendrimer using direct dyes. Prog Color Color Coatings 11:21–32

Sadeghi-Kiakhani M, Safapour S (2015) Salt-free reactive dyeing of the cotton fabric modified with chitosan-poly(propylene imine) dendrimer hybrid. Fibers Polym 16:1075–1081. https://doi.org/10.1007/s12221-015-1075-9

Said B, Souad MR, Ahmed EH (2020) A review on classifications, recent synthesis and applications of textile dyes. Inorg Chem Commun 3:107891

Samanta AK, Kar TR, Mukhopadhyay A et al (2016) Eco-friendly salt-free reactive dyeing of cotton (muslin) fabric after cationization with amino acid from soya. Text Res J 86:2179–2192. https://doi.org/10.1177/0040517515621135

Samanta AK, Kar TR, Mukhopadhyay A et al (2015) Studies on dyeing process variables for salt free reactive dyeing of glycine modified cationized cotton muslin fabric. J Inst Eng Ser E 96:31–44. https://doi.org/10.1007/s40034-015-0062-4

Sun D, Zhang X, Du H et al (2017) Application of liquid organic salt to cotton dyeing process with reactive dyes. Fibers Polym 18:1969–1974. https://doi.org/10.1007/s12221-017-1241-3

Uddin F (2021) Environmental hazard in textile dyeing wastewater from local textile industry. Cellulose 28:10715–10739. https://doi.org/10.1007/s10570-021-04228-4

Vishnu G, Palanisamy S, Joseph K (2008) Assessment of fieldscale zero liquid discharge treatment systems for recovery of water and salt from textile effluents. J Clean Prod 16:1081–1089. https://doi.org/10.1016/j.jclepro.2007.06.005

Wang L, Ma W, Zhang S et al (2009) Preparation of cationic cotton with two-bath pad-bake process and its application in salt-free dyeing. Carbohydr Polym 78:602–608. https://doi.org/10.1016/j.carbpol.2009.05.022

Wolela AD (2019) An overview on surface modification of cotton using cationic reagents for salt-free or low salt dyeing. Curr Trends Fash Technol Text Eng 5:37–46. https://doi.org/10.19080/CTFTTE.2019.05.555655

Zhang Z, Wang H, Sun J, Guo K (2021) Cotton fabrics modified with Si@ hyperbranched poly(amidoamine): their salt-free dyeing properties and thermal behaviors. Cellulose 28:565–579. https://doi.org/10.1007/s10570-020-03540-9

Chapter 10
Micelle Dyeing: Green Chemistry for Modern Coloration

Shailendra Yadav, Kanha Singh Tiwari, Manoj Kumar Sharma, Pukhraj Meena, Nahid Usmani, and Laxmi Agnihotri

Abstract Dyes and pigments make the world beautiful and are used as biological agents, signaling agents, corrosion inhibitors, and dye-sensitized photocells. Nowadays, all industries, including the textile industry, face environmental issues regarding hazardous dyeing materials and hazardous methods for their production and applications. Green chemistry helps to solve these environmental issues by adopting cost-effective natural or less hazardous synthetic dyes with their green production and applications. In the textile industry, various reactive dyes obtain the desired color for cellulosic materials and fibers. The conventional fixation of reactive dyes on cellulose is low. Green and sustainable techniques such as plasma technology, nanotechnology, and UV technology have been adopted to minimize excessive water, chemical additives, and other hazardous auxiliaries used in dyeing and for minimum energy requirements. Besides benign environmental methods, fabric dyeing is also done using surfactants that form reactive dye carrier micelles in solvents. Surfactants have amphiphilic properties, and when they are present above the critical micelle concentration (CMC), they form micelles in the aqueous phase and in the non-aqueous phase, which can also retain a small amount of water and act as a reactive dye carrier for dyeing. The micelle dyeing method is greener when environment-benign surfactants and non-aqueous solvents are used for micelle formation with dye solutions.

Keywords Reactive dyes · Dye carrier micelle · Surfactant · Greener method · Green chemistry

S. Yadav (✉) · K. S. Tiwari · M. K. Sharma · N. Usmani · L. Agnihotri
Department of Chemistry, Basic Science, AKS University, Satna-485001, (M.P.), India
e-mail: syshailendra5@gmail.com

P. Meena
Department of Food Technology, AKS University, Satna-485001, (M.P.), India

S. Maiti et al. (eds.), *Sustainable Coloration Techniques in Textiles*,
https://doi.org/10.1007/978-981-96-4975-4_10

10.1 Introduction

Green chemistry is defined as the "design of chemical products and processes to reduce or eliminate the use and generation of hazardous substances" (Anastas et al. 1996; Anastas and Warner 2000). The concept and idea of Green chemistry first emerged in the early 1990s (Collins 2011). Since then, there has been widespread global acceptance of Green chemistry, leading to the establishment of several programs and governmental efforts. The early pioneering projects were mostly based in the United States, the United Kingdom, and Italy (Anastas 2003). These have played a crucial role in providing information for sustainable development (McDonough et al. 2003; Chauhan et al. 2021). Micelle dyeing represents a transformative approach in the field of textile coloration, merging the principles of green chemistry with innovative dyeing techniques. Traditional dyeing processes often rely on significant amounts of water, energy, and chemicals, leading to environmental concerns and sustainability challenges.

Micelle dyeing, however, employs surfactants to create micellar structures that facilitate dye uptake in a more environmentally friendly manner (Tang et al. 2024). At the core of micelle dyeing is the formation of micelles—aggregates of surfactant molecules that encapsulate dye particles. When introduced into a dye bath, these micelles significantly enhance the solubility of hydrophobic dyes in water. This process reduces the reliance on harsh solvents and minimizes water usage, aligning with the principles of green chemistry. By harnessing the natural affinity of micelles for dye molecules, the process promotes efficient color fixation on textile fibers, ensuring vibrant hues while minimizing environmental impact. One of the primary advantages of micelle dyeing is its reduction of water consumption. Conventional dyeing processes can require hundreds of liters of water per kilogram of fabric, contributing to water scarcity and pollution. In contrast, micelle dyeing often uses significantly less water, decreasing the environmental footprint. Additionally, the lower temperatures required for micelle dyeing translate into reduced energy consumption, making it a more sustainable option for textile manufacturers (Tang et al. 2023a). The use of biodegradable surfactants further enhances the eco-friendliness of this method. By selecting surfactants derived from renewable resources, the micelle dyeing process minimizes toxic waste and enhances the biodegradability of the final product. This is particularly crucial in an era where consumers are increasingly demanding transparency and sustainability in the fashion industry. Micelle dyeing is not only environmentally friendly but also versatile. It can be applied to various textile substrates, including natural fibers like cotton and synthetic fibers like polyester. The adaptability of this method allows for the dyeing of complex blends, which are common in modern textiles. Micelle dyeing can achieve a broad spectrum of colors and effects, catering to the diverse demands of the fashion industry. The technique also allows for the incorporation of functional properties into textiles. By integrating antimicrobial agents or UV protectants into the micellar system, manufacturers can produce fabrics that offer additional benefits beyond mere coloration. This multifunctionality aligns well with current trends in the textile market, where consumers seek not only aesthetics but also performance (Tang et al. 2023b). Despite its numerous advantages, micelle dyeing is not without challenges. The need for specific surfactant formulations can complicate the scalability of the process. The cost of some biodegradable surfactants may hinder widespread adoption among manufacturers focused on cost-effectiveness. Addressing these challenges will be crucial for the future success of micelle dyeing. Research into alternative surfactant materials and improved formulations is ongoing, with a focus on enhancing dye uptake and fixation efficiency while maintaining environmental benefits. Collaborations between academia and industry could drive innovation, developing more cost-effective and sustainable micelle dyeing solutions (Tang et al. 2023c). Micelle dyeing embodies the principles of green chemistry, offering a sustainable and efficient alternative to traditional textile dyeing methods. By reducing water and energy consumption, utilizing biodegradable materials, and providing versatility in application, micelle dyeing has the potential to significantly impact the textile industry. As the demand for sustainable practices grows, this innovative approach may pave the way for a new era of environmentally responsible coloration in textiles. The future of micelle dyeing looks promising, with ongoing research and development poised to enhance its efficacy and accessibility, ultimately benefiting both the industry and the planet.

10.2 Green Chemistry and Its Principles

Green chemistry revolves around twelve principles that seek to minimize or eliminate hazardous substances from the synthesis, production, and execution of chemical products. This, in turn, reduces or eliminates the use of materials that pose risks to human health and the environment (Fig. 10.1). The notion of ecological chemistry was first developed in the early 1990s. The 12 principles provide design guidelines that assist chemists in effectively pursuing the deliberate objective of sustainability. Green chemistry is defined by the meticulous organization of chemical synthesis and molecular design to minimize negative outcomes. Effective design may generate synergistic outcomes rather than settling for mere trade-offs (Anastas and Eghbali 2010).

The strategy of green chemistry focuses on designing molecules, reactions, materials, and processes that prioritize environmental and human health safety. Spanning various fields like organic, inorganic, and life sciences, green chemistry aims to achieve environmental conservation and economic benefits through trends like catalysis, bio-catalysis, and the use of safe alternatives such as water-based solutions and renewable raw materials (Ivanković et al. 2017). This chapter describes how green chemistry lessens the harm chemical processes and technology do to the environment.

Fig. 10.1 Principles of green chemistry

The modern imperative demands balancing resource utilization, economic growth, and environmental preservation. Over the past few decades, this heightened awareness underscores the urgent need for environmentally sustainable practices. Consequently, there's a notable shift towards developing and adopting eco-friendly technologies. The aim is to protect ecosystems from harmful substances, with the chemical community prioritizing green chemistry principles to devise safer processes and products (Fig. 10.2).

10.3 Enhancing Sustainability in Modern Dyeing Processes Through Green Chemistry

Green chemistry makes modern dyeing processes more sustainable than older methods by reducing environmental impact and resource use while enhancing safety. Here are several key ways green chemistry principles contribute to more sustainable dyeing processes:

Fig. 10.2 Key Reductions Achieved by Green Chemistry

- **Reduced Toxicity**: Traditional dyeing processes often rely on toxic chemicals and heavy metals that can harm the environment and pose health risks to workers and consumers. Green chemistry promotes the use of safer, less toxic dyes and chemicals, reducing the risk of environmental contamination and improving worker safety (Roy Choudhury 2013).
- **Reduced Water Use**: Traditional dyeing methods typically require large volumes of water, leading to significant wastewater generation. Green chemistry encourages dyeing techniques that use less water or employ water recycling systems, reducing water consumption and the environmental impact associated with dyeing (Sharma et al. 2015).
- **Reduced Energy Consumption**: Old dyeing processes often require high temperatures and long processing times, resulting in considerable energy use. It promotes energy-efficient processes, such as low-temperature dyeing, reducing the carbon footprint of dyeing operations (Hasanbeigi and Price 2015).
- **Waste Minimization**: Conventional dyeing can produce significant waste, including excess dyes, salts, and other chemicals. Green chemistry emphasizes waste prevention, encouraging processes that use fewer chemicals and produce less waste. Techniques like sustainable micelle dyeing or closed-loop systems help minimize waste (Yousaf and Aqsa 2023).
- **Biodegradability**: Green chemistry supports the development of dyes and chemicals that are biodegradable, ensuring that any waste produced is less harmful to the environment. This contrasts with older dyeing processes, where non-biodegradable chemicals could persist in the environment (Islam et al. 2022).

- **Innovative Dyeing Techniques**: Green chemistry encourages innovative dyeing techniques that are more environmentally friendly. Techniques like waterless dyeing, digital printing, or supercritical fluid dyeing use fewer resources and produce less waste than traditional dyeing methods (Maiti et al. 2022). Green chemistry advocates for using renewable resources instead of non-renewable chemicals derived from petrochemicals. This can include using plant-based dyes or biodegradable chemicals, which are more sustainable and reduce reliance on fossil fuels.

10.4 Greener Approaches for Modern Coloration

Embracing more environmentally friendly ways is crucial in mitigating the environmental effects of textile dyeing processes, especially in current coloration techniques like micelle dyeing. One approach is using sustainable raw materials and environmentally friendly solvents, which decreases reliance on non-renewable energy sources and reduces harmful waste production (Rana and Rana 2014). Researchers are now investigating the use of natural dyes obtained from plants (Saxena and Raja 2014), insects (Shahid and Mohammad 2013), and microorganisms (Tuli et al. 2015). These dyes not only provide a wide range of brilliant colors but also help to decrease the negative impact on the environment caused by the manufacture of synthetic dyes. Furthermore, researchers are now exploring eco-friendly solvents, including $scCO_2$ (supercritical carbon dioxide) and water-based dyeing systems as substitutes for conventional organic solvents. These alternatives aim to improve the sustainability of the dyeing process (Agrawal 2015). Optimizing dyeing processes to minimize water and energy consumption is integral to green chemistry in modern coloration.

Advanced dyeing technology with precise controls and automation minimizes dye and energy wastage. Techniques like micelle dyeing enhance dye penetration and fixation, reducing water usage compared to traditional methods. Process intensification methods such as ultrasonic and microwave-assisted dyeing decrease dyeing durations and energy requirements (Khalil et al. 2023). Adopting environmentally friendly dye carriers represents a significant stride in sustainable dyeing. Biodegradable polymers like cyclodextrins and enzyme-based treatments are eco-friendly alternatives, promoting dye adhesion to fibers while minimizing environmental harm. These developments align with green chemistry principles, advocating for non-toxic materials and reducing harmful chemical discharge. Studies by Jiang et al. (2022) and Mudhoo et al. (2019) underscore the importance of green chemistry in textile dyeing, emphasizing eco-friendly methods and wastewater treatment. These efforts reflect the textile industry's commitment to sustainability across the dyeing process.

10.5 Overview of Dye Surfactant Interaction

The interactions between dyes and surfactants are a crucial feature of contemporary coloring methods. These interactions affect the solubility of dyes, their adsorption onto surfaces, and the overall efficiency of the dyeing process. Surfactants are molecules that have both hydrophilic and hydrophobic parts. They are important for helping colors spread and stick to textile fibers (Berg 2002). Based on the electronic spectra, Fig. 10.3 shows several possible configurations that affect the localization of the dye molecules in the ternary system. The left side of the diagram illustrates the identical charged dye/surfactant pairings, whereas the right side displays both polarized dye/surfactant pairs and dye/nonionic surfactant pairs. In systems where water is the dominating phase ($R > 1$), micelles of the direct type (O/W) form on the top side. In systems where oil is the dominating phase ($R < 1$), reversible micelles in the form of water-in-oil (W/O) emulsions appear at the lower end. If the dye and surfactant have the same charge, and the ratio of the amount of surfactant to the dye concentration is less than 1, then both the dye and surfactant are present in their molecular form. The dye is located inside the chain of hydrocarbons of the surfactants and the hydrophilic head groups, especially within the micelle shell (Petcu et al. 2016). Suppose a nonionic surfactant is used with a dye/surfactant that has an opposite charge, and the value of R is less than 1. In that case, the dye is located inside the oxyethylenic head groups near the center of the micelle core. In phase systems where water is dominant ($R > 1$), the dye is located between the two polar heads but on the outermost layer of the micelle shells, independent of the specific dye/surfactant combinations (Petcu et al. 2016).

Several parameters, such as the composition of the dye, the concentration of the surfactant, the pH of the solution, the temperature, and the characteristics of the substrate, influence the relationship between dyes and surfactants. Surfactants can dissolve hydrophobic dyes by creating micellar aggregates (Rathod et al. 2021). In these aggregates, the hydrophobic dye molecules are enclosed inside the hydrophobic center of the micelles, while the hydrophilic bonds interact with the surrounding water. The solubilization process improves the dye's dispersibility and inhibits its

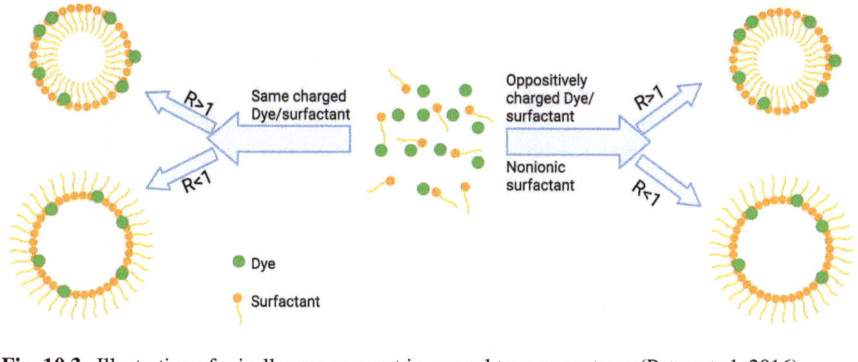

Fig. 10.3 Illustration of micelle arrangement in several ternary systems (Petcu et al. 2016)

aggregation, resulting in better penetration of the color into textile fibers during the dyeing process. Surfactant dye associations play a crucial role in the dyeing process and dye separation techniques like micellar-enhanced ultrafiltration or cloud point extraction (MEUF & CPE) (Yadav et al. 2021, 2023; Singh et al. 2021; Purkait et al. 2004a, b). The surfactant selection is critical in dye separation using MEUF since it relies on the reciprocal relationship between dye and surfactant particles. Various methods were used to describe the interactions qualitatively and quantitatively between dyes and surfactants, including potentiometry (Simončič and Špan 1998; Simončič and Kovac 1999), conductometry (Bračko and Špan 2000; Khan and Sarwar 2006), and ion-selective electrodes (Ghoreishi et al. 2007; Kuiper et al. 2001). Spectrophotometry is the most often used technique for studying dye-surfactant interactions (Yang 2004; Behera et al. 2005; Jana and Rajavenii 2004). When surfactants are present, the electronic absorption spectra of many dyes may show additional bands. The extent of the shift depends on the strength of the interaction between the dye and surfactant. The variations in the visual absorption spectra of dye, caused by the presence of surfactant at varying concentrations, are due to the balance between individual surfactant molecules, clusters of surfactant molecules (micelles), aggregates of dye molecules, complexes formed between dye and surfactant before micelle formation, and dye particles that are integrated into the micelles (Dutta and Bhat 1996; Pereira and Gehlen 2005).

Research on the interactions between dyes and surfactants may benefit chemical studies, industrial applications, and dye separation procedures (Purkait et al. 2004a, b; Bilski et al. 1997). The interactions between dyes and surfactants mostly depend on the chemical composition of the molecules (Petcu et al. 2016; Kartal and Akbaş 2005). Hydrophobic interactions, electrostatic interactions, hydrogen bonding, π-stacking, and van der Waals forces are typical examples of intermolecular forces that mostly regulate the interactions between dye molecules and surfactant aggregates (Naeem et al. 2000; Dezhampanah and Firouzi 2012; Karukstis et al. 2010).

Figure 10.4 depicts the phenomenon of "stimuli aggregation," illustrating that various causes may result in the grouping of molecules or particles upon exposure to varied external stimuli. This is especially pertinent in scientific domains like biochemistry and materials science, where comprehending aggregation is vital for various uses. The figure illustrates the concept of "Stimuli aggregation" via a visual representation of four particles converging, signifying the aggregation process. Temperature may influence the speed of molecular interactions, which may result in the formation of aggregates when there are temperature changes. pH is a measure of acidity or alkalinity that may cause alterations in molecular bonding and encourage the formation of aggregates. Light may induce photochemical processes that trigger particles to cluster by specifically targeting certain wavelengths. Hydrogen bonding is a phenomenon where molecules having polar covalent connections make bonds with hydrogen atoms, leading to the aggregation of these molecules. Hydrophobic interactions refer to the phenomenon where non-polar compounds tend to group to avoid contact with polar solvents such as water. This results in the formation of aggregates. Electrostatic interactions refer to the phenomenon where particles with opposing charges are attracted to one another, leading to their aggregation. Host–guest interactions refer to certain molecular interactions in which one molecule functions as a host for another, resulting in the creation and aggregation of complexes. Dipole–dipole interactions demonstrate the attractive forces between polar molecules that possess permanent dipoles, hence promoting the process of aggregation. The concept of magnetic forces demonstrates how magnetic particles may be persuaded to come together and form aggregates when subjected to external magnetic fields. An electric field represents how charged or polarizable particles behave and come together in reaction to electric fields. The term "Magnetic Field" refers to the overall magnetic forces that affect the behavior and grouping of particles. Several crucial inputs that impact aggregation revolve around this basic principle. The graphic presents various environmental and chemical elements, encompassing the following: During the pre-micellar phase, surfactant molecules and dye molecules form complexes by ion interaction. During the post-micellar phase, there is a higher probability for the dye molecules to accumulate on the micelles' surface. While numerous investigations have explored dyed-surfactant systems and the dynamics of surfactant-dye blends, the exact mechanism governing the interactions between dyes and surfactants within surface films and micelles remains elusive (Ghosh 2001; UlHaq et al. 2014; Chakraborty et al. 2011; Muntaha and Khan 2014; Kert and Simončič 2008). These findings may be used in the solubilization (Mathis and Zana 2002) and photography industries (Simončič and Kert 2002), where enhancing the dye concentration inside the micelles is essential. In addition, they may be used in dye-sensitized solar cells (DSSCs) (Guo et al. 2013), nonlinear optical materials (NLOMs) (Cyprych et al. 2015), and photosensitizers (Prakash et al. 2023), to position the chromophore inside the hydrophilic outer layer of the surfactant micelles. Furthermore, the use of water-in-oil systems is prevalent in eliminating colors from wastewater by applying complexes generated by dyes and surfactants (Fleancu et al. 2013).

The interactions among three unique surfactants and cationic and anionic dyes since these interactions are environmentally and economically important for the removal of colors from wastewater. The two common dyes such as Methyl Orange (MO) and Crystal Violet are frequently found in textile industry effluents. It examined three types of surfactants (anionic, nonionic, and cationic) to comprehensively understand dye-surfactant interactions, encompassing both similar and opposing charges (Shah et al. 2013; Rahimi et al. 2010; Alqaragully 2014).In a previous study, the researchers utilized the Ostwald stalagmometry method to analyze surfactant and dye-surfactant aggregations and UV–Vis spectroscopy to observe dye-surfactant interactions (Muhammad and Khan 2018). Another study calculated the standard free energy change for dye transfer into the micellar phase, equilibrium constants, and partition coefficients between surfactants and water. The study further explored how ethyl acetate (EtAc) affects dye positioning within micelles in a ternary water/surfactant/oil system, a technology known for effectively removing various contaminants (Fleancu et al. 2013).

The presence of oil was found to influence the microenvironment of dyes within micelles (Mihaly et al. 2014). Much research has examined the use of surfactants in dyeing procedures involving diverse types of fabrics and dye categorizations. Tang

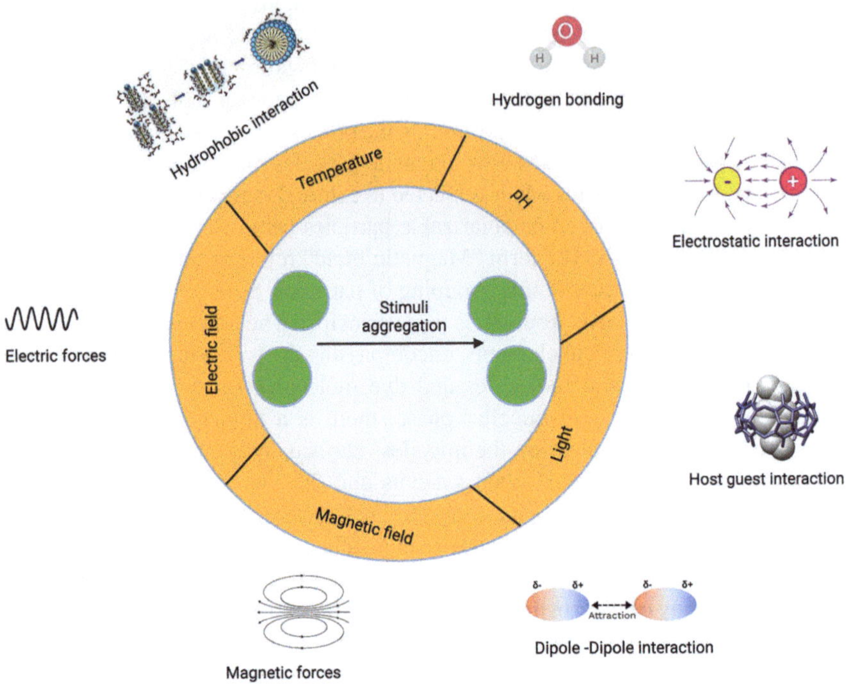

Fig. 10.4 Stimuli aggregation

et al. (Tang et al. 2023a, b, c) investigated the influence of several surfactants on the dyeing procedure of reactive dyes on cotton fabrics. Their research emphasized the need to select a suitable surfactant to improve dye fixation and color creation. The research conducted by Rashid et al. (2020) and Bezerra et al. (2020) examined how the structure and concentration of surfactants affect the dyeing behavior of acid and immersion dyes on synthetic fibers. These tests highlighted the complex interaction between colors and surfactants. The relationship between dyes and surfactants is an intricate phenomenon that substantially impacts the effectiveness and longevity of coloring processes in the textile industry. To expedite the development of dyeing methods that are both ecologically sustainable and highly effective, researchers may augment their comprehension of the underlying principles governing the interactions between dyes and surfactants. In addition, they may enhance the efficiency of surfactant use (Rashid et al. 2020).

10.6 Micelle Formation with Different Dyes and Dyeing

Micelle production is essential in dyeing processes since it directly affects the dispersion, solubility, and adsorption of colors on textile surfaces. The interplay between diverse dyes and surfactants results in the creation of micelles and their consequences for dyeing applications on varied surfaces Table 10.1.

Surfactants are molecules that have both hydrophilic and hydrophobic parts. When they are in water, they arrange themselves into micelles until the concentration reaches a particular point called the critical micelle concentration (CMC). Micelle formation occurs by the aggregation of surfactant molecules, where the hydrophobic tails are directed towards the center of the micelle while the hydrophilic heads interact with the surrounding solvent molecules. The micellar structure creates a small environment that repels water and is suitable for dissolving hydrophobic compounds like dyes, making them more evenly spread in water-based solution (Ghosh et al. 2020). The creation of micelles resulting from the interaction between dyes and surfactants is affected by several parameters, including the structure of the dye, the kind of surfactant, the concentration, and the conditions of the solution. Hydrophobic

Table 10.1 Different types of dyes and micelle formation

Dye type	Dye structure	Micelle formation	Dyeing process	References
Synthetic dyes	Azo dyes, Anthraquinone	Dyes partition into micelles formed by surfactants in aqueous solution	Conventional dyeing methods involve immersion of textile substrates into dye solutions followed by fixation	Banchero (2013)
Natural dyes	Indigo, Cochineal	Interaction with micelles varies based on dye polarity and hydrophobicity	Utilization of eco-friendly dyeing techniques like vat dyeing or natural dyeing, often involving fermentation processes	Alsantali et al. (2022), Yadav et al. (2022)
Reactive dyes	Reactive groups for covalent bonding with substrates	Covalent attachment to reactive groups on the surface of micelles	Application of reactive dyes in a dye bath with alkali and fixation agents, followed by washing and rinsing steps	Faisal (2013)
Acid dyes	Anionic dyes with acidic groups	Interaction driven by electrostatic forces with oppositely charged micelles	Acid dyeing processes involving acidic pH conditions for optimum dye adsorption onto textile substrates	Rashid et al. (2020)
Disperse dyes	Hydrophobic dyes for non-polar substrates	Solubilization within the hydrophobic core of micelles	High-temperature dyeing methods where dyes are dispersed in water with the aid of dispersing agents	Tehrani-Bagha and Holmberg (2013)

dyes, which have low solubility in water, may be dissolved in the hydrophobic center of surfactant micelles. This increases their ability to dissolve in water and makes it easier to spread out. Micelle-mediated solubilization of dyes avoids their aggregation and precipitation, facilitating equal distribution and penetration into textile fibers during the dyeing process (Koh 2011). Various studies have examined the function of micelle production in dyeing applications involving various dye classes and textile surfaces. Wang et al. (2018) researched the solubilization behavior of dispersion dyes in nonionic surfactant micelles for polyester dyeing. Their study emphasized the significance of micellar solubilization in improving dye absorption and color intensity. Two separate studies conducted by Liu et al. (2019) and Sharma et al. (2020) examined how cationic surfactants affect the process of the dissolution and adsorption of reactive dyes for cotton and wool dyeing. These studies emphasized the importance of micelle formation in enhancing dye fixation and improving fastness properties. Micelle production is an important process that explains how dyes dissolve and disperse in water, making them more effective in coloring textiles. Researchers can enhance dyeing processes by comprehending the elements influencing dye-surfactant interactions and micelle generation (Sarkar 2022). This allows them to obtain desired colored outputs while reducing the environmental impact.

10.7 Current Scenario of Micelle Dyeing and Challenges

Micelle dyeing is a potential method for coloring textiles that offers many benefits, including lower water and energy use, increased dye effectiveness, and better color durability (Khattab et al. 2020). Micelle dyeing is a process that utilizes surfactant micelles to dissolve and disperse colors in water-based solutions, making it easier for the dyes to be absorbed into textile materials. This procedure optimizes dye absorption and retention while reducing dye loss and environmental harm, which aligns with sustainability objectives in the textile sector. Nevertheless, micelle dyeing has many obstacles that restrict its extensive implementation and refinement. An important obstacle is the intricate interaction among solution conditions, textile substrates, and dye-surfactant interactions, which may impact the effectiveness of dyeing and the results of coloring. Obtaining consistent and thorough dye dispersion and penetration into textile fibers is still difficult, especially for hydrophobic or bulky dyes, since they may have low solubility or difficulty spreading inside micelles (Banchero 2013). The choice of surfactants and dye carriers plays a crucial role in micelle dyeing procedures. This is because various kinds and concentrations of surfactants may impact micelles' formation, stability, and capability to dissolve dyes. It is crucial to maintain an adequate surfactant concentration to facilitate the creation of micelles without creating excessive foam or deposition of surfactant on textile surfaces. This is necessary to provide efficient and high-quality dyeing. It is crucial to meticulously evaluate the compatibility of surfactants, dyes, and textile substrates to avoid any unwanted interactions or negative impacts on the quality of the fabric (Singh 2021). Transitioning from small-scale laboratory studies to large-scale commercial production of micelle dyeing methods poses significant logistical and technological obstacles. To achieve consistent dyeing outcomes on a wide scale while ensuring efficient processes and cost-effectiveness, developing inventive equipment designs, implementing effective process control systems, and optimizing dye compositions is necessary. To tackle these issues, textile scientists, chemists, engineers, and industry stakeholders must work together across disciplines. Their goal should be to create strong micelle dyeing procedures and production solutions that can be easily scaled up. To summarize, while micelle dyeing shows great promise for environmentally friendly textile coloring, it is crucial to tackle issues concerning the interactions between dyes and surfactants, optimize the process, and ensure its implementation on a larger scale to fully reap its advantages in industrial settings.

10.8 Significance

Sustainable micelle dyeing represents a significant advancement in textile dyeing technology, aligning with green chemistry principles by offering an eco-friendly alternative to conventional dyeing methods. Traditional dyeing often requires large volumes of water, hazardous chemicals, and high energy consumption, contributing to water pollution, toxic waste, and excessive carbon emissions. In contrast, micelle dyeing utilizes the self-assembling nature of micelles, spherical aggregates of surfactant molecules to encapsulate dyes, facilitating efficient dye transfer to fabrics while requiring less water and lower temperatures. This approach significantly reduces the ecological footprint of the textile industry, decreasing wastewater production, toxic effluent discharge, and energy use. Moreover, micelle dyeing can improve the quality and durability of dyed textiles, as the micelles ensure even dye distribution, resulting in vibrant colors and consistent patterns. This method's efficiency reduces dye wastage, aligning with the principles of waste prevention and energy efficiency.

10.9 Conclusions and Future Perspectives

Micelle dyeing holds promise as a sustainable textile coloring method, offering reduced water usage, improved dye effectiveness, and potential compatibility with eco-friendly dye carriers. However, challenges still need to be addressed, including limited access to environmentally friendly surfactants and the need for scalability and economic viability. Future research should focus on developing or modifying surfactants for improved environmental performance and enhancing the efficiency of dyeing processes on an industrial scale. Integrating modern technologies like nanotechnology and green chemistry concepts could further enhance the sustainability of micelle dyeing, contributing to a greener future for the textile industry.

Declaration

The authors declare that they have no known competing financial interests or personal relationships that could have appeared to influence the work reported in this chapter. Generative AI and AI-assisted technologies have been used in the writing process to improve the readability and language of the work only. After using this tool, the authors carefully reviewed and revised the content to ensure accuracy and quality. The authors take full responsibility for the final version of this publication.

References

Agrawal BJ (2015) Supercritical carbon-dioxide assisted dyeing of textiles: an environmentally benign waterless dyeing process. Int J Innov Res Creative Technol 1(2):201–205

Alqaragully MB (2014) Removal of textile dyes (maxilon blue and methyl orange) by date stones activated carbon. Int J Adv Res Chem Sci 1(1):48–59

Alsantali RI, Raja QA, Alzahrani AY, Sadiq A, Naeem N, Mughal EU, Al-Rooqi MM, El Guesmi N, Moussa Z, Ahmed SA (2022). Miscellaneous azo dyes: a comprehensive review on recent advancements in biological and industrial applications. Dyes Pigment 199:110050

Anastas PT (2003) Meeting the challenges to sustainability through green chemistry. Green Chem 5(2):G29–G34

Anastas P, Eghbali N (2010) Green chemistry: principles and practice. Chem Soc Rev 39(1):301–312

Anastas PT, Warner JC (2000) Green chemistry: theory and practice. Oxford University Press

Anastas PT, Williamson TC, (eds) (1996) Green chemistry: designing chemistry for the environment. Am Chem Soc

Banchero M (2013) Supercritical fluid dyeing of synthetic and natural textiles–a review. Color Technol 129(1):2–17

Behera PK, Mohapatra S, Patel S, Mishra BK (2005) Dye–surfactant interaction: solubilization of styryl pyridinium dyes of varying alkyl chain in alfa-olefinic sulfonate and linear alkyl benzene sulfonate solutions. J Photochem Photobiol A 169(3):253–260

Berg JC (2002) The role of surfactants. In: Textile science and technology, vol 13. Elsevier, pp 149–198

Bezerra FM, Lis MJ, Firmino HB, Dias da Silva JG, Curto Valle RDCS, Borges Valle JA, Scacchetti FA, Tessaro AL (2020) The role of β-cyclodextrin in the textile industry. Molecules 25(16):3624

Bilski P, Holt RN, Chignell CF (1997) Premicellar aggregates of Rose Bengal with cationic and zwitterionic surfactants. J Photochem Photobiol, A 110(1):67–74

Bračko S, Špan J (2000) Conductometric investigation of dye–surfactant ion pair formation in aqueous solution. Dyes Pigment 45(2):97–102

Chakraborty T, Chakraborty I, Ghosh S (2011) The methods of determination of critical micellar concentrations of the amphiphilic systems in aqueous medium. Arab J Chem 4(3):265–270

Chauhan DS, Yadav S, Quraishi MA (2021) Natural products as environmentally safe and green approach to combat Covid-19. Current Res Green Sustain Chem 4(100114):2666–865

Collins TJ (2011) The green evolution. Letters to a young chemist. John Wiley and Sons, Hobocken, NJ, pp 77–93

Cyprych K, Sznitko L, Morawski O, Miniewicz A, Rau I, Mysliwiec J (2015) Spontaneous crystalization and aggregation of DCNP pyrazoline-based organic dye as a way to tailor random lasers. J Phys D Appl Phys 48(19):195101

Dezhampanah H, Firouzi R (2012) Thermodynamic investigation of the interaction between anionic dye and cationic surfactant in aqueous solution. Int J Res Phys Chem 2(4):45–48

Dutta RK, Bhat SN (1996) Interaction of phenazinium dyes and methyl orange with micelles of various charge types. Colloids Surf A 106(2–3):127–134

Faisal S (2013) Synthesis of multifunctional reactive dyes and their application onto wool fabric by inkjet printing. PhD Thesis, University of Leeds

Fleancu MC, Olteanu NL, Rogozea AE, Crisciu AV, Pincovschi I, Mihaly M (2013) Physical-chemical parameters promoting phase changes in nonionic environmental-friendly microemulsions. Fluid Phase Equilib 337:18–25

Ghoreishi SM, Behpour M, Farsani AG (2007) Study of interaction between a cationic surfactant and two anionic azo dyes by ion-selective electrode technique and spectrophotometry. Dyes Pigment 74(3):585–589

Ghosh S (2001) Surface chemical and micellar properties of binary and ternary surfactant mixtures (cetyl pyridinium chloride, Tween-40, and Brij-56) in an aqueous medium. J Colloid Interface Sci 244(1):128–138

Ghosh S, Ray A, Pramanik N (2020) Self-assembly of surfactants: an overview on general aspects of amphiphiles. Biophys Chem 265:106429

Guo X, Yi P, Yang Y, Cui J, Xiao S, Wang W (2013) Effects of surfactants on agarose-based magnetic polymer electrolyte for dye-sensitized solar cells. Electrochim Acta 90:524–529

Hasanbeigi A, Price L (2015) A technical review of emerging technologies for energy and water efficiency and pollution reduction in the textile industry. J Clean Prod 95:30–44

Islam MT, Islam T, Islam T, Repon MR (2022) Synthetic dyes for textile colouration: process, factors and environmental impact. Textile Leather Rev 5:327–373

Ivanković A, Dronjić A, Bevanda AM, Talić S (2017) Review of 12 principles of green chemistry in practice. Int J Sustain Green Energy 6(3):39–48

Jana AK, Rajavenii S (2004) Studies on the molecular interaction of phenazine dyes with Triton X-100. Spectrochim Acta Part A Mol Biomol Spectrosc 60(8–9):2093–2097

Jiang H, Guo R, Mia R, Zhang H, Lü S, Yang F, Mahmud S, Liu H (2022) Eco-friendly dyeing and finishing of organic cotton fabric using natural dye (gardenia yellow) reduced-stabilized nanosilver: full factorial design. Cellulose 29(4):2663–2679

Kartal Ç, Akbaş H (2005) Study on the interaction of anionic dye–nonionic surfactants in a mixture of anionic and nonionic surfactants by absorption spectroscopy. Dyes Pigment 65(3):191–195

Karukstis KK, Litz JP, Garber MB, Angell LM, Korir GK (2010) A spectral approach to determine location and orientation of azo dyes within surfactant aggregates. Spectrochim Acta Part A Mol Biomol Spectrosc 75(4):1354–1361

Kert M, Simončič B (2008) The influence of nonionic surfactant structure on the thermodynamics of anionic dye–cationic surfactant interactions in ternary mixtures. Dyes Pigment 79(1):59–68

Khalil E, Sarkar J, Rahman MM, Shamsuzzaman M, Das D (2023) Advanced technology in textile dyeing. In: Advanced technology in textiles: fibre to apparel. Springer Nature Singapore, Singapore, pp 97–138

Khan MN, Sarwar A (2006) Study of dye–surfactant interaction: aggregation and dissolution of yellowish in N-dodecyl pyridinum chloride. Fluid Phase Equilib 239(2):166–171

Khattab TA, Abdelrahman MS, Rehan M (2020) Textile dyeing industry: environmental impacts and remediation. Environ Sci Pollut Res 27(4):3803–3818

Koh J (2011) Dyeing with disperse dyes. Textile Dyeing 10:195–220

Kuiper JM, Buwalda RT, Hulst R, Engberts JB (2001) Novel pyridinium surfactants with unsaturated alkyl chains: aggregation behavior and interactions with methyl orange in aqueous solution. Langmuir 17(17):5216–5224

Liu Y, Luo J, Zhang H, Liu W, Yu H (2019) Study on solubilization of reactive dyes in cationic surfactant micelles and their dyeing properties on cotton fabric. J Surfactants Deterg 22(2):397–404

Maiti S, Panda A, Madiwale P, Adivarekar RV (2022) Textile coloration technologies: today and tomorrow. In: Emerging technologies for textile coloration. CRC Press, pp 1–12

Mathis A, Zana R (2002) Study of complexes between lysozyme and sodium alkyl sulfates in the solid state. Colloid Polym Sci 280:968–971

McDonough W, Braungart M, Anastas PT, Zimmerman JB (2003) Peer reviewed: applying the principles of green engineering to cradle-to-cradle design. Environ Sci Technol 37(23):434A-441A

Mihaly M, Lacatusu I, Olteanu NL, Meghea A (2014) A systematic methodology to design silica templates: Silica microemulsion formulation and nanodroplet type and size estimation. ComptesRendus. Chimie 17(4):342–351

Mudhoo A, Gautam RK, Ncibi MC, Zhao F, Garg VK, Sillanpää M (2019) Green synthesis, activation and functionalization of adsorbents for dye sequestration. Environ Chem Lett 17:157–193

Muhammad MT, Khan MN (2018) Oppositely charged dye surfactant interactions: extent and selectivity of ion pair formation. J Mol Liq 266:591–596

Muntaha ST, Khan MN (2014) Study of changes in conductivity and spectral behaviour before and after micelle formation in the dye-surfactant system. J Mol Liq 197:191–196

Naeem K, Shah SS, Shah SW, Laghari GM (2000) Solubilization of cationic hemicyaninedyesin anionic surfactant micelles: a partitioning study. Monatshefte Für Chemie/Chem Monthly 131:761–767

Pereira RV, Gehlen MH (2005) Fluorescence of acridinic dyes in anionic surfactant solution. Spectrochim Acta Part A Mol Biomol Spectrosc 61(13–14):2926–2932

Petcu AR, Rogozea EA, Lazar CA, Olteanu NL, Meghea A, Mihaly M (2016) Specific interactions within micelle microenvironment in different charged dye/surfactant systems. Arab J Chem 9(1):9–17

Prakash J, Kumar P, Saxena N, Pu Z, Chen Z, Tyagi A, Zhang G, Sun S (2023) CdS-based 3D nano/micro-architectures: formation mechanism, tailoring of visible light activities and emerging applications in photocatalytic H2 production, CO2 reduction and organic pollutant degradation. J Mater Chem A 11:10015–10064

Purkait MK, DasGupta S, De S (2004a) Removal of dye from wastewater using micellar-enhanced ultrafiltration and recovery of surfactant. Sep Purif Technol 37(1):81–92

Purkait MK, DasGupta S, De S (2004b) Resistance in series model for micellar enhanced ultrafiltration of eosin dye. J Colloid Interface Sci 270(2):496–506

Rahimi R, Kerdari H, Rabbani M (2010) Adsorptive removal of crystal violet (CV), a carcinogenic textile dye, from aqueous solution by conducting polyaniline/hollow manganese ferrite nanocomposites. In: Proceedings of the ECSOC-14: the 14th international electronic conference on synthetic organic chemistry, Basel, Switzerland, pp 1–30

Rana KK, Rana S (2014) Fundamentals, representative applications and future perspectives of green chemistry: a short review. Open Access Library J 1(5):1–16

Rashid TU, Kabir SF, Biswas MC, Bhuiyan MR (2020) Sustainable wastewater treatment via dye–surfactant interaction: a critical review. Ind Eng Chem Res 59(21):9719–9745

Rathod S, Joshi A, Ray D, Aswal VK, Verma G, Bahadur P, Tiwari S (2021) Changes in aggregation properties of TPGS micelles in the presence of sodium cholate. Colloids Surf A 610:125938

Roy Choudhury AK (2013) Green chemistry and the textile industry. Text Prog 45(1):3–143

Sarkar B (2022) Micellar enhanced ultrafiltration in the treatment of dye wastewater: fundamentals, state-of-the-art and future perspectives. Groundw Sustain Dev 17:100730

Saxena S, Raja ASM (2014) Natural dyes: sources, chemistry, application and sustainability issues. In: Roadmap to sustainable textiles and clothing: eco-friendly raw materials, technologies, and processing methods. Springer Singapore, Singapore, pp 37–80

Shah MP, Patel KA, Nair SS, Darji AM (2013) Microbial decolourization of methyl orange dye by Pseudomonas spp. OA Biotechnol 2(1):10

Shahid M, Mohammad F (2013) Recent advancements in natural dye applications: a review. J Clean Prod 53:310–331

Sharma RK, Das A, Singh DP, Patel RK (2020) Effect of cationic surfactants on solubilization and adsorption of reactive dyes: a potential tool for wool dyeing. J Nat Fibers 17(1):27–36

Sharma SK (ed) (2015) Green chemistry for dyes removal from waste water: research trends and applications, Scrivener Publishing LLC

Simončič B, Kovac F (1999) A study of dye–surfactant interactions. Part 2. The effect of purity of a commercial cationic azo dye on dye–surfactant complex formation. Dyes Pigment 40(1):1–9

Simončič B, Špan J (1998) A study of dye-surfactant interactions. Part 1. Effect of chemical structure of acid dyes and surfactants on the complex formation. Dyes Pigment 36(1):1–14

Simončič B, Kert M (2002) A study of anionic dye–cationic surfactant interactions in mixtures of cationic and nonionic surfactants. Dyes Pigment 54(3):221–237

Singh MK (2021) Textiles functionalization-a review of materials, processes, and assessment. In Textiles for Functional Applications. http://dx.doi.org/10.5772/intechopen.96936

Singh J, Alauddin S, Yadav S (2021) Arsenic contamination removal from groundwater utilizing natural adsorbent. J Appl Geochem 23(2):132–143. (removal of dye from water or water purification by micelle)

Tang YLA, Lee CH, Wang Y, Kan C (2023a) Comparison of water-based dyeing system and alkyl polyglucoside (APG) surfactant-based reverse micellar dyeing system with reactive dyes for cotton. Cellulose 30(6):4011–4023

Tang YLA, Jin S, Lee CH, Law HS, Yu J, Wang Y, Kan CW (2023b) Reverse micellar dyeing of cotton fabric with reactive dye using biodegradable non-ionic surfactant as nanoscale carrier: an optimisation study by one-factor-at-one-time approach. Polymers 15(20):4175

Tang AYL, Lee CH, Wang Y, Kan C (2023c) Rhamnolipid (RL) microbial biosurfactant-based reverse micellar dyeing of cotton fabric with reactive dyes: a salt-free and alkali-free one-bath one-step approach. Colloids Surfaces A Physicochem Eng Asp 658:130725

Tang YLA, Yu J, Lee CH, Wang Y, Kan CW (2024) Limonene biosolvent-based non-aqueous dyeing medium for salt-free and alkali-free rhamnolipid (RL) biosurfactant reverse micellar dyeing of cotton fabric with reactive dyes. Sustain Mater Technol 42:e01137

Tehrani-Bagha AR, Holmberg K (2013) Solubilization of hydrophobic dyes in surfactant solutions. Materials 6(2):580–608

Tuli HS, Chaudhary P, Beniwal V, Sharma AK (2015) Microbial pigments as natural color sources: current trends and future perspectives. J Food Sci Technol 52:4669–4678

UlHaq N, Usman M, Mansha A, Rashid MA, Munir M, Rana UA (2014) Solubilization of reactive blue 19 by the micelles of cationic surfactant Cetyltrimethyl ammonium bromide (CTAB). J Mol Liq 196:264–269

Wang Q, Zhang J, Wang Y, Li Y (2018) Solubilization behaviour of dispersed dyes in nonionic surfactant micelles and its application in polyester dyeing. J Dispers Sci Technol 39(4):549–556

Yadav S et al (2023) A brief review on natural dyes, pigments: recent advances and future perspectives. Results Chem 5:100733

Yadav S et al (2021) A concise overview on heterocyclic compounds exhibiting pesticidal activities. Int J Adv Res 9(8):989–1004

Yadav S et al (2023) Physical, chemical, and biological methods of heavy metal management. In heavy metals in the environment: Mmnagement strategies for global pollution. Am Chem Soc pp. 247–259

Yang J (2004) Interaction of surfactants and aminoindophenol dye. J Colloid Interface Sci 274(1):237–243

Yousaf AM, Aqsa R (2023) Integrating circular economy, SBTI, digital LCA, and ESG benchmarks for sustainable textile dyeing: a critical review of industrial textile practices. Glob. NEST J 25:39–51

Chapter 11
Bio-Derived Dyes and Mordants for Greener Textile Coloration

Tithi Mitra and Sankar Roy Maulik

Abstract Synthetic colorants and auxiliaries offer various advantages in terms of durability, color range, and improved functionality to meet the diverse demands of consumers. However, this industry contributes to pollution and environmental degradation and has been criticised for its significant environmental footprint. On the contrary, bio-derived dyes and auxiliaries obtained from renewable resources, such as plant extracts, microorganisms, and agricultural by-products, offer a sustainable alternative and eco-friendly approach in the textile industry. The biodegradability of the colorants derived from renewable resources aims to introduce natural compounds that are both environmentally friendly and exhibit unique properties. This characteristic aligns with the growing demand for sustainable practices in the textile industry, promoting circularity and reducing the environmental burden of textile waste. Bio-based alternatives are generally non-toxic and pose fewer health risks than their synthetic counterparts, creating a safer working environment for textile industry workers. Furthermore, the utilisation of bio-derived compounds in textile processes promotes agricultural diversification; thus, the textile industry can contribute to the economic viability of agricultural sectors. A comprehensive case study is conducted using colorants obtained from renewable resources along with bio-mordant and innocuous metallic salts to dye and print cotton, wool, silk, and hemp-cotton union fabric for product development. The results show positive outcomes, particularly in the case of printing with bio-mordant and also with clay. Using clay as a resisting material is an environmentally friendly approach, as it is sourced from nature and does not involve harmful chemicals. However, challenges like consistent results and cost-effectiveness need to be addressed further.

Keywords Auxiliaries · Bio-mordant · Dyeing · Finishing · Natural Dyes · Sustainability · Textile

T. Mitra · S. Roy Maulik (✉)
Department of Silpa-Sadana, Visva-Bharati (A Central University), Sriniketan, India
e-mail: sankar.raymaulik@visva-bharati.ac.in

S. Maiti et al. (eds.), *Sustainable Coloration Techniques in Textiles*,
https://doi.org/10.1007/978-981-96-4975-4_11

11.1 Introduction

The fiber production, manufacturing, usage, and disposal of textiles and clothes, all are part of a vast supply chain. In the textile industry, synthetic colorants and auxiliaries have acquired popularity due to their ability to meet the diverse expectations of consumers. The vibrant hues and improved properties imparted by synthetic dyes to textiles influenced their widespread adoption. However, the common chemicals, auxiliaries, and dyestuffs popularly used by the manufacturers sometimes have mutagenic and carcinogenic effects (Kishor et al. 2021). The textile industry contributes to the environmental burden daily by emitting harmful chemicals and highly colored wastewater into water bodies, contaminating soil and polluting the atmosphere. The adverse effects go beyond the environmental issues since these chemicals affect human health and safety, especially for those who come into direct contact with them. Numerous energy-intensive chemical processes involving petrochemical-based color intermediates are required to manufacture synthetic dyes, and these reactions release toxic compounds into the environment (Christie 2007). The dyestuffs discharged to the wastewater are subsequently released into the environment and 95% of textile wastewater originates from coloring, whereas 5% from washing (Novotný et al. 2006). Heavy metals found in textile waste have the potential to become carcinogenic if accumulate in the human body. Wastewater also generates BOD, COD, and TSS, and increases pH values beyond the norm, if not treated properly (Elsahida et al. 2019). Azo dyes are one of the major classes of colorants used in the dyeing of textiles. Some bacteria metabolised these azo dyes to form colorless aromatic amines, some of which are carcinogenic. The structures and carcinogenicity of azo dyes have been studied and reported (Chen 2006). Reactive dyes form a covalent bond with the primary hydroxyl group of cellulose and are extensively used in textile industries due to their known advantages viz., wide range of shades, high wet fastness, ease of application, etc. However, under typical dyeing conditions, 20–50% of the unfixed dye, called hydrolysed dye remains in the effluent (Lee and Pavlostathis 2004).

The researchers are now focusing on reducing harmful chemicals in the manufacturing stages, implementing effective wastewater treatment, and exploring nature-based alternatives to reduce these environmental issues caused by using harmful chemicals in manufacturing. Using natural colorants and auxiliaries is one of the key strategies for promoting a more effective workplace for humans and the environment, as they are eco-friendly and skin-friendly.

Before the advent of synthetic dyes in 1856, textiles were dyed with colorants obtained from different natural resources, viz., plants, animals, and minerals. However, certain limitations, which include poor substantivity, especially for cellulosic substrates, colorfastness, limited range of shades, etc. forced natural dye into the back seat after the discovery of synthetic dyes. The use of natural dyes receded due to the ease of the application process, reproducibility of shades, superior colorfastness properties, and low cost of production of synthetic dyes (Periyasamy et al. 2020). Natural dyes can be obtained from different sources, such as plants, animals, minerals, microorganisms, agricultural and forestry waste, and food waste (skin, shells, etc.) (Elsahida et al. 2019). These dyes can be classified based on their origin, chemical structure, solubility, and mordanting method. The chemical composition, color, and dyeing technique of each dye obtained from various sources are distinct. Plant-based dyes derived from indigo and madder roots offer wide shades of blue and red respectively. Minerals such as ochre and lapis lazuli produce earthy hues. Presently only 1% of global requirements can be met by natural dyes (Elsahida et al. 2019). The substantivity of natural dyed textiles can be improved by using mordants, which are generally inorganic metal salts. However, heavy metals used in the mordanting process seriously affect the environment.

Nowadays, eco-conscious consumers prefer bio-inspired dyes for their soft and natural hues, medicinal properties, and ability to harmonise with nature. However, these dyes have certain limitations in terms of color value and fastness properties, which can be improved by mordants. Metal-based mordants are popularly used along with natural dyes for dyeing and printing/painting purposes without considering their toxicity effect. These mordants, if not used judiciously can cause environmental and health issues. After dyeing, the effluent contains these free metal salts and subsequently leaches from the textiles during domestic washing.

Thus, researchers are trying to replace these heavy metals with environment-friendly bio-inspired mordants in natural color dyeing, which may be a good alternative for developing a greener approach (Repon et al. 2024). The use of bio-inspired mordant, a safe alternative to natural color dyeing, greatly impacts the research on natural dyes and dyeing processes (Rani et al. 2020; Vankar et al. 2008a). Moreover, implementing bio-derived substances into textile processing fosters agricultural diversification, which may support economic growth. It is essential to understand that replacing synthetic dyes and auxiliaries with natural ones would be challenging. To successfully use natural dyes commercially for textile applications, one of the crucial steps includes standardisation of the process, starting from the extraction, mordanting, and application to the fiber. Also, identifying the key components in the dyes and analysing their bioactive properties is crucial for imparting additional functions to the fabric. The commercial success of natural dyes also depends on standardised shade cards and various color combinations through compound shades. Different researchers are exploring eco-friendly alternatives to inorganic salt-based mordants in the dyeing and printing of textiles with natural dyes.

Green and environmentally friendly wool dyeing with myrobalan extracts as green mordants has been studied and reported. According to the researchers, myrobalan can be a potential bio-mordant for eco-friendly dyeing (Hosseinnezhad et al. 2022). Bio-mordants and microwave-assisted extraction of natural dyes under mild conditions are gaining popularity worldwide. The tannin isolated from neem (*Azadirachta indica*) bark is used to dye silk fabric under microwave radiation in the presence of turmeric (*Curcuma longa*), henna (*Lawsonia inermis*), etc., and acts as bio-mordants. The microwave treatment enhanced the extraction of tannin and also improved the color characteristics (Zuber et al. 2020). Studies have reported microwave-assisted extraction of natural dye from cochineal insects for imparting color on bio-mordanted (henna leaves, acacia bark, and pomegranate peels used as bio-mordants) silk fabric. It is revealed that bio-mordanted silk fabric dyed with the color extracted under microwave irradiation shows significant improvements in dyeing behaviour and color characteristics (Amin et al. 2020). Sustainable dyes derived from different plants are in demand nowadays among the global communities. Cinnamon bark is a source of yellow natural colorant and can be used for dyeing silk in the presence of different bio-mordants, producing improved fastness properties (Adeel et al. 2020a). The dyeing of silk fabric using safflower (*Carthamus tinctorius* L.) in microwave irradiation for 1–5 min. in the presence of both bio-mordants and chemical mordants has been reported. The microwave treatment of the extract for 3 min is suitable to get darker shades and acceptable color characteristics (Adeel et al. 2020b). Rose flowers contain anthocyanin pigment that can be used for dyeing wool at optimal conditions in the presence of chemical and bio-mordant to produce a variety of shades with appreciable fastness properties along with antioxidant, anti-inflammatory, and antibacterial activity (Adeel et al. 2021). The colorant obtained from cochineal insects through microwave-assisted extraction techniques in different mediums is used for dyeing bio-mordanted silk. Using bio-mordants in optimised conditions makes the process greener and more sustainable. The acid-solubilised extract containing Glauber's salt as an exhausting agent produces high color strength on microwave-treated silk fabric, and the use of bio-mordants further improves the color depth and fastness properties (Amin et al. 2020).

11.2 Mordants

Natural dyes possess less affinity for mainly cellulosic fiber in terms of color value and fastness properties and hence it is essential to add mordant during dyeing and printing/painting. Metal mordants or inorganic salts viz., alum, ferrous sulphate, zinc sulphate, cupric sulphate, potassium dichromate, etc., are among the common mordants used by textile practitioners. Varying the concentration of mordants leads to more intense and wide color variations with improved colorfastness properties. Although metal mordants can develop a broad range of shades, it is crucial to note that the use of heavy metal can be hazardous to human health and the environment. Waterbodies, soil, and ecosystems can be contaminated by heavy metal mordants, harming aquatic life and disrupting natural balance. There are three types of mordant application processes: pre-, post-, and simultaneous (Elsahida et al. 2019). As the name indicates, pre-mordanting involves applying the mordant to the textile substrate before the application of dye. Simultaneous mordanting involves the application of dye and mordant together and post-mordanting is the application of mordant after the addition of dye on the substrate (Elsahida et al. 2019). Nowadays, researchers also try to adopt contemporary processes rather than traditional ones to save labour, chemical, and process expenses (Periyasamy 2022). These contemporary methods include extraction aided by microwave, ultrasonic, gamma, ultraviolet, and plasma. Out of these, ultrasonic stands out as an effective choice as it is easily scalable and gained recognition as a green technology (Periyasamy 2022).

11.3 Bio-Mordant

The mordants commonly used in natural color dyeing are inorganic salts and some of them are toxic and non-environment friendly, thus the sustainability of the natural dye is under question. In contrast, bio-mordants are increasingly drawing the attention of researchers as a substitute for heavy metals due to their biodegradability and low environmental impact. A lot of research on the use of bio-mordants with high tannin or metal content as a safe substitute for chemical-based inorganic salts is reported in the literature (Guesmi et al. 2013; Vankar et al. 2008b). Bio-mordants from different natural origins are also used during dyeing with natural dyes. *Terminalia chebula, Punica granatum*, tannin, tannic acid, tartaric acid, guava, and banana leaves ash are some examples of bio-mordants (Aminoddin 2010). To improve the affinity of natural dyes for textile substrates, researchers have explored the use of tannins as a mordanting agent. The polyphenolic structure in the tannins plays a vital role in forming the crosslink with dye and fiber. Plants high in tannins act as mordants and some of the promising sources include oak bark and wood (*Quercus infectoria*), and cutch (*Acacia catechu*), etc. (Benli 2024). Researchers have also explored the possibility of using banana midrib as a natural colorant for textile dyeing, which also possesses antimicrobial, anti-inflammatory, anti-allergic, and antioxidant properties on dyed substrates. *Acacia nilotica* as a natural mordant is used to increase the affinity of the dye for pineapple fabric to achieve improved colorfastness, along with some functional properties such as antibacterial and UV protection. Thus, banana midrib and *Acacia nilotica* emerge as promising textile colorants and bio-mordants with multifaceted benefits (Kushwaha et al. 2024). Researchers have experimented with the possibility of dyeing wool with papaya (*Carica Papaya L.*) leaf in the presence of bio-mordants such as orange peel, pomegranate rind, amla, and harda (myrobalan) powder following pre-, post-, and simultaneous mordanting methods. Their outcomes reveal that the bio-mordants provided good colorfastness properties in addition to appreciable functional properties viz., antimicrobial and UV protection (Rani et al. 2020). The carpenter sawdust waste from *Millettia laurantii* can be used as a sustainable source to dye wool fabric, with improved biological and functional properties with the possibility of upscaling carpenter waste as a natural dye source (Safapour et al. 2023a). The mugwort herbs, rue herbs, black cherry stems and hops flowers can be used as bio-mordants, highlighting their potential benefits and environmental sustainability. The atomic absorption spectroscopy (AAS) analysis reveals a substantial amount of copper and ferrous content in hope flower and black cherry stems. These bio-based mordants have excellent rubbing and washing fastness properties when applied to cotton fabric (Thakker and Sun 2022). The classification of mordants based on their properties and mechanisms has been reported, focusing on their contribution to environmental degradation. Researchers have also studied the effects of dye uptake and optimisation of mordant usage and concentrations (İşmal and Yıldırım 2019). The effect of hydrolyzable and non-hydrolyzable tannins as bio-mordant on cotton fabric dyed with *Acacia nilotica* has been reported. The bio-mordant containing hydrolyzable tannin produces a light-yellow color, whereas, with non-hydrolyzable tannin, a medium to deep brown color is produced (Dhanania et al. 2021). Nowadays, bio-dyes are considered a promising sustainable alternative for cleaner dyeing and functional finishing of textiles. Wool fiber dyed with flavonoid-based bio-dyes extracted from lemon balm plant in the presence of bio-mordants imparts antimicrobial and antioxidant properties (Safapour et al. 2023b). The dragon's blood resin extract is used to dye nylon fabric, and different process variables are optimized using response surface methodology. The colorfastness properties of the dyed substrates in the presence of different bio-mordants viz., peppermint, mugwort (Artemisia), *Dorema ammoniacum* gum, and pomegranate rind are assessed and found comparable to the metallic mordants (Haji et al. 2023). Turmeric extract is used to dye cotton fabric in combination with bio-mordants such as citrus lemon and *Colocasia esculenta.* The result shows a promising approach toward sustainable textile coloration. Pre-treated samples with bio-mordants exhibit enhanced color strength and improved fastness properties compared to metallic mordants (Hosen et al. 2021). Researchers have reviewed several studies to find effective bio-mordants instead of metallic mordants to facilitate the natural dyeing process (Benli 2024) and reported that most bio-mordants have various functional properties. Table 11.1 shows different plants that can be used as bio-mordants (Benli 2024).

Researchers employed a natural dye extracted from *Syzygium cumini* fruit (commonly known as *Jamun)* and soy powder as a bio-mordant to dye cotton fabric. The finding showed that cotton fabric dyed and mordanted using bio-derived sources through ultrasonic dyeing and mordanting techniques over traditional techniques provides higher surface color strength (Periyasamy 2022). Researchers have investigated and utilized bio-mordants in their study, which resulted in enhanced colorfastness of dyed cotton fabrics. The bio-mordants produced excellent light, wash, and dry or wet rubbing fastness. Based on their findings, researchers recommended the

Table 11.1 Plants and their parts that can be utilized as bio-mordants

Plant name	Botanical name	Parts used	Used as
Khayer	*Acacia catechu*	Wood	Dye and Mordant
Pomegranate	*Punica granatum*	Rind	Dye and Mordant
Myrobalan	*Terminalia chebula Retz*	Fruit	Dye and Mordant
Aloe vera	*Aloe barbadensis Miller*	Leaves	Mordant
Banana	*Musa paradisiaca*	Pseudostem sap	Mordant
Lemon	*Citrus lemon L*	Fruit	Mordant
Amla	*Emblica ofcinalis G*	Fruit	Mordant
Babool	*Acacia nilotica*	Bark	Mordant
Cashew	*Anacardium occidental*	Leaves	Mordant
Amaltas/Golden shower	*Cassia fistula*	Bark	Mordant
Eucalyptus	*Eucalyptus spp*	Bark and wood	Mordant
Tamarind	*Tamarindus indica*	Fruit	Mordant
Guava	*Psidium guajava L*	Leaves	Mordant

incorporation of bio-mordants in the natural dyeing process (Shahmoradi Ghaheh et al. 2021).

Although bio-mordants are more environment-friendly than metallic salts, they have certain limitations in producing a range of colors on textiles and reproducibility of shades. Hence, the judicious use of metallic salts, which do not affect the environment, can be used along with bio-mordants.

11.4 Case Study

In the present case study, natural dyes extracted from vegetable resources were used to dye and print/paint various natural-based fiber/fabrics to produce environmentally friendly products in the presence of bio-mordants and innocuous inorganic salts. It is observed that most of the researchers used bio-mordants for dyeing purposes. The present study reports using bio and innocuous mordants in dyeing and printing/painting with natural dyes.

11.4.1 Materials

11.4.1.1 Fabrics and Yarn

The cotton fabric of areal density 100 gsm, 2/28 NM wool yarn, and 2/20 Ne organic cotton yarn, hemp-cotton union fabric, and ramie-cotton union fabric were used in the study.

11.4.1.2 Natural Dyes

Indian Madder (*Rubia cordifolia* L.), Indian teak leaf (*Tectona grandis*), Lac (*Laccifer lacca*), Khayer (*Acacia Catechu*), Pomegranate rind (*Punica granatum*), Myrobalan (*Terminalia chebula*), Alizarin and Natural indigo (*Indigofera tinctoria*) were used for dyeing and printing purposes of various textile substrates.

11.4.1.3 Mordants

Alumn, iron rust, myrobalan, citric acid, and clay were used as mordants as and when required.

11.4.1.4 Chemicals

Laboratory reagent (LR) grade sodium hydroxide, sodium hydrosulphite, sodium meta-silicate, hydrogen peroxide (30% w/v), sodium carbonate, and acetic acid obtained from Loba Chemie Pvt. Ltd, Mumbai and Emplura, Merck Life Science Pvt. Ltd, Mumbai, and T.R. Oil and non-ionic detergent of the commercial grade obtained from the local market were used in the present study. Locally available red soil, wheat powder, gum arabic, and calcium oxide were also used during mud-resist printing.

11.4.2 Methods

11.4.2.1 Aqueous Extraction of Natural Dyes

Aqueous extraction is one of the earliest and most simple practices of natural dye extraction used worldwide. In this method, all the vegetable resources were first chopped into fine pieces or ground to fine powder, then separately boiled for 45 min and filtered or strained, except for natural indigo, which comes in powder form.

11.4.2.2 Dyeing of Wool and Cellulosic Fabric

Natural dye was applied to wool yarn, ramie-cotton union fabric, and hemp-cotton union fabric following the exhaust application method. In this case, textile substrates were first treated with various mordants viz., alum, iron rust, and citric acid at 70 °C for 20 min, followed by an aqueous extract of specified natural colorants for 30 min at 85 °C temperature. Different shades of wool yarn, ramie-cotton union fabric, and hemp-cotton union fabric were developed at 40% dye concentration in the presence of citric acid, alum, and iron rust with varying concentrations. Mordanting and dyeing were performed at a material-to-liquor ratio of 1:30. Tables 11.2 and 11.3 show the recipe used for dyeing and some of the developed shades, respectively.

The application of natural dye on fabrics to be used for product development was done through the pad-dry-steam method. This method provides better uniformity of shades on the fabric. In the padding process, after mordanting and dyeing, the fabrics were dried in the air and steamed for 45 min at 102 °C. All the dyed samples were thoroughly washed using non-ionic detergent.

11.4.2.3 Mud-Resist Printing

Patterns can be created on the fabric without using toxic or hazardous chemicals. One specific printing method is mud-resist, which uses mud as a mechanical resisting agent to prevent the dye from penetrating certain areas of the fabric to create beautiful

Table 11.2 Dyeing recipe for shade development

Dyes used	Shade depth (%) (o.w.f.)	Mordants used	Concentration (g/l)		
			Ramie-cotton union fabrics	Hemp-cotton union fabrics	Wool yarn
Rubia cordifolia L	40	Alum	10	10	10
Punica granatum	40	Iron rust	10	10	2
Laccifer lacca	40	Citric acid	10	10	10
Acacia catechu	40	Iron rust	10	10	
Tectona grandis	40	Citric acid	Nil	Nil	10

and desired patterns. This age-old textile printing process originates in the Indian state of Rajasthan and is colloquially known as *Dabu*.

In this study, clay is sourced from the Birbhum district of West Bengal. The other ingredients used in the study are insect-eaten wheat powder, gum arabic, and calcium oxide. Paintbrushes were used to apply the mud paste to the fabric to achieve bold designs. Figure 11.1 portrays the application process utilizing a paintbrush as a transferring medium. It is important to sprinkle sawdust on the fabric after applying mud to speed up drying. Once completely dried, the fabric was dipped in an indigo dye bath previously prepared, followed by oxidation and washing. The indigo bath was prepared with sodium hydrosulphite and sodium hydroxide and dyeing was carried out at room temperature.

11.4.2.4 Printing with Clay as Bio-Mordant

In another process, the local clay of Birbhum district was employed as a mordant to produce surface ornamented fabrics. The process resembles mud-resist printing, where the significant step is the clay preparation, which involves the homogenous mixing of clay with other ingredients (e.g., gum and insect-eaten wheat powder). Here, the printing process was initiated with mordanting in two phases. In the first phase, a bio-mordant, myrobalan (*Terminalia chebula*), was applied to obtain the ground color on the scoured cotton fabric following the pad-dry-steam method. The process was performed in a miniature laboratory model hand-driven padding mangle at an expression of 100%. In the second phase, prepared clay was applied only in the areas where designs were desired on darker shades with the help of a wooden block (Fig. 11.2). After the mud was applied, sawdust was sprinkled to facilitate the drying of the mud (Fig. 11.3). Once the fabric was completely dried, it was then dyed with Alizarine.

Table 11.3 Developed shades

Sl No	Developed shade	Natural dye	Mordant	Fabric/Yarn
1		*Rubia cordifolia* L	Citric acid	Hemp-cotton union fabric
2		*Acacia catechu*	Iron rust	Hemp-cotton union fabric
3		*Punica granatum*	Alum	Ramie-cotton union fabric
4		*Rubia cordifolia* L	Alum	Ramie-cotton union fabric

(continued)

Table 11.3 (continued)

Sl No	Developed shade	Natural dye	Mordant	Fabric/Yarn
5		*Acacia catechu*	Iron rust	Ramie-cotton union fabric
6		*Rubia cordifolia* L	Alum	Wool yarn

(continued)

Table 11.3 (continued)

Sl No	Developed shade	Natural dye	Mordant	Fabric/Yarn
7		*Tectona grandis*	Citric acid	Wool yarn
8		*Laccifer lacca*	Alum	Wool yarn

11.4.2.5 Block Printing with Natural Dyes

Printing on fabric with wooden blocks is an age-old technique where designs are carved out on the block using hand tools; the process is quite intricate and time-consuming. In this exploration, silk fabric was dyed with an aqueous extract of *Laccifer lacca,* i.e., lac dye, in the presence of alum used as mordant. The application of both mordant and dye was performed following the padding process. After padding, the fabric was dried at room temperature and then printed with blocks. The dark shade of the print paste was produced from the extraction of pomegranate rind and myrobalan and by the addition of iron rust as a mordant. In the print paste, gum indulka was added to increase its viscosity. After printing, the fabric was dried and steamed for about 45 min to fix the natural dye on the fabric, followed by washing with non-ionic detergent. Figure 11.4 shows the block printing process on silk fabric.

Fig. 11.1 Application of mud paste on cotton fabric using a paintbrush

Fig. 11.2 Application of mud paste on myrobalan-treated cotton fabric using wood block

11.4.3 Results and Discussion

11.4.3.1 Mud Resist and Print Development

The viscosity of the mud paste plays a crucial role during the printing process in facilitating the proper transfer of the mud paste onto blocks and from blocks to the fabric. The approach of using mud to create prints is an environment-friendly process

Fig. 11.3 Drying in direct sunlight

Fig. 11.4 Printing on silk fabric

since it comes from nature and is readily available in the surroundings. The process does not involve the use of any harmful chemicals. This printing method depends only on manual labor and sunlight. Since this process consumes less energy, it reduces the environmental damage and thus stands out as a greener approach.

11.4.3.2 Block Printing with Natural Dyes

Figure 11.5, 11.6 and 11.7 shows various products developed using natural dyes and different innocuous and bio-mordants.

Fig. 11.5 Mud resists printed cotton stole

Fig. 11.6 Clay mordanted and developed dress

Fig. 11.7 Block printed saree with natural dyes

11.5 Conclusion

Sustainability lies in our society's choices, including the suitable materials for every product, from our clothes to our daily life products. The choices of utilizing dyes and their auxiliaries in the textile industry affect the environment in numerous ways. Natural colorants over toxic synthetic dyes can become a matter of choice for professionals. With time, research on this matter is increasing, and numerous alternatives are developing, such as bio-mordanting in textile dyeing with natural dye. One of the most exciting parts is that one can utilize many natural wastes viz., pomegranate rind, onion peel, dried leaves, etc., as a dye and also bio-mordants, which would otherwise be discarded. Designing fabrics that adhere to the concept is essential in the sustainability journey. This study portrayed different utilizations of natural dyes. Textiles dyed with specific colorants extracted from renewable natural resources require inorganic salts as mordants to improve the dye affinity and fastness properties. The shift toward bio-derived mordants and finishing agents represents an ethical approach to dyeing practices, promoting sustainability and environmental growth. However, there are some challenges for adopting these bio-derived alternatives, including consistent results, cost-effectiveness, etc. Researchers are trying to overcome these hurdles and optimize the performance of bio-based colorants in industrial applications.

References

Adeel S, Habib N, Arif S et al (2020a) Microwave-assisted eco-dyeing of bio-mordanted silk fabric using cinnamon bark (*Cinnamomum Verum*) based yellow natural dye. Sustain Chem Pharm 17:100306. https://doi.org/10.1016/j.scp.2020.100306

Adeel S, Naseer K, Javed S et al (2020b) Microwave-assisted improvement in dyeing behavior of chemical and bio-mordanted silk fabric using safflower (*Carthamus tinctorius* L) extract. J Nat Fibers 17:55–65. https://doi.org/10.1080/15440478.2018.1465877

Adeel S, Salman M, Usama M et al (2021) Sustainable isolation and application of rose petals-based anthocyanin natural dye for coloration of bio-mordanted wool fabric: short title: dyeing of bio mordanted wool with rose petal extract. J Nat Fibers 19:6089–6103. https://doi.org/10.1080/15440478.2021.1904480

Amin N, Rehman F, Adeel S et al (2020) Sustainable application of cochineal-based anthraquinone dye for the coloration of bio-mordanted silk fabric. Environ Sci Pollut Res 27:6851–6860. https://doi.org/10.1007/s11356-019-06868-3

Aminoddin H (2010) Functional dyeing of wool with natural dye extracted from *Berberis vulgaris* wood and *Rumex hymenosepolus* root as bio-mordant. Iran J Chem Chem Eng 29:55–60

Benli H (2024) Bio-mordants: a review. Environ Sci Pollut Res 31:20714–20771. https://doi.org/10.1007/s11356-024-32174-8

Chen H (2006) Recent advances in azo dye degrading enzyme research. Curr Protein Pept Sci 7:101–111. https://doi.org/10.2174/138920306776359786

Christie RM (2007) Environmental aspects of textile dyeing. CRC Press [u.a.], Boca Raton, Fla

Dhanania Y, Singhee D, Samanta AK (2021) Optimization of dyeing process variables for eco-friendly dyeing of cotton fabric with babul bark extract as a natural dye and gallnut extract as a bio-mordant. J Nat Fibers 19(13):5478–5495. https://doi.org/10.1080/15440478.2021.1875955

Elsahida K, Fauzi AM, Sailah I, Siregar IZ (2019) Sustainability of the use of natural dyes in the textile industry. IOP Conf Ser Earth Environ Sci 399:012065. https://doi.org/10.1088/1755-1315/399/1/012065

Guesmi A, Ladhari N, Hamadi NB, Msaddek M, Sakli F (2013) First application of chlorophyll-a as bio-mordant: sonicator dyeing of wool with betanin dye. J Clean Prod 39:97–104

Haji A, Shahmoradi Ghaheh F, Mohammadi L (2023) Dyeing of polyamide 6 fabric with new bio-colorant and bio-mordants. Environ Sci Pollut Res 30:37981–37996. https://doi.org/10.1007/s11356-022-24862-0

Hosen MdD, Rabbi MdF, Raihan MdA, Al Mamun MdA (2021) Effect of turmeric dye and bio-mordants on knitted cotton fabric coloration: A promising alternative to metallic mordanting. Clean Eng Technol 3:100124. https://doi.org/10.1016/j.clet.2021.100124

Hosseinnezhad M, Gharanjig K, Imani H, Razani N (2022) Green dyeing of wool yarns with yellow and black myrobalan extract as bio-mordant with natural dyes. J Nat Fibers 19:3893–3915. https://doi.org/10.1080/15440478.2020.1848735

İşmal ÖE, Yıldırım L (2019) Metal mordants and bio-mordants. In: The impact and prospects of green chemistry for textile technology. Elsevier, pp 57–82

Kishor R, Purchase D, Saratale GD, Saratale RG, Ferreira LFR, Bilal M, Chandra R, Bharagava RN (2021) Ecotoxicological and health concerns of persistent coloring pollutants of textile industry wastewater and treatment approaches for environmental safety. J Environ Chem Eng 9:105012. https://doi.org/10.1016/j.jece.2020.105012

Kushwaha A, Singh S, Chaudhary K (2024) Cleaner dyeing of pineapple fabric using musa acuminata natural dye and *acacia nilotica* bio-mordant with enhanced functionality. J Inst Eng India Ser E. https://doi.org/10.1007/s40034-023-00284-6

Lee YH, Pavlostathis SG (2004) Decolorization and toxicity of reactive anthraquinone textile dyes under methanogenic conditions. Water Res 38:1838–1852. https://doi.org/10.1016/j.watres.2003.12.028

Novotný Č, Dias N, Kapanen A, Malachová K, Vándrovcová M, Itävaara M, Lima N (2006) Comparative use of bacterial, algal and protozoan tests to study the toxicity of azo- and anthraquinone dyes. Chemosphere 63:1436–1442. https://doi.org/10.1016/j.chemosphere.2005.10.002

Periyasamy AP (2022) Natural dyeing of cellulose fibers using *syzygium cumini* fruit extracts and a bio-mordant: A step toward sustainable dyeing. Sustain Mater Technol 33:e00472. https://doi.org/10.1016/j.susmat.2022.e00472

Periyasamy AP, Militky J (2020) Sustainability in textile dyeing: recent developments. In: Muthu SS, Gardetti MA (eds) Sustainability in the Textile and Apparel Industries. Springer International Publishing, Cham, pp 37–79

Rani N, Jajpura L, Butola BS (2020) Ecological dyeing of protein fabrics with carica papaya L. leaf natural extract in the presence of bio-mordants as an alternative copartner to metal mordants. J Inst Eng India Ser E 101:19–31. https://doi.org/10.1007/s40034-020-00158-1

Repon MR, Dev B, Rahman MA et al (2024) Textile dyeing using natural mordants and dyes: a review. Environ Chem Lett 22:1473–1520. https://doi.org/10.1007/s10311-024-01716-4

Safapour S, Rather LJ, Mir SS, Dar QF (2023a) Upscaling Millettia laurentii carpentry sawdust into natural dyes: imparting antimicrobial, antioxidant, and UV-protective finish to wool yarns through an ecological and sustainable natural dyeing process. Biomass Convers Biorefinery. https://doi.org/10.1007/s13399-023-04184-y

Safapour S, Rather LJ, Moradnejad J et al (2023b) Functional and colorful wool textiles through ecological dyeing with lemon balm bio-dyes and mordants. Fibers Polym 24:4357–4370. https://doi.org/10.1007/s12221-023-00397-0

Shahmoradi Ghaheh F, Moghaddam MK, Tehrani M (2021) Comparison of the effect of metal mordants and bio-mordants on the colorimetric and antibacterial properties of natural dyes on cotton fabric. Color Technol 137:689–698. https://doi.org/10.1111/cote.12569

Thakker AM, Sun D (2022) Innovative plant-based mordants and colorants for application on cotton fabric. J Nat Fibers 19:14346–14364. https://doi.org/10.1080/15440478.2022.2064391

Vankar PS, Shanker R, Mahanta D, Tiwari SC (2008a) Ecofriendly sonicator dyeing of cotton with Rubia cordifolia Linn. using biomordant. Dyes Pigments 76:207–212. https://doi.org/10.1016/j.dyepig.2006.08.023
Vankar PS, Shanker R, Mahanta D, Tiwari SC (2008) Ecofriendly sonicator dyeing of cotton with *Rubia cordifolia* Linn. using bio-mordant. Dyes Pigments 76, 207–212
Zuber M, Adeel S, Rehman FU et al (2020) Influence of microwave radiation on dyeing of bio-mordanted silk fabric using neem bark (*Azadirachta indica*)-Based tannin natural dye. J Nat Fibers 17:1410–1422. https://doi.org/10.1080/15440478.2019.1576569

Chapter 12
Biosurfactants: Eco-friendly Substitutes to Synthetic Auxiliaries for Sustainable Wet Processing of Textiles

Sk. Mohammad Raafi

Abstract Wet processing of textiles involves the intense application of widespread auxiliary chemicals mostly in the pretreatment, coloration, and finishing stages to improve the process performance. Within the range of auxiliaries, synthetic surfactants are dominantly used in water softening, wetting, washing, dispersing, dye leveling, fixing, and so many other operations. Due to the high levels of toxicity and very little biodegradability, the petroleum-based, chemically synthesized surfactants ultimately pose adverse environmental effects when discharged as effluents. A green approach to resolve the challenges could be the application of bioderived surfactants or 'biosurfactants'. Biosurfactants are natural surface-active compounds obtained directly from the cell walls or membranes of microorganisms. The scope of the identity can be extended to certain biomolecules of plant and animal origin. Biosurfactants are highly biocompatible, less toxic, and efficiently functional even in extreme conditions of the processing environment. However, the industrial applications of biosurfactants are currently not on the full scale, mostly because of the high production costs, relatively low yield, and incompetent methodology. The current chapter addresses the key aspects of biosurfactants, including their sources, types, properties, and synthesis. The shift of focus from petroleum derivatives to bioderivatives for surfactant production has been discussed. Special emphasis was placed on exploring the potential of biosurfactants at various wet processing stages of textiles. Challenges and prospects were reported in the concluding remarks.

Keywords Biosurfactants · Synthetic surfactants · Microorganisms · Eco-friendly · Textile auxiliaries · Sustainability

Sk. M. Raafi (✉)
Department of Wet Process Engineering, Bangladesh University of Textiles, Tejgaon Dhaka-1208, Bangladesh
e-mail: raafi@wpe.butex.edu.bd

S. Maiti et al. (eds.), *Sustainable Coloration Techniques in Textiles*,
https://doi.org/10.1007/978-981-96-4975-4_12

12.1 Introduction

Chemically, a surfactant possesses a pair of hydrophilic and hydrophobic portions suitable to reduce the surface tension and interfacial tensions between molecules in solid–liquid, liquid–liquid, and liquid–gas phases. Surfactants of different classes, such as anionic, cationic, nonionic, and amphoteric, are available considering the presence and nature of the charge of the hydrophilic moieties. Besides lowering the surface tension, the surfactants possess other qualities, enabling them to be used as emulsifiers, foaming agents, dispersants, etc., in various industrial applications (Alwadani and Fatehi 2018). Surfactants of synthetic origin hold the major proportion of the current market. Attributed to high toxicity levels and lack of biodegradability, these compounds disrupt aquatic microbial populations and wipe out fishes and other marine life. The high foaming power of the surfactants reduces the oxygen level of water and thus deteriorates the aquatic ecosystems. An estimated 60% of more than 15 million tons of global consumption of surfactants eventually enter the aquatic environment (Johnson et al. 2021). Atmospheric pollutants (nitrogen oxides, carbon dioxide, sulfur dioxide, hydrocarbons), waterborne effluents, and solid wastes, generated during the synthesis of both petrochemical and oleo chemical-based surfactants can cause eutrophication and acidification of freshwater reservoirs (Rebello et al. 2014). Environmental regulations and governmental provisions concerning the use of toxic detergents, as well as increased awareness of consumers regarding the use of sustainable materials, have motivated the quest for natural, biodegradable compounds as possible alternatives to synthetic surfactants (Farias et al. 2021).

In the widest sense, surfactants acquired from natural sources, including plants, animals, and microbes, are commonly known as 'green surfactants' or 'biosurfactants' (Nagtode et al. 2023). Their capability of demonstrating surface activities originates from the polar lipids, fatty acid esters, or amides, that resemble the hydrophilic head-hydrophobic tail combination (Fiechter 1992). Microbial biosurfactants differ from bio-based surfactants originating from plants and animals in such a sense that bacteria and yeasts mostly contribute solely to the production of microbial surfactants without the intervention of organic synthesis, whereas bio-based surfactants are typically produced by chemical synthesis but incorporate surface-active ingredients from renewable sources structurally (Moldes et al. 2021).

Biosurfactants are superior to their synthetic counterparts as they are manufactured from renewable sources, less toxic, more biocompatible, biodegradable, functional over expansive temperatures and pH values as well as exhibit antimicrobial activity (Nguyen et al. 2010; Kumari 2022). Relating to the efficacy of application, certain biosurfactants can achieve a similar lessening in the surface tension of water at the expense of a significantly smaller number of molecules because of their 10- to 40-fold lower CMC (Critical Micelle Concentration) value than that of the synthetic surfactants (Desai and Banat 1997). Modification of the composition and structure of the microbe-derived natural surfactants also offers customization to suit a particular application (Fiechter 1992). The international market of green surfactants stood at a value of almost USD 2.84 billion in 2023, which is further envisioned to rise over the forecast period of 2024–2032 to attain a value of USD 3.56 billion by 2026 (at a CAGR of 5.70%) and USD 3.98 billion by 2032 (at a CAGR of 5.80%) (Nagtode et al. 2023; Expert Market Research 2024).

12.2 Sources of Biosurfactants

12.2.1 Biosurfactants from Plants

Plant-derived biosurfactants are widespread in nature and are collected from roots, stems, leaves, seeds, fruits, etc. parts of plants. The surface-active molecules commonly found in plants are structurally categorized as phospholipids, proteins, protein hydrolysates, and saponins (Bezerra et al. 2021). Phospholipid-based potential emulsifiers are found in lecithins derived from soya beans, sunflower kernels, rapeseed, and many more oil-carrying seeds (van Nieuwenhuyzen 2010). Proteins of vegetable origin have high molecular weight and comprise different proportions of polar and non-polar groups spread irregularly throughout the structure (Jahan et al. 2020). Proteins are more useful as emulsifiers rather than wetting agents, as they produce more stable emulsions and foams. In the case of protein hydrolysates, the structure is chemically, thermally, or enzymatically modified affecting their constitution and size to enhance solubility, water and oil retention, foaming and stability, emulsification, viscosity, gelation, and other functional properties (Moure et al. 2006).

Saponins belong to a diverse group of plant metabolites where the hydrophobic part composed of either a triterpenoid or an alkaloid steroid is linked with hydrophilic sugar chains via ether or ester bonds (Tucker et al. 2020). Triterpenoid saponins are abundantly found in dicotyledonous plants of families like Araliaceae, Fabaceae, and Caryophyllaceae. In contrast, steroidal saponins are derived from monocotyledonous plants of families belonging to Agavaceae, Dioscoreaceae, and Liliaceae (Rai et al. 2021). The critical micelle concentrations of weakly acidic saponin, extracted from *Sapindus mukorossi* are 0.13 and 0.105% at 20 °C and 30 °C respectively (Kommalapatu et al. 1997; Muntaha and Khan 2015).

12.2.2 Biosurfactants from Animals

Biosurfactants of animal origin usually cover lecithin, gelatin, casein, wool fat, cholesterol, and wax. Lecithin is generally obtained from egg yolk. Gelatin is manufactured via the dilute acid- or base-induced partial hydrolysis of collagen. Being a fairly poor protein surfactant, the emulsifying properties of gelatin can be improved by attaching hydrophobic side chains via enzymatic treatment or by partnering with other surface-active agents. Casein is a heterogeneous phosphoprotein finding its use as an emulsifier, thickener, and gelling agent in the food industry. Albumin from egg, bovine serum, and human serum are popular among protein-based biosurfactants. Bile acids and pulmonary surfactants are physiologically important. The manufacturing and application of animal-derived surfactants are challenging due to expensive animal feedstock, deviations in bulk reproducibility, religious constraints, and stringent acts of government (Xu et al. 2011).

12.2.3 Biosurfactants from Microbes

Microbial biosurfactants are inherently procured from bacteria and yeast belonging to some major species such as *Bacillus subtilis*, *Pseudomonas aeruginosa*, *Acinetobacter calcoaceticus*, *Candida lipolytica*, and *Starmerella (Candida) bombicola*. Long-chain fatty acids and hydroxylated fatty acids frequently comprise the hydrophobic part while carbohydrates, amino acids, carboxylic acids, cyclic peptides, phosphate, or alcohol constitute the hydrophilic portion (Drakontis and Amin 2020; Farias et al. 2021).

12.3 Types of Microbial Biosurfactants

The majority of the biosurfactants are anionic or nonionic. Applications of anionic biosurfactants are popular in cleaning and oil removal processes, while cationic biosurfactants are effective as emulsifiers of skin care commodities, and non-ionic biosurfactants are utilized as low-temperature detergents (Selva Filho et al. 2023). The average molar mass of a biosurfactant varies from 500 to 1500 Da (Santos et al. 2016).

The structures of microbial biosurfactants are diversified. Low molecular weight biosurfactants are competent in decreasing the surface tension and the interfacial tension, whereas high molecular weight biosurfactants are successful in maintaining emulsion stability (Bognolo 1999; Drakontis and Amin 2020).

12.3.1 Glycolipids

Carbohydrates bonded to long, open-chained acids or hydroxyl aliphatic acids by ether and ester linkages constitute glycolipids. A few renowned glycolipids are-rhamnolipids, trehalolipids, and sophorolipids (Sharma et al. 2021).

12.3.2 Lipopeptides and Lipoproteins

Fatty acid chains containing cyclic peptides form the backbone of this class of biosurfactants. For instance, surfactin produced from *Bacillus subtilis,* is formulated of seven amino acids connected to a fatty acid chain via lactone bridge (Sharma et al. 2021). Iturin, another biosurfactant from *Bacillus subtilis,* has high surfactant and antimicrobial properties also like surfactin. Moreover, iturin was found to sustain its functionality throughout a storage span of 6 months at −18 °C and regardless of autoclaving at weak acidic to moderate alkaline conditions. Lichenysins, obtained from *Bacillus licheniformis* showcase great stability under a broad range of temperatures, pH, and salt conditions, analogous to surfactin (Rawat et al. 2020).

12.3.3 Fatty Acids, Phospholipids, and Neutral Lipids

A good quantity of fatty acids and phospholipid biosurfactants are achievable through the growth of some bacteria and yeast on linear alkanes. *Acinetobacter* spp. yields phosphatidylethanolamine-rich vesicles, which emulsify alkanes in water (Santos et al. 2016). Spiculisporic acid produced from the yeast *Penicillium spiculisporum,* consists of a single lactone ring and two carboxyl groups (Sharma et al. 2021).

12.3.4 Polymeric Biosurfactants

Best-known polymeric biosurfactants include alasan, emulsan, liposan, and lipomannan. *Acinetobacter radioresistens* contributes to a strong emulsifier 'alasan', which is a complex of alanine, proteins, and polysaccharides. Emulsan from *A. calcoaceticus* has an average molecular weight of around 1,000 kDa. It is capable of emulsifying hydrocarbons effectively in water, even if applied at a low dosage of 0.001–0.01% (Rawat et al. 2020). Liposan from *Candida lipolytica* is an extracellular water-soluble emulsifier containing 83% carbohydrate and 17% protein in its composition. Good stability of emulsions with a variety of oils and hydrocarbons along with antimicrobial properties can be brought by mannoproteins sourced from *Acinetobacter* sp., *Saccharomyces cerevisiae,* etc. (Sharma et al. 2021).

12.3.5 Particulate Biosurfactants

Particulate biosurfactants make partitions of extracellular membrane vesicles to form a microemulsion, capable of influencing the alkane uptake by microbial cells. Such vesicles were reported in the cells of *Acinetobacter* sp. HO1-N (Vijayakumar and

Saravanan 2015). In comparison to the outer membrane of the same microbe, the purified vesicles exhibit an increase in phospholipids and polysaccharides by 5 times and 360 times respectively. Generally, hydrocarbon-degrading microbes like cyanobacteria and non-pathogenic bacteria own this biosurfactant activity (Rawat et al. 2020).

A summary of derivatives of microbial biosurfactants based on their molecular weight, origin, and chemical characteristics is provided in Table 12.1 and the chemical structures of a few important microbial biosurfactants are illustrated in Fig. 12.1 (Banat et al. 2010).

12.4 Properties of Biosurfactants

The significant properties of biosurfactants, making them competent to substitute synthetic surfactants are discussed as mentioned below:

12.4.1 Surface and Interfacial Activity

The addition of biosurfactants in very low concentrations can efficiently reduce different solutions' surface and interfacial tensions because of their lower CMC values. The lower CMC value corresponds to the much smaller number of molecules, sufficient to wrap the entire interface. Since the molecular size of the biosurfactants is large containing numerous branching chains, the maximum number of molecules adsorbed onto the interface becomes less (Johnson et al. 2021).

12.4.2 Stability to pH and Temperature

The functionality of many biosurfactants remains unaffected by the change in pH and temperature of the surroundings. For instance, lichenysin can survive a pH of 4.5–9.0 and temperatures up to 50 °C. Likewise, a lipopeptide originating from *Bacillus subtilis* LB5a is a good emulsifier withstanding pH 4–10, temperatures up to 120 °C, and a salty environment containing sodium chloride up to 10% (w/v) (Purwasena et al. 2019; Gayathiri et al. 2022).

12.4.3 Biodegradability and Low Toxicity

Unlike synthetic surfactants, microbe-derived surfactants are easy to degrade and mainly suitable for environmental practices such as bioremediation and dispersion

Table 12.1 Types of microbial biosurfactants based on molecular weight, origin, and chemical characteristics

Types of biosurfactants			Microbial origin	Chemical characteristics		References
				Surface tension (mN/m)	Critical micelle concentration CMC (mg/L)	
Low molecular weight	Glycolipids	Rhamnolipids	*Pseudomonas aeruginosa*	25–38	0.1–20	Desai and Banat (1997), Nikolova and Gutierrez (2021), Sundaram et al. (2024)
		Sophorolipids	*Candida bombicola*			
		Trehalolipids	*Rhodococcus erythropolis*			
	Lipopeptides	Iturin	*Bacillus subtilis*	26.5–33	12–160	Desai and Banat (1997), Nikolova and Gutierrez (2021), Nagtode et al. (2023)
		Viscosin	*Pseudomonas fluorescens*			
		Surfactin	*Bacillus subtilis, Bacillus nealsonii*			
		Lichenysin	*Bacillus lichenformis*			
	Fatty acids		*Corynebacterium lepus*	30–32	150	Beebe' and Umbreit (1971), Cooper et al. (1979), Macdonald et al. (1981), Desai and Banat (1997), Sharma et al. (2022)
	Phospholipids		*Thiobacillus thiooxidans, Klebsiella pneumoniae* WME02			
	Neutral lipids		*Nocardia erythropolis*			
High molecular weight	Polymeric	Emulsan	*Acinetobacter calcoaceticus* RAG-1	29–32	20–58	Nikolova and Gutierrez (2021), Sharma et al. (2022), Baccile and Poirier (2023)
		Liposan	*Candida lipolytica*			
		Alasan	*Acinetobacter radioresistens* KA53			

(continued)

Table 12.1 (continued)

Types of biosurfactants			Microbial origin	Chemical characteristics		References
				Surface tension (mN/m)	Critical micelle concentration CMC (mg/L)	
	Particulate	Vesicles	*Acinetobacter calcoaceticus*	Diameter: 20–50 nm; Buoyant density 1.158 g/cm^3		Käppeli and Finnerty (1979), De et al. (2015)
		Whole cells	Variety of bacteria			

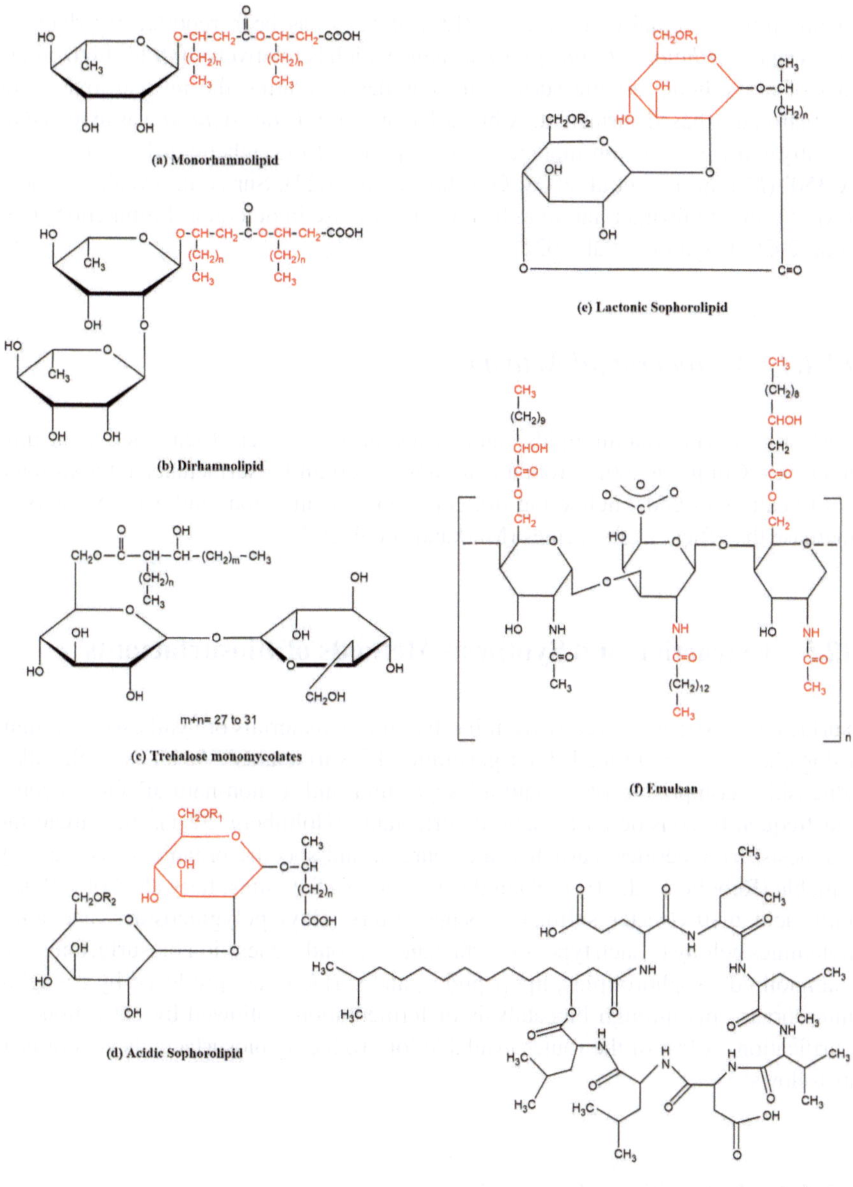

Fig. 12.1 Chemical structures of a few microbial biosurfactants (Adapted from the reference Banat et al. 2010 © Springer-Verlag)

of oil spills. A lethal concentration (LC) of 50% has been reported for rhamnolipids against *Photobacterium phosphoreum*, which is relatively 10 folds higher than that of a synthetic anionic surfactant, thus demonstrating the intense toxicity of the petroleum-based surfactant. A biosurfactant from *P. aeruginosa* was assessed as slightly non-toxic and non-mutagenic in comparison to a synthetic surfactant 'Marlon A-350' (Muthusamy et al. 2008; Gayathiri et al. 2022). Surfactin revealed its non-toxicity, and nonirritant nature indicating its safe use in detergent formulations (Fei et al. 2020; Gayathiri et al. 2022).

12.4.4 Antimicrobial Activity

Surfactin, iturin, rhamnolipids, and fengysin have been found active against microbes. Common antimicrobial activities of certain biosurfactants are associated with healing wounds, delivering drugs to spots of infection, and even boosters to increase the efficacy of vaccines (Markande et al. 2021).

12.5 Extraction and Synthesis Methods of Biosurfactants

Surfactants, extracted directly from bio-based raw materials or synthesized straight using chemicals are termed 'First-generation Biosurfactants'. As an example, alkyl glucosides comprising of a 'natural' sugar unit and a 'non-natural' fatty alcohol are frequently considered as natural surfactants (Holmberg 2001). To initiate the synthesis, triglycerides, carbohydrate sources, and specific organic acids are very suitable (Rebello et al. 2014; Nagtode et al. 2023). Saponins, fatty alcohol sulfates, fatty acid methyl ester sulfonates, sugar esters, alkyl polyglucosides, and alkanolamines belong to such types of surfactants. Second-generation biosurfactants like rhamnolipids, sophorolipids, lipopeptides, and surfactin are produced by engaging microorganisms through biocatalysis or fermentation, followed by extraction and purification. A few of the routes available for producing biosurfactants are outlined as follows:

12.5.1 Extraction of Saponins

Saponin can be obtained conventionally through simple techniques like maceration, Soxhlet extraction, and reflux extraction. Maceration employs water as the extraction solvent. The seed-free outer pericarps of soapnut from the plant *Sapindus mukorossi* are oven-dried at 50 °C for about 2 days before grinding. The ground pericarps are sieved through 840 μm-sized pores and added to deionized water (for example, 1 g per 100 ml water to make a 1% solution) under continuous magnetic stirring at room temperature for 3 h. The supernatant found after the centrifugation at 10,000 rpm for 45 min, is filtered through a 0.44 mm pre-filter followed by a 0.45 μm membrane filter (Kommalapatu et al. 1997).

In the cases of Soxhlet extraction and Reflux extraction, the powdered plant materials are defatted by oleophilic solvents like petroleum ether or n-hexane first. Methanol or ethanol is then used for the extraction. Next, the alcoholic crude extract is evaporated using a rotary evaporator to produce the aqueous solution containing concentrated saponin. Via a separating funnel, the solution is then extracted with n-butanol in a 1:3 ratio. Finally, drying of the butanol fraction yields the active saponin extract. A few advanced methods utilizing ultrasound, microwave, and high-pressure technologies are also in practice to ensure high extraction efficiency in less time and at less consumption of solvents (Rai et al. 2021).

12.5.2 Biosynthesis

The polar and non-polar components of biosurfactants are synthesized by enzyme-catalyzed pathways of hydrocarbon and carbohydrate metabolisms. According to researchers, four possibilities are there for the biosynthesis of hydrophilic (HL) and hydrophobic (HB) parts of biosurfactants (De et al. 2015; Nagtode et al. 2023):

- Both HL and HB parts are manufactured *de novo* by two independent routes.
- The HL part is prepared *de novo* and the HB part is induced by substrate.
- The HB part is synthesized *de novo* and the HL part is dependent on substrate.
- The synthesis of both HL and HB parts is substrate-dependent.

12.5.3 Fermentation

Microorganisms acquire carbon and energy while converting complex organic substrates into important products like ethanol, amino acids, polysaccharides, and vitamins through fermentation. Based on the nature of microorganisms, submerged (SmF) and solid-state fermentations (SSF) can produce cost-efficient biosurfactants by utilizing byproducts, wastes, or effluents from multiple industries such as agro, starch, dairy, and sugar (Pardhi et al. 2022). Figure 12.2 illustrates the synthesis of microbial biosurfactants from plant biomass in biorefinery contexts (Barbosa et al. 2022).

SmF is suitable for the manufacturing of biosurfactants from bacteria and yeasts because of their requirement of water for optimal growth. Though the purification of biosurfactants thus produced is pretty easy, there is a risk of losing some important compounds from the liquid fragment during recovery. On the other hand, SSF usually utilizes solid renewable wastes including banana peel, cassava bagasse, molasses, rice husk, wheat bran, etc., as a substrate. The economic recovery and downstream processes are crucial to verify the commercial feasibility of a bioprocess since those

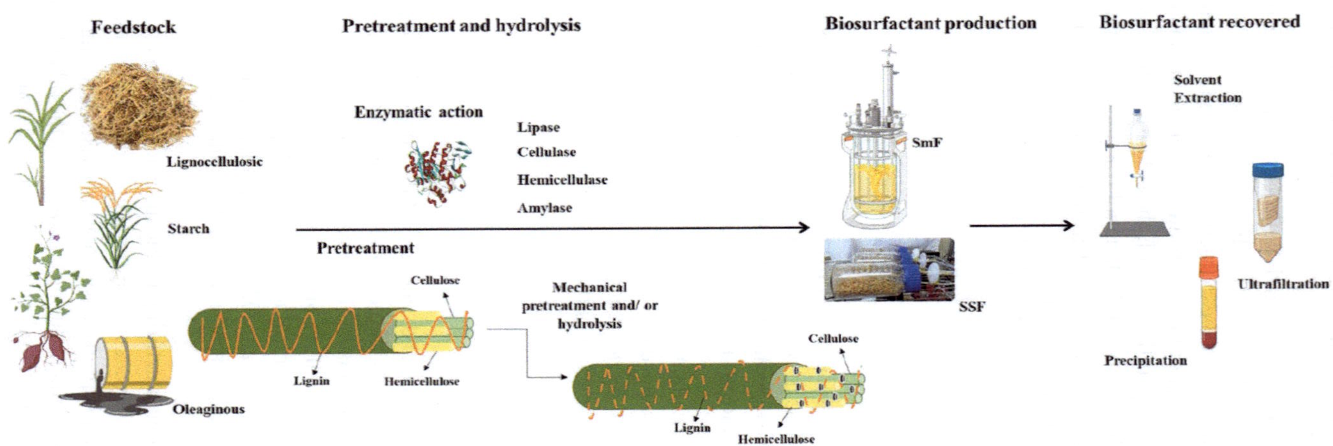

Fig. 12.2 Synthesis route of microbial biosurfactants in biorefinery using plant biomasses and agro-industrial byproducts (Reprinted from the reference Barbosa et al. 2022 © Sociedade Brasileira de Quimica)

induce more than 50% of total production costs. The quality of the recovered products at various stages is improved through multi-step recovery strategies with a set of purification and concentration steps (Pardhi et al. 2022).

12.5.4 Enzymatic Methods

The most popular example of a biocatalytic route is using lipases to produce sugar esters. Glycolipid surfactants are based on a sugar moiety and a fatty acid tail, obtained from renewable substrates. Fungi-based extracellular lipase enzymes (EC 3.1.1.3), are considered most prospective to hydrolyze lipids in aqueous media. However, in the presence of a moderate amount of water, ester bonds are synthesized (Barbosa et al. 2022). Amino acid-based esters and amine biosurfactants can be exclusively synthesized from other lipases or enzymes (Nagtode et al. 2023).

12.5.5 Ultrasonication

Ultrasound belongs to sounds, out of earshot of humans. It can be used in stimulating and controlling microorganism activity as well as other processes like fermentation, through probes or bath ultrasound systems. Ultrasound of low intensity (power < 1 W/cm^2 and frequency > 100 kHz) causes enhanced mass transfer through the boundary layer and the thickness thus gets narrowed. This results in altering the active sites of enzymes, eventually changing the enzymatic activity. In contrast, high-intensity ultrasound (power > 10 W/cm^2 and frequency < 100 kHz) is responsible for disrupting cellular membranes, triggering damage to vital macromolecules. Hence, the application of mild ultrasound can increase efficiency and productivity (Behzadnia et al. 2020; Nagtode et al. 2023).

12.5.6 Hydrodynamic Cavitation

The theory of generating vapor-filled cavities in this method is based on the variation of local pressure within a liquid. The cavities are formed and grown in a low-pressure region. The persistent reduction in pressure eventually collapses the cavities releasing high-pressure (>100 MPa) shock waves and highly reactive free radicals with extreme temperatures up to 10,000 K. Thus, hydrodynamic cavitation could change the target matrix physically and chemically (Panda et al. 2020; Ranade et al. 2022). In comparison to ultrasonication, a significantly greater amount of lipid can be extracted from wet microalgae via venturi-type hydrodynamic cavitation (Nagtode et al. 2023).

12.5.7 Microwave-Assisted Extraction

In microwave-assisted extraction (MAE), microwaves for heating are combined with typical solvent extraction. The exposure of the target compounds and cells to the solvent causes extraction by rupturing the cells. Microwaves assist in expanding the cells through localized heating of the free water molecules available in the desired compounds and cell matrix. Subsequently, the expected metabolites easily flow out of the cells (Nagtode et al. 2023).

12.6 Applications of Biosurfactants in Textiles

12.6.1 Wetting

Surfactants reduce the interfacial tension at the solid–liquid interface and enhance the wetting behavior (Rai et al. 2021). In an investigation on bleached cotton Poplin fabric of 89.29 g/m^2, saponin derived from *Sapindus mukorossi* (Ritha) was able to lower the surface tension to 35.30 mN/m along with a good wetting time (Pradhan and Bhattacharyya 2017).

12.6.2 Emulsification

The adsorption of surfactants at the oil–water interface incurs repulsive interactions between the droplets and generates an energy barrier against rupture to form a stable emulsion. At higher concentrations, soapnut solution is found to form very stable emulsions that remain unseparated for almost 120 min. Further increase in the surfactant concentration decreased the emulsion stability, perhaps due to rapid droplet coalescence (Pradhan and Bhattacharyya 2017). Another study of the natural surfactant, derived from *Sapindus laurifolius* revealed the emulsion stability for more than 180 min when used near the CMC (Saxena et al. 2018).

12.6.3 Cleaning

In order to eliminate natural polymers, mineral oils, natural oils (fatty acid-triglycerides), fatty acid esters, ethoxylated substrates, etc., desizing, bleaching, and washing processes are performed prior to coloration and finishing. Biosurfactants have the potential to degrade fiber preparations such as lubricants ecologically and further mineralize to CO_2 and H_2O (Kesting et al. 1996). They can emulsify the hydrophobic impurities from the fabric surface and suspend them in the aqueous media to facilitate the cleaning.

Sustainable scouring of cotton fabric separately with saponins extracted from soapnut (*Sapindus mukorossi*) and shikakai (*Acacia concinna*) was commented as cleaning-efficient due to the associated rates of water absorption of 65.62% and 65.02% respectively, compared to that of 75.30% from conventional scouring using sodium hydroxide (Thakker 2021). Enzymatic scouring coupled with low concentrations of soapnut extract was found to confer good hydrophilicity denoted by an average wetting time of 4.86 s, sacrificing 3.1%–3.8% weight. Moreover, no significant difference between the alkali-scoured and the bioscoured fabrics was observed after the subsequent dyeing (Raafi et al. 2023). Upon testing against simulated dirt prepared from coconut oil, paraffin wax, and hexane, *Sapindus mukorossi* (soapnut) extract and synthetic cleaning agent displayed notable cleanability at a high concentration while the efficiency observed for *Zephyranthes carinata* (Pyagi Phool) extract was at lower concentrations probably due to the early onset of micelles (Pradhan and Bhattacharyya 2017). Saponin was proven efficient in lowering the residual ash and grease on the wool, as well as making the wool brighter and less yellowish which were not statistically inconsistent from those of wool scoured commercially using NPEO (nonylphenol ethoxylates) (Leighs et al. 2019). Saponification of seed-derived oil from neem (*Azadirachta indica*) plant with sodium hydroxide solution (30% w/v) was used to extract a natural surfactant characterized with alkaline pH, foaming of 1.5 cm height, CMC of nearly 0.12 g/mL, HLB value higher than 10 and weak-to-moderate antimicrobial activity (Siqueira de Azevedo Sá et al. 2022).

Rhodococcus globerulus-derived surfactant has been found to degrade various pure compounds like triglycerides, fatty acid esters, modified fatty acid esters (spinning oil), and polyoxyethylene esters. Glycolipids-surfactant from *Rhodococcus erythropolis* has been proven to remove a greater portion of hydrophobic substrates (coconut oil) from polyester fabric compared to that involving a surfactant-free nutrient system as a washing agent (Kesting et al. 1996).

Combined application of a rhamnolipid (RhL) biosurfactant with pectinase in cotton fabric scouring was determined as a successful competitor to alkali used in traditional scouring. Moreover, the fabric texture gets damaged in conventional scouring but the bioscouring was gentle to remove the required hydrophobic impurities without deteriorating the substrate. Within the pectinase concentration of 4–6 g/L, RhL of 100 mg/L resulted in 2.8–3.0% loss in weight, CIE whiteness of 27.00–30.81, and water wettability of less than 1 s (Raza et al. 2014).

In textile washing, biosurfactants segregated from green and food urban remnants were characterized with similar efficiency as selective commercial anionic and nonionic surfactants when used above the relatively low CMC. The suffering of those biosurfactants from water hardness and turning the fabrics yellowish could be improved by using the mixture of 1:1 w/w biosurfactants and chemical surfactants (Savarino et al. 2010).

12.6.4 Leveling

Humic acid-like biosurfactants (cHAL) derived from compost have been conveniently used in dyeing Nylon 6 microfiber by acidic and dispersive colorants individually. The mechanism governing the rate and the yield of the water-soluble dye uptake by nylon can be associated with affinity and/or the ability of the surfactants in the medium to accumulate in micellar form. In the second case, the free dye concentration in the medium is controlled by the diffusion of the dye into the micellar core of cHAL and the foundation of a complex by polar and/or hydrophobic interactions. The weakly acidic, anionic cHAL was expected to work according to the later mechanism since its concentration was quite low and was capable of yielding micelles at a lower CMC of 0.403 g/L, compared to those of the other two synthetic surfactants used in the experiment. Due to the poor aqueous solubility, water-insoluble dyes are distributed non-homogenously all over the fabric's solid phase. Diffusion of dye into the micellar core of cHAL can confer enhanced water solubility. Excellent color levelness along with superior color intensity was obtained from cHAL at more suitable operational conditions than that achieved with the experimental synthetic surfactants (Savarino et al. 2007; Montoneri et al. 2008).

12.6.5 Solubilization

Bio-based surfactants from different biomass sources were found with no significant performance differences in the disperse dyeing of cellulose acetate fabrics, compared to that containing the synthetic surfactants. Since the biosurfactants cooperate in solubilizing the dyes at the molecular level, high-quality products from dyeing can be achieved. Upon reaching the CMC and at the beginning of forming the final micellar aggregate, no more dye is solubilized with the increment of biosurfactant concentration. The dye solubilization resumes after forming the final micellar aggregate and building up a critical number of micelles (Savarino et al. 2009).

RhL surfactant incorporated disperse dyeing of polyester fabric was found to yield comparable shade evenness, excellent tensile and colorfastness properties without damaging the fabric surface. At the CMC of 45 mg/L, the produced biodispersant was reported to reduce the tension across the water/oil (n-hexadecane) interface from ~50 mN/m to ~1 mN/m at room temperature. Upon increasing the concentration of the dye, Rhl at 1.5 times the CMC (Above CMC or A-CMC) contributed to better penetration of dye molecules into the amorphous regions of polyester. Therefore, the reduction in inter-fiber spaces resulted in greater tensile strength of the dyed fabric. This evidence of better dispersibility was also supported by the corresponding increase in the K/S value observed while increasing the Rhl concentration from CMC to A-CMC (Khubaib et al. 2021).

12.6.6 Dye Carrier

Conventional reactive dyeing consumes massive quantities of water and chemicals and poses catastrophic effects on the environment by generating a significant number of effluents. In recent days, concerns have been focused on inventing alternate dyeing systems by shifting the dyeing medium from a water-based system to a solvent-based non-aqueous system. Reverse micellar dyeing of textiles is one of the most prominent non-aqueous techniques. As a uniform carrier of dye within a nonpolar oil phase, reverse micelles structurally resemble nano-dimensional droplets of a polar liquid phase, confined by a unilayer of surfactant molecules (Tang and Kan 2020). Utilization of RhL reverse micelles in octane solvent was explored as a one-bath, one-step approach to obtain a better color yield, good to excellent levelness, washing, and crocking fastness of reactive dyed cotton fabric with adding neither salt nor alkali. Moreover, the fabric surface was hardly damaged by the chemicals (Tang et al. 2023a). Upon investigating the influence of solvent medium in RhL-based reverse micellar reactive dyeing, a lack of levelness was observed for cotton fabrics maintained in decamethylcyclopentasiloxane (D5), and paraffin liquid than that of heptane and nonane solvent systems (Tang et al. 2023b).

12.6.7 Wastewater Treatment

Surfactants can solubilize organic non-polar dyes in their hydrophobic core. Surfactant micelles are able to solubilize and bind pollutants in the aqueous system, facilitating their removal via Micellar-enhanced Ultrafiltration (MEUF). In an ultrafiltration membrane process, the hydrodynamic size of pollutant solutes is increased by dosing surfactant above the critical micelle concentration. Thus the large-sized pollutants get rejected by membrane pores, which could ultimately be discarded by a low molecular weight cut-off nanofiltration or reverse osmosis membranes. Although the use of chemically synthesized surfactants in this field causes secondary pollution due to the slow degradation process, persistency, and chronic toxicity, the biosurfactants are both functionally and ecologically adaptable. Saponin from *Sapindus mukorossi* was noticeable in solubilizing acid and basic dyes (Samal et al. 2017). Rhamnolipid has fast sorption rates toward the dye Direct Orange 2GL from wastewater. Theoretically, all the dyes could be removed from the effluent by using higher mass values of the biosurfactant because the adsorption was proportionate to the mass of the RhL (da Silva et al. 2021). Experimental results revealed the supremacy of rhamnolipid over sodium dodecyl sulfate (SDS) in removing the basic dye crystal violet from an aqueous solution with technical, economical, and ecological advantages (Verma et al. 2020).

12.7 Conclusions

The growing interest in sustainability has fostered the exploration of biosurfactants in versatile industrial applications including textiles. Their bio-friendly nature has been proven to avoid the problems caused by petroleum-based, chemically synthesized surfactants. However, the current challenges for biosurfactants are their economic viability, large-scale production, and improvement of yields. Moreover, the ready biodegradability of the bio-based surfactants might risk reactivity at the point of use due to the relatively short shelf life. Continuous research and developments in the quest for suitable biosurfactants satisfying the superiority of performance and the improvements to the addressed bottlenecks could help in attaining sustainability in textile chemistry.

Declaration

The author declare that he has no known competing financial interests or personal relationships that could have appeared to influence the work reported in this chapter. Generative AI and AI-assisted technologies have been used in the writing process to improve the readability and language of the work only. After using this tool, the author carefully reviewed and revised the content to ensure accuracy and quality. The author take full responsibility for the final version of this publication.

Acknowledgements The author is thankful to Dr. Md. Rashedul Islam of the Department of Wet Process Engineering, BUTEX, for the invitation and inspiration to contribute to this book and the cooperation in providing access to multiple scientific articles.

References

Alwadani N, Fatehi P (2018) Synthetic and lignin-based surfactants: challenges and opportunities. Carbon Res Convers 1:126–138. https://doi.org/10.1016/j.crcon.2018.07.006

Baccile N, Poirier A (2023) Microbial bio-based amphiphiles (biosurfactants): general aspects on critical micelle concentration, surface tension, and phase behavior. In: Biosurfactants. Elsevier, pp 3–31

Banat IM, Franzetti A, Gandolfi I et al (2010) Microbial biosurfactants production, applications and future potential. Appl Microbiol Biotechnol 87:427–444. https://doi.org/10.1007/s00253-010-2589-0

Barbosa FG, Ribeaux DR, Rocha T et al (2022) Biosurfactants: sustainable and versatile molecules. J Braz Chem Soc 33:870–893. https://doi.org/10.21577/0103-5053.20220074

Beebe' JL, Umbreit WW (1971) Extracellular Lipid of Thiobacillus thiooxidans. J Bacteriol 108:612–614

Behzadnia A, Moosavi-Nasab M, Ojha S, Tiwari BK (2020) Exploitation of ultrasound technique for enhancement of microbial metabolites production. Molecules 25. https://doi.org/10.3390/MOLECULES25225473

Bezerra KGO, Silva IGS, Almeida FCG et al (2021) Plant-derived biosurfactants: extraction, characteristics and properties for application in cosmetics. Biocatal Agric Biotechnol 34. https://doi.org/10.1016/j.bcab.2021.102036

Bognolo G (1999) Biosurfactants as emulsifying agents for hydrocarbons. Colloids Surf A Physicochem Eng Asp 152:41–52
Cooper DG, Zajic JE, Gerson DF (1979) Production of surface-active lipids by corynebacterium lepus. Appl Environ Microbiol 37:4–10
da Silva VL, Dilarri G, Mendes CR et al (2021) Rhamnolipid from Pseudomonas aeruginosa can improve the removal of Direct Orange 2GL in textile dye industry effluents. J Mol Liq 321. https://doi.org/10.1016/j.molliq.2020.114753
De S, Malik S, Ghosh A et al (2015) A review on natural surfactants. RSC Adv 5:65757–65767. https://doi.org/10.1039/c5ra11101c
Desai JD, Banat IM (1997) Microbial production of surfactants and their commercial potential. Microbiol Mol Biol Rev 61:47–64
Drakontis CE, Amin S (2020) Biosurfactants: formulations, properties, and applications. Curr Opin Colloid Interface Sci 48:77–90. https://doi.org/10.1016/j.cocis.2020.03.013
Expert Market Research (2024) Global green surfactants market report and forecast 2024-2032. https://www.expertmarketresearch.com/pressrelease/global-green-surfactants-market. Accessed 11 Jun 2024
Farias CBB, Almeida FCG, Silva IA et al (2021) Production of green surfactants: market prospects. Electron J Biotechnol 51:28–39. https://doi.org/10.1016/j.ejbt.2021.02.002
Fei D, Zhou GW, Yu ZQ et al (2020) Low-toxic and nonirritant biosurfactant surfactin and its performances in detergent formulations. J Surfactants Deterg 23:109–118. https://doi.org/10.1002/jsde.12356
Fiechter A (1992) Biosurfactants: moving towards industrial application. Trends Biotechnol 10:208–217. https://doi.org/10.1016/0167-7799(92)90215-H
Gayathiri E, Prakash P, Karmegam N et al (2022) Biosurfactants: potential and eco-friendly material for sustainable agriculture and environmental safety—a review. Agronomy 12. https://doi.org/10.3390/agronomy12030662
Holmberg K (2001) Natural surfactants. Curr Opin Colloid Interface Sci 6:148–159
Jahan R, Bodratti AM, Tsianou M, Alexandridis P (2020) Biosurfactants, natural alternatives to synthetic surfactants: physicochemical properties and applications. Adv Colloid Interface Sci 275. https://doi.org/10.1016/j.cis.2019.102061
Johnson P, Trybala A, Starov V, Pinfield VJ (2021) Effect of synthetic surfactants on the environment and the potential for substitution by biosurfactants. Adv Colloid Interface Sci 288. https://doi.org/10.1016/j.cis.2020.102340
Käppeli O, Finnerty WR (1979) Partition of alkane by an extracellular vesicle derived from hexadecane-grown Acinetobacter. J Bacteriol 140:707–712
Kesting W, Tummuscheit M, Schacht H, Schollmeyer E (1996) Ecological washing of textiles with microbial surfactants. Interfaces, surfactants and colloids in engineering 125–130
Khubaib MA, Raza ZA, Abid S et al (2021) Cell-free culture broth of pseudomonas aeruginosa—an alternative source of biodispersant to synthetic surfactants for dyeing the polyester fabric. J Surfactants Deterg 24:343–355. https://doi.org/10.1002/jsde.12485
Kommalapatu RR, Valsaraj KT, Constant WD, Roy D (1997) Aqueous solubility enhancement and desorption of hexachlorobenzene from soil using a plant-based surfactant. Water Res 31:2161–2170
Kumari V (2022) Biosurfactant as an antimicrobial and biodegradable agent a review. In: Microbial surfactants. CRC Press, pp 139–157
Leighs SJ, McNeil SJ, Ranford SL (2019) The application of biosurfactants for scouring wool. Color Technol 135:48–52. https://doi.org/10.1111/cote.12370
Macdonald CR, Cooper DG, Zajic JE (1981) Surface-active lipids from Nocardia erythropolis grown on hydrocarbons. Appl Environ Microbiol 117–123
Markande AR, Patel D, Varjani S (2021) A review on biosurfactants: properties, applications and current developments. Bioresour Technol 330. https://doi.org/10.1016/j.biortech.2021.124963

Moldes AB, Rodríguez-López L, Rincón-Fontán M et al (2021) Synthetic and bio-derived surfactants versus microbial biosurfactants in the cosmetic industry: an overview. Int J Mol Sci 22:1–23. https://doi.org/10.3390/ijms22052371

Montoneri E, Savarino P, Bottigliengo S et al (2008) Humic acid-like matter isolated from green urban wastes. Part II: Performance in chemical and environmental technologies. Bioresources 3:217–233. https://doi.org/10.15376/biores.3.1.217-233

Moure A, Sineiro J, Domínguez H, Parajó JC (2006) Functionality of oilseed protein products: a review. Food Res Int 39:945–963. https://doi.org/10.1016/j.foodres.2006.07.002

Muntaha ST, Khan MN (2015) Natural surfactant extracted from Sapindus mukurossi as an eco-friendly alternate to synthetic surfactant-a dye surfactant interaction study. J Clean Prod 93:145–150. https://doi.org/10.1016/j.jclepro.2015.01.023

Muthusamy K, Gopalakrishnan S, Kochupappy Ravi T, Sivachidambaram P (2008) Biosurfactants: properties, commercial production and application. Curr Sci 94:736–747

Nagtode VS, Cardoza C, Yasin HKA et al (2023) Green Surfactants (Biosurfactants): a petroleum-free substitute for sustainability–comparison, applications, market, and future prospects. ACS Omega 8:11674–11699. https://doi.org/10.1021/acsomega.3c00591

Nguyen TTL, Edelen A, Neighbors B, Sabatini DA (2010) Biocompatible lecithin-based microemulsions with rhamnolipid and sophorolipid biosurfactants: formulation and potential applications. J Colloid Interface Sci 348:498–504. https://doi.org/10.1016/j.jcis.2010.04.053

Nikolova C, Gutierrez T (2021) Biosurfactants and their applications in the oil and gas industry: current state of knowledge and future perspectives. Front Bioeng Biotechnol 9. https://doi.org/10.3389/fbioe.2021.626639

Panda D, Saharan VK, Manickam S (2020) Controlled hydrodynamic cavitation: a review of recent advances and perspectives for greener processing. Processes 8. https://doi.org/10.3390/pr8020220

Pardhi DS, Panchal RR, Raval VH et al (2022) Microbial surfactants: a journey from fundamentals to recent advances. Front Microbiol 13. https://doi.org/10.3389/fmicb.2022.982603

Pradhan A, Bhattacharyya A (2017) Quest for an eco-friendly alternative surfactant: surface and foam characteristics of natural surfactants. J Clean Prod 150:127–134. https://doi.org/10.1016/j.jclepro.2017.03.013

Purwasena IA, Astuti DI, Syukron M et al (2019) Stability test of biosurfactant produced by Bacillus licheniformis DS1 using experimental design and its application for MEOR. J Pet Sci Eng 183. https://doi.org/10.1016/j.petrol.2019.106383

Raafi SM, Arju SN, Asaduzzaman M et al (2023) Eco-friendly scouring of cotton knit fabrics with enzyme and soapnut: an alternative to conventional NaOH and synthetic surfactant based scouring. Heliyon 9. https://doi.org/10.1016/j.heliyon.2023.e15236

Rai S, Acharya-Siwakoti E, Kafle A et al (2021) Plant-derived saponins: a review of their surfactant properties and applications. Science 3. https://doi.org/10.3390/sci3040044

Ranade VV, Bhandari VM, Nagarajan S et al (2022) Hydrodynamic cavitation. In: Hydrodynamic cavitation: devices, design, and applications, 1st edn. John Wiley & Sons, pp 3–20

Rawat G, Dhasmana A, Kumar V (2020) Biosurfactants: the next generation biomolecules for diverse applications. Environ Sustain 3:353–369. https://doi.org/10.1007/s42398-020-00128-8

Raza ZA, Rehman A, Hussain MT et al (2014) Production of rhamnolipid surfactant and its application in bioscouring of cotton fabric. Carbohydr Res 391:97–105. https://doi.org/10.1016/j.carres.2014.03.009

Rebello S, Asok AK, Mundayoor S, Jisha MS (2014) Surfactants: toxicity, remediation and green surfactants. Environ Chem Lett 12:275–287. https://doi.org/10.1007/s10311-014-0466-2

Samal K, Das C, Mohanty K (2017) Eco-friendly biosurfactant saponin for the solubilization of cationic and anionic dyes in aqueous system. Dyes Pigm 140:100–108. https://doi.org/10.1016/j.dyepig.2017.01.031

Santos DKF, Rufino RD, Luna JM et al (2016) Biosurfactants: multifunctional biomolecules of the 21st century. Int J Mol Sci 17. https://doi.org/10.3390/ijms17030401

Savarino P, Montoneri E, Biasizzo M et al (2007) Upgrading biomass wastes in chemical technology. Humic acid-like matter isolated from compost as chemical auxiliary for textile dyeing. J Chem Technol Biotechnol 82:939–948. https://doi.org/10.1002/jctb.1767

Savarino P, Montoneri E, Bottigliengo S et al (2009) Biosurfactants from urban wastes as auxiliaries for textile dyeing. Ind Eng Chem Res 48:3738–3748. https://doi.org/10.1021/ie801853x

Savarino P, Montoneri E, Musso G, Boffa V (2010) Biosurfactants from urban wastes for detergent formulation: Surface activity and washing performance. J Surfactants Deterg 13:59–68. https://doi.org/10.1007/s11743-009-1133-7

Saxena N, Pal N, Ojha K et al (2018) Synthesis, characterization, physical and thermodynamic properties of a novel anionic surfactant derived from: sapindus laurifolius. RSC Adv 8:24485–24499. https://doi.org/10.1039/c8ra03888k

Selva Filho AAP, Converti A, Soares da Silva R de CF, Sarubbo LA (2023) Biosurfactants as multifunctional remediation agents of environmental pollutants generated by the petroleum industry. Energies (Basel) 16. https://doi.org/10.3390/en16031209

Sharma J, Sundar D, Srivastava P (2021) Biosurfactants: potential agents for controlling cellular communication, motility, and antagonism. Front Mol Biosci 8. https://doi.org/10.3389/fmolb.2021.727070

Sharma N, Lavania M, Lal B (2022) Biosurfactant: A next-generation tool for sustainable remediation of organic pollutants. Front Microbiol 12. https://doi.org/10.3389/fmicb.2021.821531

Siqueira de Azevedo Sá C, Ladchumananandasivam R, Rossi CGFT et al (2022) Characterization of a natural surfactant from an essential oil from neem (Azadirachta indica A. Juss) for textile industry applications. Textile Res J 92:2643–2650. https://doi.org/10.1177/00405175211007518

Sundaram T, Govindarajan RK, Vinayagam S et al (2024) Advancements in biosurfactant production using agro-industrial waste for industrial and environmental applications. Front Microbiol 15. https://doi.org/10.3389/fmicb.2024.1357302

Tang AYL, Kan C (2020) Non-aqueous dyeing of cotton fibre with reactive dyes: a review. Color Technol 136:214–223. https://doi.org/10.1111/cote.12459

Tang AYL, Lee CH, Wang Y, Kan C (2023a) Rhamnolipid (RL) microbial biosurfactant-based reverse micellar dyeing of cotton fabric with reactive dyes: a salt-free and alkali-free one-bath one-step approach. Colloids Surf A Physicochem Eng Asp 658. https://doi.org/10.1016/j.colsurfa.2022.130725

Tang YLA, Lee CH, Wang Y, Kan C (2023b) Effect of solvent dyeing medium on dye encapsulation and non-aqueous dyeing of cotton fabric with Rhamnolipid (RL) microbial biosurfactant-based reactive dye encapsulated reverse micelles. Fibers Polym 24:4323–4335. https://doi.org/10.1007/s12221-023-00371-w

Thakker AM (2021) Sustainable processing of cotton fabrics with plant-based biomaterials Sapindus mukorossi and Acacia concinna for health-care applications. J Text Inst 112:718–726. https://doi.org/10.1080/00405000.2020.1776537

Tucker IM, Burley A, Petkova RE et al (2020) Surfactant/biosurfactant mixing: adsorption of saponin/nonionic surfactant mixtures at the air-water interface. J Colloid Interface Sci 574:385–392. https://doi.org/10.1016/j.jcis.2020.04.061

van Nieuwenhuyzen W (2010) Lecithin and other phospholipids. In: Surfactants from renew able resources. Wiley, pp 191–212

Verma SP, Mallela NR, Sarkar B (2020) An efficient removal of crystal violet from aqueous solution using rhamnolipid micellar solubilization followed by ultrafiltration and modeling of flux decline. J Environ Chem Eng 8. https://doi.org/10.1016/j.jece.2020.104443

Vijayakumar S, Saravanan V (2015) Biosurfactants-types, sources and applications. Res J Microbiol 10:181–192

Xu Q, Nakajima M, Liu Z, Shiina T (2011) Biosurfactants for microbubble preparation and application. Int J Mol Sci 12:462–475. https://doi.org/10.3390/ijms12010462

Chapter 13
Recent Advances in Textile Printing Techniques

Md. Belal Uddin Rabbi, Tabassum Ferdous, Rasheda Begum Dina, and Md. Rashedul Islam

Abstract The textile industry has witnessed significant advancements in printing technologies, reshaping traditional processes and enabling new creative possibilities. This chapter provides a comprehensive overview of recent innovations in textile printing techniques, focusing on both digital and traditional methods. It explores the integration of digital technologies, such as inkjet and 3D printing, which have revolutionized design flexibility, precision, and sustainability. Additionally, advancements in eco-friendly practices, including water-based inks and sustainable materials, are discussed in relation to their environmental impact. The chapter also discusses the potential of these cutting-edge techniques to transform the future of textile production and design.

Keywords Printing · 3D printing · Inkjet printing · Printed-textile · Screen printing

13.1 Introduction

Textile printing is a very versatile and innovative technique for fabric coloration. It enhances the visual appeal and imparts a contemporary ambiance to textiles. This method employs colorants (dye, pigment) for printing on designated surfaces (front, back) of fabrics utilizing appropriate printing techniques (block, screen, roller printing methods). Traditional printing methods such as block, screen, and roller are used on fabric for their detailed design, high durability, and color consistency from ancient times. Block printing emerged in ancient India and China, where hand-carved wooden blocks were used to transfer the design onto the fabric surface. Similarly, screen printing introduced stencils and mesh screens for vibrant color, dimensional stability, uniform thickness, surface smoothness, and high durability on the fabric surface. However, this printing technique has less adaptation, customization, and

Md. B. U. Rabbi · T. Ferdous · R. B. Dina · Md. R. Islam (✉)
Department of Wet Process Engineering, Bangladesh University of Textiles (BUTEX), Dhaka, Bangladesh
e-mail: rashedul.islam@wpe.butex.edu.bd

S. Maiti et al. (eds.), *Sustainable Coloration Techniques in Textiles*,
https://doi.org/10.1007/978-981-96-4975-4_13

modification and can't create a complex design on the fabric surface. Those traditional printing techniques are not suitable for large-scale production in the industry (Broadbent 2001; Miles 1981; Chatterjee and Ghosh 2020).

As a result, traditional printing industries have shifted to the modern printing industry for sustainability and green fashion. Traditional printing techniques often require large amounts of colorant (Dye, pigment), water, and chemicals (Huang et al. 2017; Wang et al. 2017; Zhang et al. 2016b). Conversely, digital printing methods reduce water and colorant consumption while utilizing environmentally friendly inks. Modern techniques, including inkjet printing, sublimation, and 3D printing (3DP), have transformed the textile business through adaptability and personalization. Digital inkjet printing uses sophisticated printers to directly imprint exact patterns onto fabrics, facilitating intricate designs and vivid colors without the necessity for complex screens or blocks. Sublimation printing entails the transfer of designs onto synthetic materials using heat and pressure, yielding lasting and vivid graphics.3D printing (3DP), or additive manufacturing (AM), uses digital design and layer-by-layer (l-b-l) construction to create intricate three-dimensional structures on demand, across a wide range of scales from microns to meters, rendering it suitable for economical customization and beyond (Chatterjee and Ghosh 2020; Ngo et al. 2018; Islam et al. 2024; Özen et al. 2022). The textile printing industry is transforming due to the influences of fast fashion and increased environmental awareness. The demand for high-quality, distinctive designs produced under strict time constraints has catalyzed progress in printing technology. At the same time, increasing consumer awareness regarding sustainable practices has inspired manufacturers to implement strategies that integrate environmental responsibility with aesthetic innovation. The simultaneous focus on rapidity and ecological responsibility highlights the necessity for ongoing technological progress. Consequently, AI, shown by the development of ChatGPT, an AI-driven language model capable of developing and analyzing text using machine learning (ML), has provided enhanced solutions to address various issues (Cao et al. 2024b; Zhu et al. 2020; Joshi et al. 2020; Rahmani Dabbagh et al. 2022).

This chapter explores the extensive history of textile printing, examining its evolution from conventional techniques to contemporary advancements, as shown in Fig. 13.1. This analysis discusses the technological advancements that have influenced the industry, the environmental and cultural consequences of these developments, and the challenges and opportunities on the horizon. It also describes the relationship between tradition and technology, illustrating how textile printing is evolving to meet the challenges of a rapidly transforming environment while conserving its artistic integrity and cultural importance, as shown in Fig. 13.2.

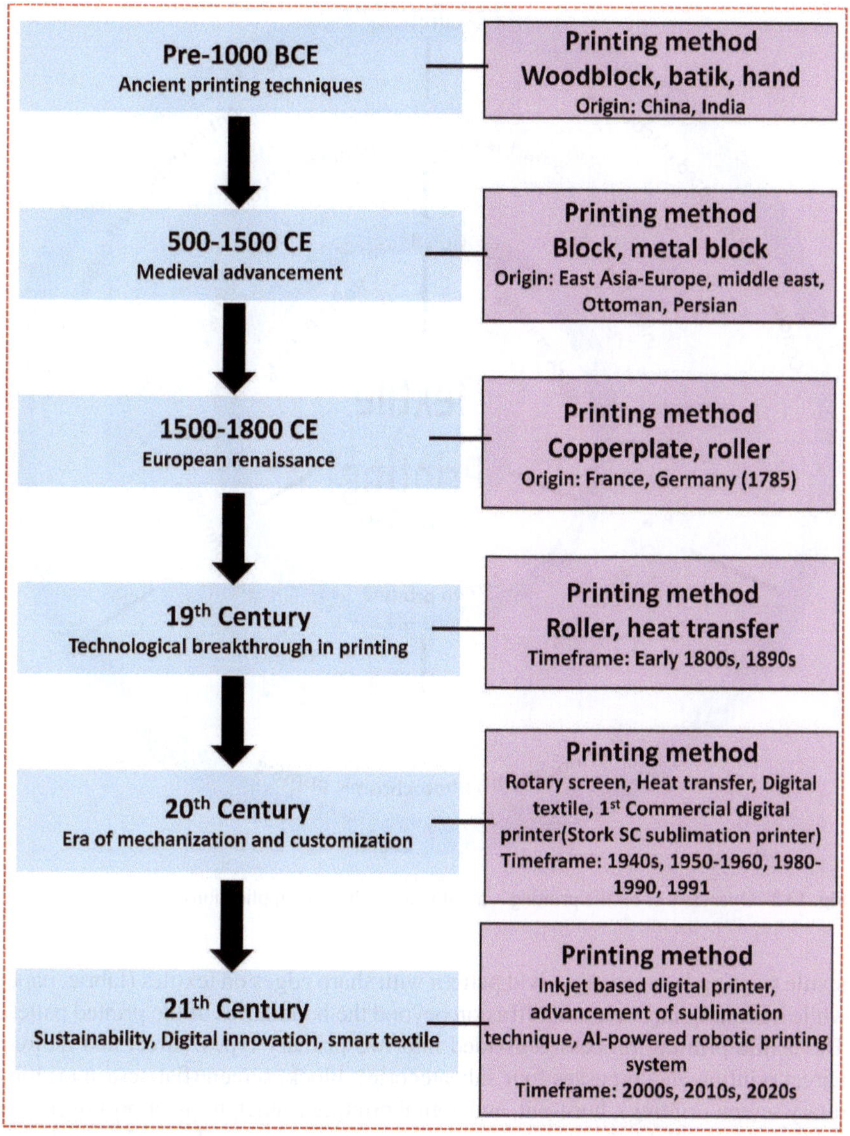

Fig. 13.1 History of textile printing—past, present, and future (Miles 1981)

13.2 Traditional Textile Printing Techniques

Textile printing is a fundamental technique in which colorants (dyes, pigments) are used to create specific patterns or motifs on the surfaces of textiles (fabric, paper) using particular printing techniques and machines. The principal objective of

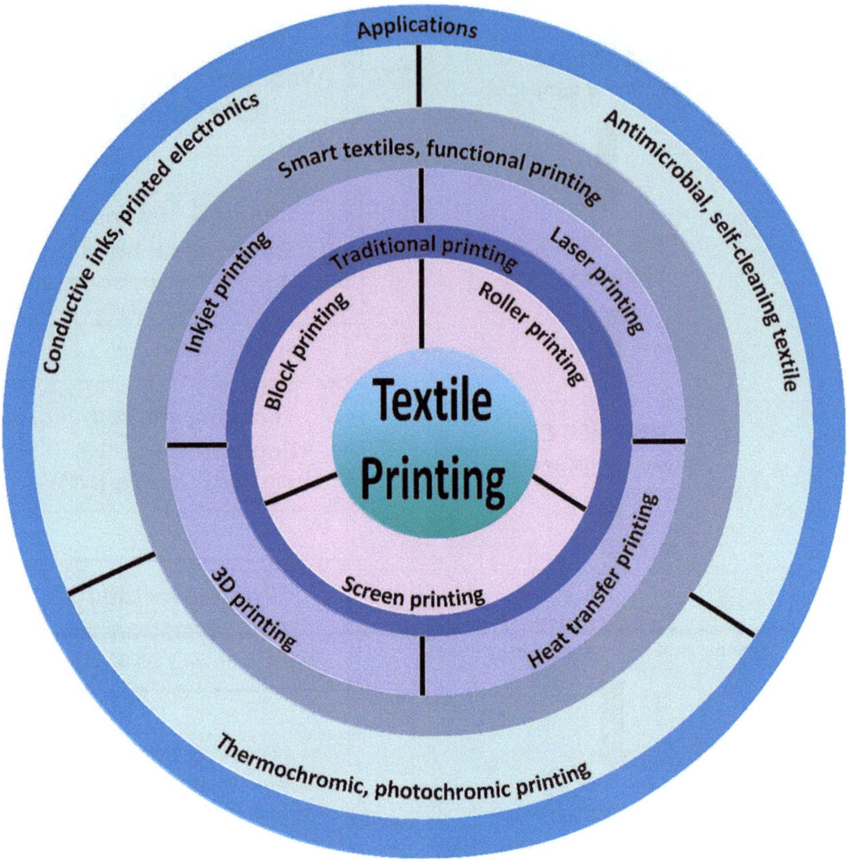

Fig. 13.2 Overview of textile printing—traditional methods to applications

textile printing is to create a vivid pattern with sharp edges on textiles (fabric, paper) while inhibiting any colorant diffusion beyond the boundaries of the printed pattern. The textile printing process is divided into two primary types: direct and indirect. Direct printing encompasses four subcategories: block, screen (flat-screen printing, rotary screen printing), burn-out, and digital printing (inkjet, transfer printing). Indirect printing is divided into two subcategories: resist (physical and chemical resist printing) and discharge printing (conventional discharge printing employing chemical substances, bio discharge printing utilizing environmentally friendly agents) (Ragab et al. 2022).

Block printing is one of the oldest textile printing methods for traditional designs on textiles (fabric, paper). In this method, wooden, metal, or rubber blocks with unique designs are coated with colorants (dyes, pigments) and pressed onto the textiles. This approach is well-suited for complex, detailed designs and offers versatility for small-scale manufacturing. This process utilizes natural dyes, making it an environmentally friendly option. Nonetheless, this process demands significant time and labor, presenting challenges in terms of scalability for extensive production efforts. The screen printing approach is well-suited for various substrates (paper, fabrics) and offers vibrant color with high durability, as shown in Fig. 13.3a. It uses a stencil on a fine mesh screen, and the design is created by blocking a specific mesh area, allowing the dye to pass through the substrate, as shown in Fig. 13.3b. This technique is appropriate for production at a medium to large scale. The roller printing process is another traditional printing process shown in Fig. 13.3c, where engraved metal rollers transfer patterns on substrates. This process is efficient for high-volume production with consistent print quality over large surface areas. However, the high initial costs of setting up engraved rollers make it less flexible for small-scale production (Majumdar et al. 2013). Table 13.1 shows the overview of block, screen, and roller printing methods.

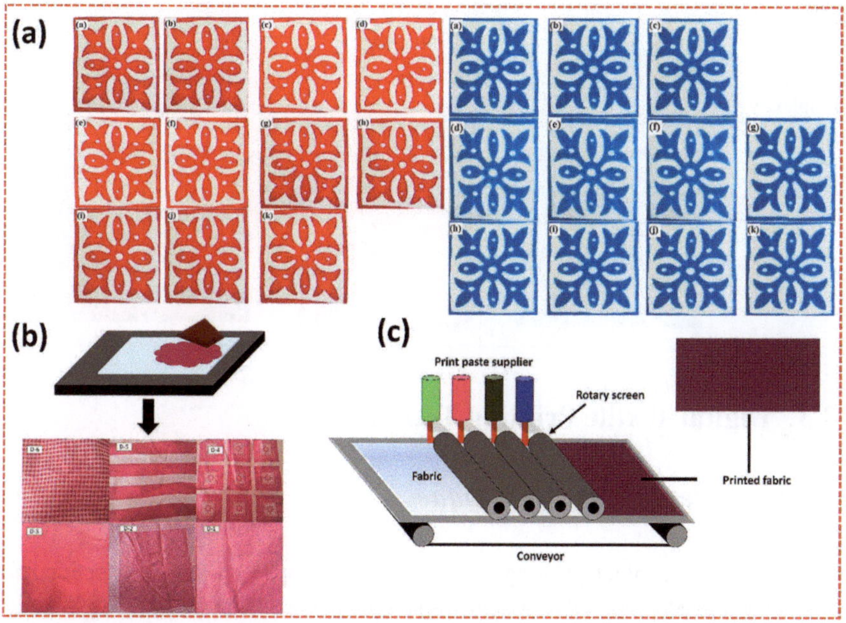

Fig. 13.3 **a** Screen-printed fabric containing reactive red dye and turquoise blue reactive dye. Reproduced (adapted) with permission. Copyright 2024, Elsevier (Akter et al. 2023). **b** Screen printing techniques and designs on cellulosic fabric utilizing azo-reactive dye. Reproduced (adapted) with permission. Copyright 2024, Nature (Siddiqua et al. 2023). **c** Roller printing process

Table 13.1 Overview of block, screen, and roller printing

	Block	Screen	Roller
Fiber	Cotton	All kinds of fibers	All kinds of fibers
Fabric construction	Woven	• Woven • Knit	• Woven • Knit
Advantage	• Detailed design • Versatility • Environmentally friendly	• High durability • Vibrant color	• Consistent print quality • Efficient for high-volume production
Disadvantages	• Time and labor-consuming	• Printing speed: Low • Printing image quality: Low (small-scale production)	• Low-level automation • High cost for engraving rollers
Production scale	Small	Medium-large	Large
Cost	Low	Medium-high	High
Design complexity	High	Medium	Low-medium
Sample		 (a)	

13.3 Digital Textile Printing Techniques

13.3.1 Inkjet Printing Technique

In manufacturing, inkjet printing is an innovative approach that is beneficial for high-volume production duties and prototyping. Diverse techniques are employed to comprehensively eject ink droplets from an inkjet printhead on the fabric's surface, resulting in the formation of text and graphics with remarkable precision. This technique can be classified into two primary categories: (i) Continuous Inkjet Technology (CIJ), (ii) Drop-on-Demand Inkjet Technology (DOD) (Cao et al. 2024a).

The textile industry employs inkjet printing techniques, including direct-to-garment, direct-to-fabric, and sublimation. The inkjet printing system operates without a squeeze mechanism, using automatic screen movement systems instead. The system is completely integrated with CAD, eliminating the need for costs associated with screens, engraving, washing, and storage. Sampling or proofing presents a straightforward and cost-effective approach for short runs, eliminating the need for a screen. As a result, all expenses associated with the printing process are eradicated.

The registration process is seamless, and an extensive array of repeat sizes exists, as all dye is effectively applied to the fabric. The process does not require thickener or paste, demonstrating low water and energy consumption. This technology is compatible with just-in-time delivery systems and facilitates mass customization (Kan and Yuen 2012). The efficacy of inkjet printing depends upon the ink's characteristics. The influence of ink spreading and penetration is significant in determining printing quality. In addition, the general printing quality is greatly affected by the ink's properties, which include pH, surface tension, viscosity, pH, and conductivity (Cao et al. 2024a). The features of inkjet printing is summarized in Table 13.2.

13.3.2 Laser Printing and 3D Printing on Textiles

Laser printing on textiles involves the application of a concentrated laser beam to engrave or cut intricate patterns directly onto fabric. This method exhibits significant versatility and compatibility with diverse materials, encompassing natural (cotton, silk) and synthetic fabric (polyester, nylon). The primary applications encompass the development of intricate designs, the enhancement of texture, and the advancement of fabric customization for fashion and interior decor. Facilitates engraving of raised and recessed patterns, effectively simulating various textures, including embossing and lace. The output consists of intricate, tactile patterns that emulate conventional embroidery or lacework; however, they are achieved with remarkable precision and efficiency (Ozguney 2007; Nourbakhsh and Ashjaran 2012; Tse and Kan 2020).

3D printing on textiles entails the sequential deposition of material to construct elevated structures or decorative elements directly onto the surfaces of fabrics. Thermoplastic materials, including 100% polyester, polyethylene terephthalate (PET), polyurethane, flat knitted surface, polylactic acid (PLA), thermoplastic polyurethane (TPU), and silicone, are frequently utilized due to their inherent flexibility and durability. Various materials used in 3D printing include wax, resins, sand, metals, and polymers. Depending on the specific printing technique, these materials can exist in multiple states, including liquid, solid, powder, or gas (Jandyal et al. 2021). Natural and synthetic fibers, like cotton, nylon polymers, and leather, are also available for 3D printing, and recent advancements in material science are expected to bring more textile fiber-based materials into 3D printing to incorporate more in the fashion industry (Chatterjee and Ghosh 2020). A new generation of innovative materials, responsive to stimuli such as heat or water, is being developed for advanced applications such as self-evolving structures and soft robotics. For example, nickel-titanium alloys and shape memory polymers enable complex, adaptive designs to be produced with high precision and functionality (Lee et al. 2017). Recent advancements in biocompatible materials, primarily natural and biocompatible synthetic polymers (modified diblock copolymers, chitosan, sodium alginate, and acrylate-based polymers) have revolutionized 3D bioprinting. While the variety of materials available is constrained, it facilitates the development of functional tissues. These

Table 13.2 Summary of inkjet printing

Printing properties	Inkjet printing	Refs.
Ink type	• Color, functional ink-functionality • Pigment, dye, reactive, coating ink-coloration method • Publication, packaging, label, ceramics, textile, PCB inks-applied industry • Textile, ceramic, PCB ink-ink-based method	Cao et al. (2024a)
Viscosity of ink	5–50 CP	Khan et al. (2020)
2D patterning	Yes	
Design optimization, flexibility	High	
Film roughness	Low	
Speed of printing	0.01–0.1 ms^{-1}	
Technology used in inkjet printing	Piezo-driven, EHD, laser-assisted, needle type, aerosol jet, acoustophoretic	Shah et al. (2021)
Use	Flexible hybrid electronics, printed sensors	
Types	(a) CIJ (b) DOD	Reproduced with permission. Copyright 2024, Elsevier (Hu et al. 2018)
	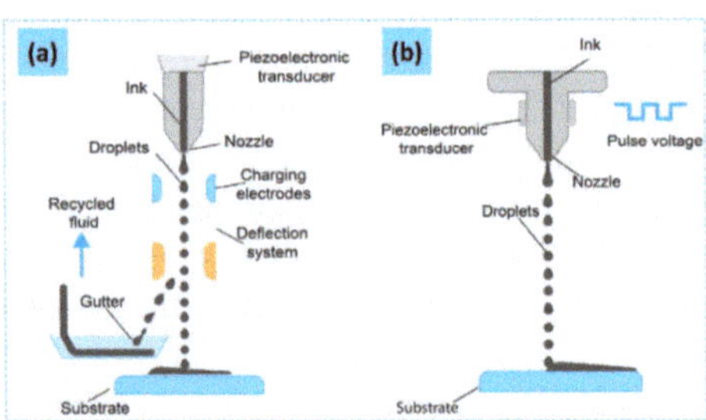	

functional living tissues are essential for regenerative medicine and meet the pressing need for organ transplantation (Lee et al. 2017). Composite materials in 3D printing combine different substances to enhance strength, flexibility, and conductivity. Examples include conductive composites like carbon black/polycaprolactone for sensors and piezoelectric composites such as barium titanate in polymers for energy applications. In fields like electronics and healthcare, these materials support innovation by offering customized mechanical performance (Lee et al. 2017).

The scope of applications includes wearable electronics, tailored fashion, and structural enhancements in functional textiles. 3D printing enables incorporating volumetric designs, including raised geometric patterns, logos, or decorative elements, applicable in luxury fashion or technical textiles that demand enhanced structural properties. 3D printing facilitates the creation of customized components or accessories, addressing specialized markets and personal preferences. The swift prototyping abilities of 3D printing make it an essential textile design instrument. Designers can efficiently generate and evaluate novel textures, patterns, or wearable prototypes, thereby expediting the development process and minimizing expenses (Chatterjee and Ghosh 2020; Dip et al. 2020; Pei et al. 2015; Peng et al. 2015). Table 13.3 shows the summary of 3D printing.

Table 13.4 shows the application of inkjet and 3D printing in textile industries.

Table 13.3 Summary of 3D printing

Printing properties	Type	Refs.
Fabric use	Natural: cotton, linen, viscose fabric (weave design), plain ivory Synthetic: 100% polyester, PET, polyurethane, Flat knitted surface	Dip et al. (2020)
Type	1. Binder jetting 2. Material extrusion 3. Direct energy deposition 4. Material jetting 5. Sheet lamination 6. Vat photopolymerization 7. Power bed fusion	
Compatible software	• 3D simulation: FDM • Structure creation and representation: Lectra modaris, CLO 3D • Design and modeling: Autodesk Fusion 360, Solid Works, AutoCAD • G-code generator: Chitobox, cubic reactor • Slicing software: Ultimaker Cura, Simplify 3D	
Process	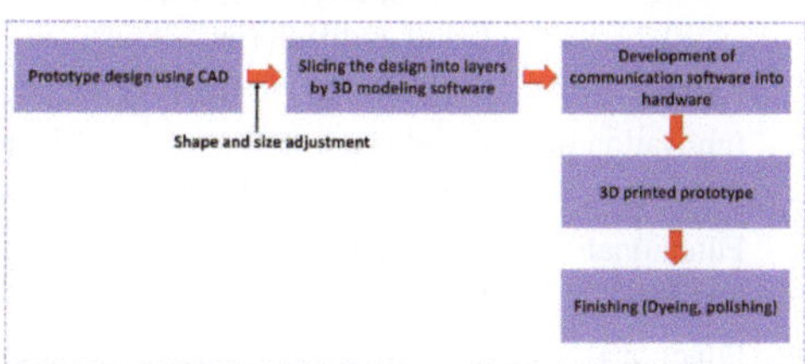	Reproduced (adapted) with permission. Copyright 2024, Elsevier (Chakraborty and Biswas 2020)
Parameters	• Printing bed temperature: 65 °C • Printing speed: 50 mm/s • Printing time: 2 h and 16 min • Layer thickness: 0.2 mm • Layer number: 9	Spahiu et al. (2020)

Table 13.4 Summary of inkjet printing, 3D printing in textile industries

Materials/fabric	Printing method	Application	Characteristics	Refs.
PET cylindrical fabric	Inkjet printing	Strain sensors	Capacitance change: 0.7% Strain: 1%	Quintero et al. (2015)
PET		Strain sensors	Gauge factor (GF): 165 Strain: 0–0.33%	Kang (2021)
Cotton		Pressure sensors	Sensitivity: 1.45/kPa Response time: 30 ms Detection limit: 150 g	Khan et al. (2022)
Taffeta fabric		Temperature sensor	Bending stability Resistance:0.3% Cycle: 1000	Kuzubasoglu et al. (2021)
Polyester sheet		Humidity sensor	Humidity range: 20–30%	Kulkarni et al. (2013)
Hosiery		ECG electrodes	SNR (dry): 12.93 ± 0.80 SNR (Gel):13.75 ± 0.26	Bihar et al. (2017)
Polyester		Green printing	High production efficiency Low energy consumption	Gao et al. (2021)
Polyester		E-textile	Sheet resistance: 0.2 ± 0.025–0.9 ± 0.02 Ω/m	Shahariar et al. (2019)
Wool		Functional textile	Anti-felting Antibacterial	Wang et al. (2023c)
Linen		Functional textile	WCA: 46.5° Wetting time: 0.9 s	Liang et al. (2021)
Cotton		Theoretical guidance for printing	Guiding: Printing Reduce: Discharge of wastewater	Zhao et al. (2022)
Cotton		Functional textile	Adjustment of the crystallinity: 58.5–73.9% Good antibacterial property	Wang et al. (2023a)
Cotton, linen		Innovation in printing	Good printing Excellent color fastness	Li et al. (2023)
Wool		Functional textile	Good antibacterial property	Wang et al. (2023b)
Cotton		Functional textile	UV protection: 134.87 Good antimicrobial property	Wang et al. (2022)
PLA filament	3D printing	Functional textile	Mechanical stability: Better Tensile strength: High	Chatterjee and Ghosh (2020)

13.4 Sustainable and Eco-friendly Printing Techniques

Waterless printing technologies have emerged as an innovative solution to address environmental concerns and reduce water use in printing technology. A significant amount of water is used during the production of textile products, especially in the printing process. A textile industry with a daily production capacity of around 8,000 kg uses 1.6 million liters of water, and nearly 25% of water is used in the whole dyeing and printing process. The US EPA states that, on average, 40 L of pure water are used to dye 1 kg of fabric, potentially increasing this requirement based on the textile material and dyeing method employed (Behera et al. 2021). If we could skip the pretreatment and do a waterless printing process, washing-free printing could be a good option, as it would save considerable water in the printing industry.

Waterless pigment printing could be a viable option to minimize water consumption and reduce wastewater production. In waterless pigment printing, the binding agent must comprise resin. A colorant and neutralizing agent will be incorporated. The printing procedure can then be executed using the lithographic method (Offsetdruckfarbe 2008). Li et al. developed a washing-free inkjet printing method using disperse dye, where water-borne polymer is used, as shown in Fig. 13.4a–j. The printing fabric showed excellent color strength and fastness against wash. The disperse dye ink was prepared according to Fig. 13.1. The printing procedure was completed using an inkjet printing process on polyester fabric and PET film. The printed samples showed excellent rubbing and wash fastness properties (Li et al. 2022). Gao et al. developed an innovative inkjet printing process for polyester fabric that does not require pretreatment or soaping by the parameters of inks, as shown in Fig. 13.4k–q. They enhanced the ink diffusion on polyester fabric by controlling the intermolecular force between ink components. Zirconia crystals (0.4–0.6 mm diameter) are combined with disperse dye, dispersing agent, deionized water, and ethylene glycol in this process, and the mixture is ground at 600 rpm for 20 h. An Epson R330 inkjet printer prints the inkjet patterns at 20 °C. The printed patterns are thermofixed in the oven (110 °C, 30 s), and then post-treatment is performed (180 °C, 1 min). The printed samples exhibited exceptional color fastness to sublimation, abrasion, and washing (Gao et al. 2021).

Eco-friendly ink could be a good option for sustainable pigment printing, as it is environmentally friendly, cheap, and uses fewer toxic chemicals. *Terminalia catappa*, also known as almond, sea almond, and tropical almond, grows mainly in the tropics of Africa, Asia, and Australia and could be a good choice for ink preparation, as shown in Fig. 13.4r. At first, *T. Catappa* leaves in three different colors such as green, yellow–brown, and reddish pink, are used where the aqueous extracted leaves solution is centrifuged for 2–4 days and then filtered to remove dust and other particles from the extracted solutions. After heating, the extracted solution turned into the pigment powder. For ink preparation, xanthan gum (10%) is mixed with pigment solutions (Pigment:solvent:binder = 1:1:1), and the developed inks are applied on paper and film. This eco-friendly ink could be used as an alternative to artificial inks on paper and film, which is cheap and eco-friendly (Madhan et al. 2024). Hyata et al.

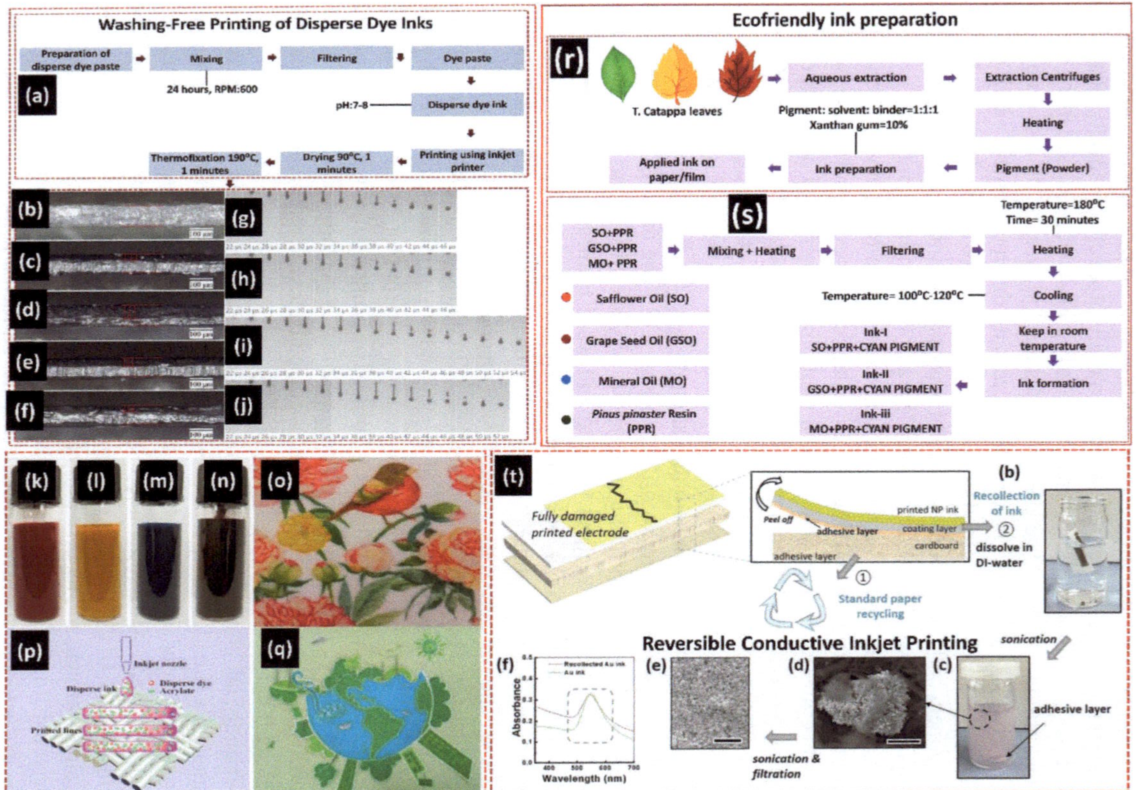

Fig. 13.4 **a** Washing-free printing using disperse dye, cross-section images of printed PET films: **b** pure PET, **c** PET@PVP, **d** PET@PVA, **e** PET@PEG, and **f** PET control sample, Ejection properties of disperse dye ink: **g** PVP, **h** PVA, **i** PEG, **j** control sample. Reproduced with permission. Copyright 2024, MDPI (Li et al. 2022), **k**–**n** Disperse ink samples, **o**, **q** Polyester fabric printed sample, **p** Disperse ink inkjet on polyester fabric. Reproduced with permission. Copyright 2024, Elsevier (Gao et al. 2021), **r**, **s** Eco-friendly ink preparation (Madhan et al. 2024; Hayta et al. 2022). **t** Reversible conductive inkjet printing. Reproduced

developed eco-friendly offset printing inks from vegetable oils (Safflower, mineral, and grape seed oil) and resin (*Pinus pinaster* resin), an appropriate utilization of renewable sources, shown in Fig. 13.4s. At first, mineral oils and resin are mixed with specific ratios (oil:resin: 4:7) and heated for proper mixing. After that, filtering removes dust and other impurities from the solutions. Then, the solutions go through heating (180 °C, 30 min) and cooling (at first temperature should be 100–120 °C and then stay at room temperature), resulting in ink forming where cyan pigment is added in a specific amount in the solution as a colorant. Colorimetric analysis, gloss measurements, and density showed similar properties of both inks compared to the international standard (Hayta et al. 2022).

The reduction of waste and chemicals is essential for sustainable and eco-friendly printing, as the dyeing and printing industries increasingly prioritize energy conservation, consumption reduction, and green practices (Huang et al. 2017; Wang et al. 2017; Zhang et al. 2016b). For this reason, the reversible conductive ink printing and improved de-inking for inkjet-printed paper using low-temperature plasma at low pressure could reduce waste and chemicals (Mauchauffé et al. 2019; Kang et al. 2020). Kang et al. first collected the fully damaged electrode and peeled the adhesive layer. After collecting the adhesive layer, it was dissolved into deionized water for the ink to be recollected, and then the sonication process was performed. Then, the ink solutions went through a sonication and filtering process, and as a result, hybrid ink was recovered, showed in Fig. 13.4t (Kang et al. 2020).

13.5 Smart Textiles and Functional Printing

13.5.1 Conductive Inks and Printed Electronics

The advancement of conductive inks for printed electronics has accelerated significantly due to the expanding range of applications in wearable textiles, sensors (strain sensor, pressure sensor, temperature sensor, humidity sensor, electrodes for ECG measurement, electrodes for EEG measurement, electrodes for EMG measurement, electrodes for EOG measurement) as shown in Fig. 13.5a–k, Table 13.5, antennae, and radio frequency identification. Consequently, inks with high conductivity need to be formulated using silver, graphene, and metal oxide nanoparticles, including copper and aluminum, due to their superior conductivity, excellent leveling characteristics, cost-effectiveness, and potential for scalability, as shown in Fig. 13.6a (Pan et al. 2018). Conductive nanoparticles are primarily utilized in inkjet, gravure, flexographic, and screen-printing techniques for fabricating printed electronics, as shown in Fig. 13.6b–c. Carbonaceous materials (Carbon dot, CNT, graphene, GO), Conductive polymers (PEDOT, PPy, PPP, PTH), and 2D (MX_2 where M-Mo, W, Ti, X-Se, S, transition metal dichalcogenides (TMDS), Te) materials are used for the printed electronics, as shown in Fig. 13.6d–g.

Table 13.5 Summary of printing techniques used in e-textiles

Type	Printing method	Substrate	Characteristics	Refs.
Strain sensor	Screen printing	PU-based stretchable fabric	• Sensing range: 130% • Cycle: >10,000 • WCA: 112°	Zhang et al. (2021a)
	Inkjet printing	Graphene-PEDOT: PSS conductive ink	• Gauge factor (GF): 165 • Strain range: 0–0.33%	Kang (2021)
Pressure sensor	Screen printing	Cotton	• High sensitivity: 81.61 kPa^{-1} • Range: 0–10 kPa • Rapid response: 6 ms • Relaxation time: 30 ms	Liu et al. (2019)
	Screen and inkjet printing combined	Cotton, polyester	• Sensitivity: 1.45/kPa • Response time ≈112 ms • Detection limit: 150 g	Khan et al. (2022)
Temperature sensor	Inkjet printing	Taffeta fabric	• Temperature range: RT-50 °C • Sensitivity:0.15%/°C (CNT) • Sensitivity: 0.41%/°C (PEDOT: PSS) • Sensitivity: 0.31%/°C (CNT/PEDOT: PSS) • Bending stability with a resistance change of 0.3% up to 1000 cycles	Kuzubasoglu et al. (2021)
Humidity sensor	Screen printing	Polyester	• Humidity range: 30–95% • Sensitivity: 2.7	Komazaki and Uemura (2019)
ECG electrodes	Brush printing	Woven, knit	• Place of operation: Cycling suit	Paiva et al. (2019)
	Screen printing	Polyester knit fabric	• SNR: 30.17 (polyester) • SNR: 33.52 dB (Ag/AgCl electrodes)	Nigusse et al. (2020)
	Inkjet printing	Woven	• SNR: 18 dB	Li et al. (2019)
EEG electrodes	Screen printing	Cotton woven	• Type: Forehead • Performance: >0.998	Islam et al. (2022)

(continued)

Table 13.5 (continued)

Type	Printing method	Substrate	Characteristics	Refs.
	Jet printing	Polyurethane nonwoven	• Type: Ears	La et al. (2018)
EMG electrodes	Screen printing	Polyester-cotton fabric	• Place: Forearm • SNR: 93%	Court and Torah (2021)
	Screen printing	Textile and a silver-plated lycra fabric (CCSM)	• Composed: Ag, PEDOT: PSS	Ohiri et al. (2022)
	Jet printing	Polyurethane nonwoven	• Place: Several body part	La et al. (2018)
EOG electrodes	Screen printing	PET	• Place: Forehead	Toral et al. (2020)
Supercapacitors	Screen printing	Cotton	• Device capacitance: 3.2 mF cm^{-2} • Energy density: 0.28–3 mW cm^{-2} • Capacitance: 95% (10,000 cycles)	Islam et al. (2022)
	Screen printing	Polyester	• Device capacitance: 16.8 mF cm^{-2} • Power density: 0.0023 mW cm^{-2} • Capacitance retention: 60°, 90°, 180° (bending angle) • Cycle: 100	Liang et al. (2020)
Personalized thermoregulation	Screen printing	Stretchable textile	• Maximum temperature: 56.1°C • Performance: 95% • Cycle: 100,000	Islam et al. (2024)
	Patternable spray coating	Stretchable fabric	• Maximum temperature: 35–55 °C	Ahn et al. (2019)
	Scalpel printing	Cotton, polyester woven, and nonwoven	• Maximum temperature: 140 °C	Arbab et al. (2019)
	Pattern printing	Stretchable LM@PDMS	• Maximum temperature: 45.26, 95.9 °C • Performance: Wearable thermotherapy	Silva et al. (2020)

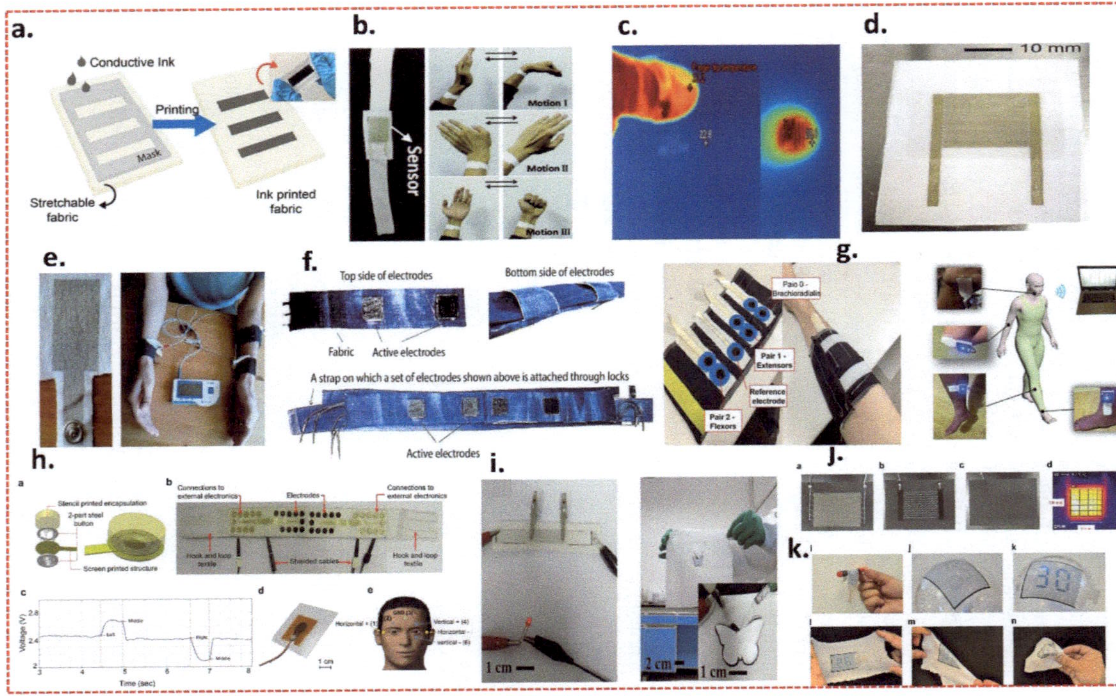

Fig. 13.5 Printed textile **a** strain sensor, **b** pressure sensor, **c** temperature sensor, **d** humidity sensor, electrodes for **e** ECG measurement, **f** EEG measurement, **g** EMG measurement, **h** EOG measurement, printed textile for **i** supercapacitors, **j** thermoregulation, **k** electrochromic devices. Reproduced (adapted) with permission.Copyright 2024, Wiley (Islam et al. 2024)

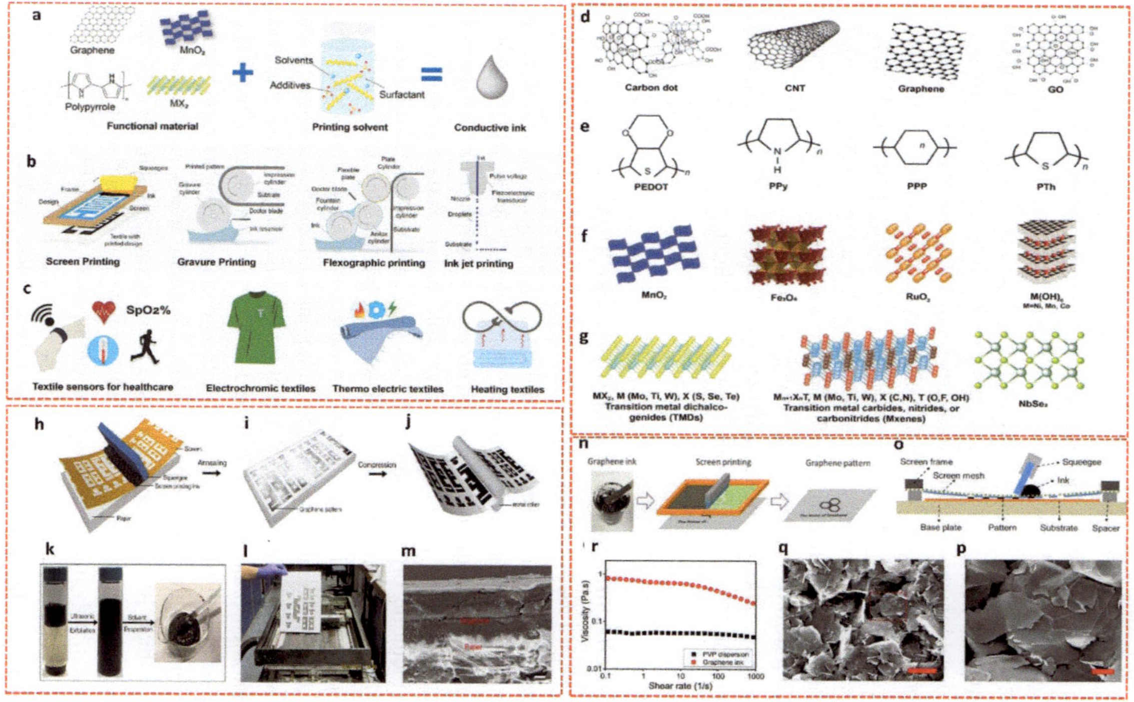

Fig. 13.6 Print ink and printing processes for e-textiles. **a** Conductive ink preparation **b** fabrication and **c** applications of printed e-textiles, Printable functional materials for e-textiles. Reproduced with permission. Copyright 2024, Wiley (Islam et al. 2024). **d** Carbonaceous materials, **e** conductive polymers, **f** metal oxides/hydroxides, and **g** 2D materials. Reproduced with permission. Copyright 2024, Wiley (Islam et al. 2024). Screen-printing steps: **h** graphene ink patterning, **i** printed patterns annealing, and **j** compressing. **k** Graphene ink and screen-printing ink. **l** Printing on A4 paper. **m** SEM of the printed graphene antenna. Reproduced with permission. Copyright 2024, Nature (Pan et al. 2018). **n** Screen-printing process (inks: graphene). **p** SEM image of printing process (p) Higher magnification image of the region outlined on **p**, scale bar = 2 μm, **q** scale bar = 10 μm. **r** Polymer solution viscosity and GNP-ink (different shear rates). Reproduced with permission. Copyright 2024, American Chemical Society (He et al. 2019)

Pan et al. showed a sustainable and environmentally friendly route for producing graphene ink suitable for screen printing technology, as shown in Fig. 13.6h–m. Dihydrolevoglucosenone (cyrene) was used as a solvent, which sped up the ink production process and reduced the cost. Using the screen-printing method, the developed inks were printed on paper, where graphene patterns were visible, and then a roller was used for compression. The graphene printed inks on paper showed very high conductivity (7.13 × 104 Sm^{-1}) devices, which allows us to produce wireless connectivity antenna operational from MHz to tens of GHz, which can be used for wireless data communication and energy harvesting, which brings us very close to the ubiquitous use of printed graphene technology for such applications (Pan et al. 2018). He et al. demonstrated the development of highly conductive, super flexible, and low-cost inks derived from graphene nanoplates for flexible electronics applications, as shown in Fig. 13.6n–p. Screen printing on various substrates, including paper and plastic, achieved this. The graphene-printed patterns were dried in an oven (80 °C, 2 h). In the case of graphene patterns involving multiple printing layers, the printed pattern underwent annealing at (80 °C, 30 min) following each printing pass. The printed pattern demonstrates remarkable conductivity and excellent flexibility, maintaining its conductivity after undergoing 1000 bending cycles without substantial loss.

13.5.2 Thermochromic and Photochromic Printing

Thermochromism refers to the reversible alteration in the color of a compound as a function of temperature. The thermochromic color change is distinctly observable, frequently exhibiting a dramatic effect, and occurs within a narrow temperature range. Photochromism refers to the reversible color change observed in various materials, including organic, inorganic, biological, and polymeric substances, when exposed to light. The fundamental processes involved in photochromic reactions include cis–trans isomerization, electrocyclization, redox reactions, and light-induced proton transfer. The absorption or photoluminescence (PL) spectra exhibit notable changes resulting from molecular structure modifications before and after light irradiation. Photochromic materials can be used to design and prepare rewritable paper owing to their switch-like molecular structural changes demonstrating stable reversibility and remarkable light responsiveness. Photochromism can be categorized into T-type, which includes spiropyrans, spirooxazines, naphthopyrans, and P-type photochromism. Conversely, thermochromic materials comprise organic and inorganic compounds, polymers, and sol–gel systems, as shown in Fig. 13.7 (Rožić et al. 2020; Ma et al. 2021).

When integrated with digital and traditional printing, they possess varied applications within the textile sector. Printing processes offer various uses. In fashion, they facilitate dynamic and personalized designs. In-home decor, they create safety gear improves additionally, and in safety gear that improves visibility under different conditions. Recent developments in dye chemistry and printing methodologies are enhancing the accessibility and durability of these materials, thereby broadening

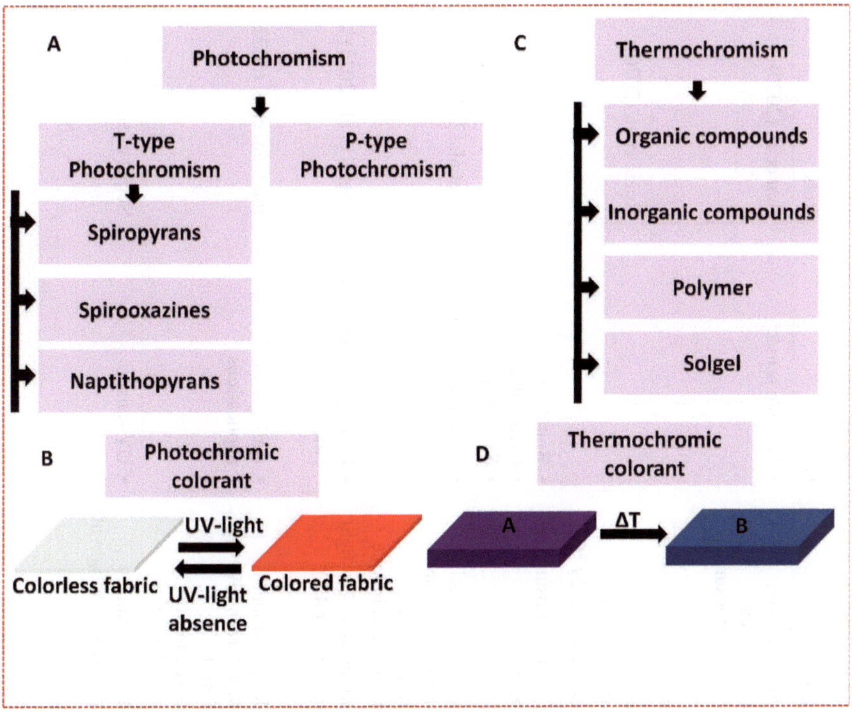

Fig. 13.7 Mechanism and type of photochromism and thermochromism

their attractiveness across various sectors. Also, using this printing method, flexible sensors, brand protection, security printing, military camouflage, and security encoding are made (Ma et al. 2021). Table 13.6 shows the summary of thermochromic and photochromic printing on textiles.

Sun et al. conducted a study on a responsive chromic textile, employing the screen-printing technique to synthesize a COP^{2+} in the cotton fabric's surface, as shown in Fig. 13.8a, b. The CCF exhibited an immediate color change (yellow-green) upon exposure to NH_3 gas, reverting to yellow once the stimulus was removed. The process of reversible color transition can be initiated through exposure to UV light (10 s) or by applying heat (110 °C), demonstrating its cyclability. This robust multi-responsive textile exhibits significant potential for application as a flexible sensor in environmental monitoring (Sun et al. 2020).

Table 13.6 Summary of thermochromic and photochromic printing

Material/fabric	Printing method	Case	Characteristics	Use	Refs.
Covalent organic polymer (COP^{2+})	Screen printing	• Photochromic • Thermochromic	• Color transition: Yellow-Green • UV light (exposure for 10 s) • Heat (exposure 110 °C)	• Flexible sensor	Sun et al. (2020)
Thermochromic offset ink	Printability tester	• Thermochromic	• Color transition: Blue-Colorless • Heat (exposure 29 °C)		Rožić et al. (2020)
SiO_2@naphthopyran Nanoparticles	Screen printing	• Photochromic	• UV light (exposure for 1 min)	• Brand protection • Camouflage • Security printing	Pinto et al. (2016)
PDMS + thermochromic microcapsules	3D printing	• Thermochromic	• System: self-heating thermochromic and infrared thermal encryption technique	• Encryption device	Li et al. (2025)
4% Polyamide and 26% elastane elastics	Screen printing	• Photochromic • Thermochromic • Hydro chromic	• Color change: UV light (exposure for 4 h)	• Display a wide range of color	Cabral et al. (2023)
Viologen-based nanocomposite ink	Screen printing	• Photochromic • Thermochromic • Vapochromic	• Color transition: Yellow-Green (399 and 585 nm) • Heat (exposure 25–95 °C)	• Anti-counterfeiting applications	Ibarhiam et al. (2022)
Photochromic dye: Reversacol Ruby Red	Inkjet printing	• Photochromic	• Color transition: Exposure to UV light at 500 nm	• UV-sensor	Seipel et al. (2018)

(continued)

Table 13.6 (continued)

Material/fabric	Printing method	Case	Characteristics	Use	Refs.
Solar-driven thermochromic fabric (SDTCF)	Screen printing	• Photochromic	• Color transition: Orange-Green • Heat (exposure 21.5–38.0 °C)	• Human body protection • Health management	Ge et al. (2021)
Pigment powder	Screen printing	• Photochromic	• Color transition: Colorless-Green • UV light (exposure 516 nm)	• Military camouflage • Security encoding • Brand protection • Smart packaging	Ahmed et al. (2023)
Thermochromic microcapsules	Screen printing	• Thermochromic	• Color transition: Green-magenta -Heat (exposure 30–80 °C) • Green-purple (humidity drops from 100 to 0%)	• Good wearability • 5.52 cm bending length • 101.0 mm/s air permeability	Yang et al. (2024)
Phosphor-binder nanocomposites	Screen printing	• Photochromic	• Color transition: Colorless-Greenish yellow • UV light (exposure 382 nm, 400 nm)	• Brand protection • Smart packaging	Ahmed et al. (2022)
PSBM-SP latex particles based on spiropyran	Screen printing	• Photochromic	• Color transition: colorless-purple • UV light (exposure 550 nm)	• Wearable textile • Flexible textile	Yang et al. (2021)
Photochromic (PC) pigments	3D printing	• Thermochromic	• Color transition: Colorless-purple • UV light (overexposure)	• Help to protect against overexposure to sunlight	Alam et al. (2024)

(continued)

Table 13.6 (continued)

Material/fabric	Printing method	Case	Characteristics	Use	Refs.
Poly (L-lactic acid) based materials	Screen printing	• Thermochromic	• Color transition: Colorless-blue • Heat (exposure 50 °C) • Thermal stability 15-months	• Thermochromic sensing material	Cruz et al. (2024)
Novel multifunctional coordination polymer	Screen printing (ink-free printing medium)	• Photochromic	• Photo-controlled fluorescence properties	• Ink-free printing medium	Zhang et al. (2021b)
PDMS composite	3D printing	• Thermochromic • Photochromic	• Color transition: Colorless-purple • UV light: 0–2500 μW/cm^2 • Skin temperatures: 0–38 °C • Sweat pH: 4.0–7.0	• Wearable electronic device • UV-sensing	Liu et al. (2022)

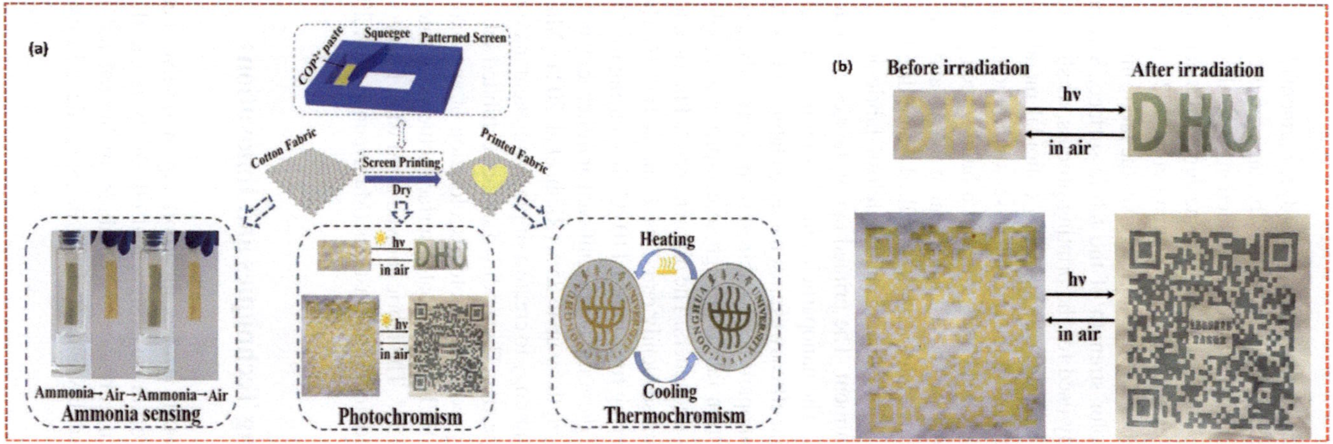

Fig. 13.8 **a** Screen-printed fabric: photochromism, thermochromism, and NH_3 sensing. **b** Screen-printed cellulosic fabric before and after UV irradiation (40 s, room temperature), UV-Vis diffuse reflectance spectra of CCF upon irradiation with UV light. Reproduced (adapted) with permission. Copyright 2024, Springer Nature (Sun et al. 2020)

13.5.3 Antimicrobial and Self-cleaning Textiles

Implementation of nanoparticles with colorant (dye, pigment) by printing method has become popular in recent years in textiles because of the functional properties of nanoparticles such as antimicrobial reduction and self-cleaning properties of silver, titanium dioxide, zinc oxide, and magnesium oxide nanoparticles (Rabbi et al. 2024; Abdelrahman et al. 2020; Manivannan et al. 2021; Uddin Rabbi et al. 2024). Combining nanoparticles with colorants gives the fabric multifunctional properties and enhances the color strength of the fabric surface. Screen printing and inkjet printing are generally used for the printing process, as shown in Table 13.7. Jabli et al. effectively printed 2, 5, 8, and 10% colloidal metal oxide nanoparticles by reducing copper, magnesium, and nickel salts with biological *Cynomorium coccineum* extract onto cotton surfaces utilizing a hand screen printing approach. Sodium alginate biopolymer served as a thickening agent, and the printed cellulosic fabrics were meticulously cleaned with distilled water, dried, and subjected to treatment (150 °C, 5 min). Ultimately, printed cellulosic fabric underwent treatment in washing and drying treatment. The printed cotton fabrics exhibited satisfactory wash and rubbing fastness. The nanoparticles in the print paste imparted the fabric with significant antibacterial efficacy against bacteria (Jabli et al. 2020). Rabbi et al. demonstrated that silver nanoparticles combined with reactive dye and pigment can prepare multifunctional cotton fabric through screen-printing, employing Forthick HV and sodium alginate as thickeners. The pigment-printed fabric undergoes drying with an air blower and is subsequently cured for 1 min at 160 °C. The reactive dye printed fabric undergoes drying for 2 min at 110 °C, then curing for 1 min at 120 °C. The pigment-printed fabric demonstrated significant antimicrobial reduction properties and increased color strength by up to 25% (Rabbi et al. 2024). Manivannan et al. utilized digital printing technology to create a single-component ink for self-cleaning and superhydrophobic properties. The modified surface exhibited self-cleaning properties, evidenced by the reproducible photo discoloration of approximately 90% of methylene blue (MB). The PET@TPS-coated surface achieved a water contact angle (WCA) of approximately 152°. The fabric showed good wash fastness against water and organic solvent and had good light fastness (Manivannan et al. 2021).

13.6 Hybrid Printing Techniques and Innovations

Textile industries continuously evolve, incorporating innovative methods to meet modern design demands, optimize design, and efficiently create the best design. Hybrid printing is a modern technique that combines traditional and digital printing techniques to improve versatility and efficiency in the textile industry.

Table 13.7 Summary of combining nanoparticles using inkjet and screen printing on textiles for antimicrobial and self-cleaning properties

Material/ fabric	Printing method	Dye/nanoparticle	Characteristics	Refs.
Cotton	Inkjet printing	• C. I. Reactive Red 218 • C. I. Reactive Blue 49 • C. I. Reactive Yellow 95 • Nanoparticle: Antireflective	• Improved coloration	Song et al. (2023)
Cotton	Inkjet printing	• Silver nanoparticle	• Good antimicrobial reduction	Soleimani-Gorgani et al. (2023)
Cotton	Inkjet printing	• Silver nanoparticle	• Superhydrophobic ability	Boumegnane et al. (2022)
Cotton polyester	Screen printing	• Silver nanoparticle	• Good antimicrobial reduction	Rabbi et al. (2024)
Cotton	Inkjet printing	• Wood-derived cellulose nanofibrils (CNFs) • Water-based cyan pigment ink	• Supercapacitor • Antenna	Nechyporchuk et al. (2017)
Wool synthetic acrylic	Screen printing	• Nanoparticles: TiO_2, MgO, ZnO • C. I. Basic Yellow 28 • C. I. Acid Red 88	• Good superhydrophobicity • Photocatalytic self-cleaning • Good antimicrobial reduction	Abdelrahman et al. (2020)
Cotton	Screen printing	• Nanoparticles: Colloid metal oxide nanoparticles	• Good antimicrobial reduction	Alam et al. (2024)
Polyethylene terephthalate fabric (PET)	Digital printing method	• Nanoparticle: Anatase TiO_2	• Self-cleaning surface • Superhydrophobic ability	Manivannan et al. (2021)

13.6.1 Combination of Traditional and Digital Techniques

Combining conventional printing techniques (roller, screen, and block printing), with advanced digital technologies (3D and inkjet printing) presents a distinctive fusion of adaptability and effectiveness. Conventional approaches, recognized for their robustness and accuracy on textile surfaces where adaptability and personalization are limited, can be enhanced by applying digital printing technologies.

Hybrid printing enables designers to use the advantages inherent in both techniques. Screen printing establishes a foundational layer or produces vivid hues, whereas digital methods contribute complex details and smooth transitions. This combination is especially advantageous for designs that demand high precision and durability, including sportswear and home textiles. Applying hybrid printing techniques facilitates the development of distinctive textures and complex, multi-layered designs. For example, one can achieve raised textures or metallic effects through screen printing, enhanced by the digital accuracy of photorealistic imagery. This method improves visual attractiveness and creates opportunities for practical uses, including UV protection and antimicrobial characteristics (Swetha et al. 2024; Ugur Koseoglu 2019).

13.6.2 Multi-process Printing Systems

The evolution of multi-process printing machinery has transformed textile manufacturing, facilitated quicker production cycles, and allowed more intricate designs. Consequently, fashion is undergoing daily evolution, characterized by integrating multi-process systems that incorporate diverse printing technologies, thereby facilitating innovative applications.

Recent advancements in multi-process printers facilitate smooth transitions among screen, digital, and printing techniques within a unified configuration. Implementing these machines enhances production efficiency by minimizing downtime and enables the concurrent application of multiple layers or techniques. This significantly benefits extensive production processes and small-scale, precision-focused design initiatives. Multi-process systems are being utilized more frequently for rapid prototyping, enabling designers to visualize and evaluate intricate designs efficiently. Manufacturers can create complex textiles that integrate vivid colors with tactile features by combining various processes. This capability primarily serves high-end fashion and technical textiles, where aesthetic and functional qualities are important (Ugur Koseoglu 2019; Tkalec et al. 2024).

13.7 Automation, AI, and Robotics in Textile Printing

13.7.1 AI and Machine Learning in Textile Printing

The advancements in automation, artificial intelligence, and robotics technology have profoundly transformed the field of textile printing, enhancing precision, optimizing pattern design efficiency, improving color matching, and promoting sustainable printing practices, as shown in Fig. 13.9. Incorporating AI improves the enhancement of perception and prediction within the printing process, alongside advancing material development and fabrication techniques. Artificial intelligence algorithms are employed to enhance pattern design, facilitating the development of seamless layouts. Artificial intelligence can perceive, adjust, and predict the conditions of the

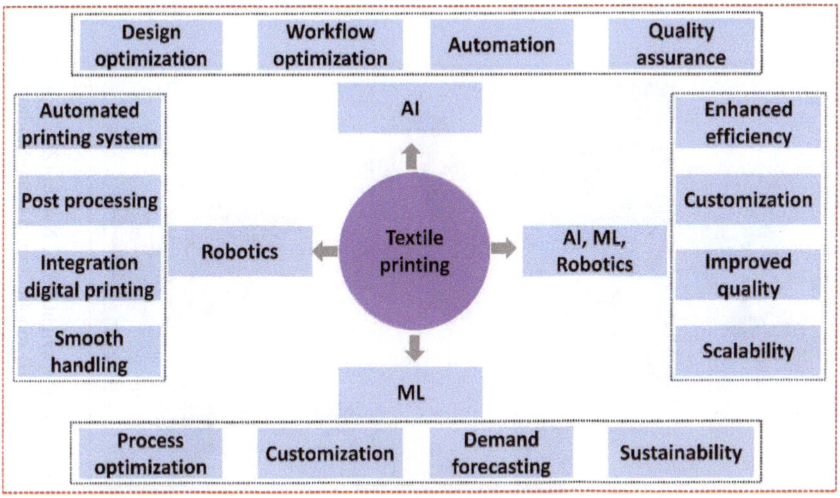

Fig. 13.9 Role of AI, ML, and robotics in textile printing

printing technique, thereby enhancing the overall quality of the textiles, as shown in Fig. 13.10a. Using automation, AI can predict the most efficient design and optimize workflow. As a result, the quality of printed textiles can be improved (Zhu et al. 2020).

Machine learning (ML) is a branch of artificial intelligence (AI) dedicated to developing statistical algorithms that identify correlations within extensive datasets, enabling predicting future outcomes. Process optimization, design optimization, demand forecasting, and sustainable printing are all facilitated by the datasets employed in machine learning. Machine learning has been widely implemented in the textile industry. Joshi et al. created a machine learning-enhanced nanoparticle-printed FINP-IµC biochip utilizing inkjet printing and microfluidic technology, as shown in Fig. 13.10b. This biochip may classify cancer subtypes by analyzing and quantifying tumor heterogeneity using the machine-learning dataset (Joshi et al. 2020). The integration of printing and machine learning technology can result in the production of multifunctional sensors. Zhu et al. created high-performance hydrogen (H_2) sensors that are both flexible and foldable. These sensors were applied to all textile substrates produced through inkjet printing with rGO. Furthermore, the triboelectric-textile and ML-enabled technology used to power IoT demonstrated that H_2 is used for wearable applications with robust mechanical performance (Zhu et al. 2021).

13.7.2 Robotic Assistance in Printing Processes

Defence, aerospace, fashion, aeronautics, healthcare, and construction are among the sectors in which robot-assisted printing has been effectively implemented

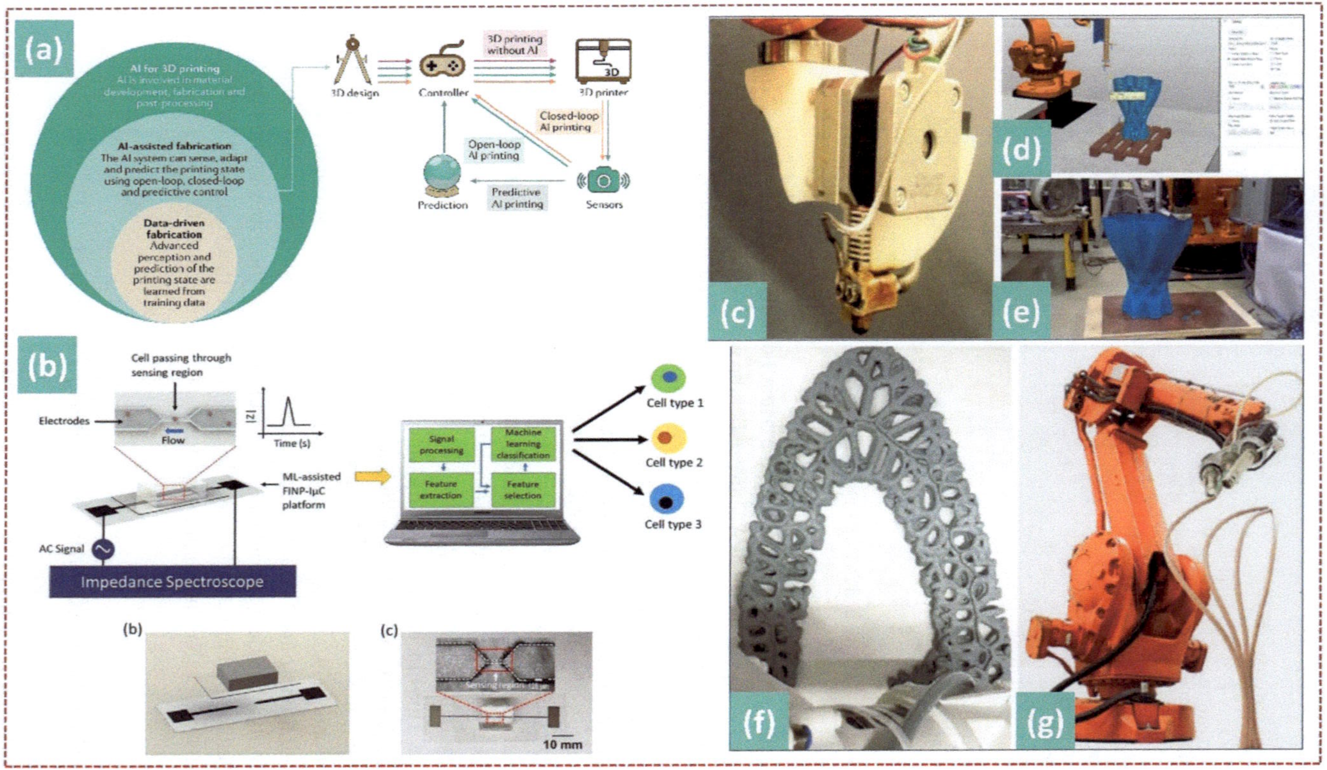

Fig. 13.10 **a** AI can be involved in 3D printing and fabrication in different ways. Reproduced with permission. Copyright 2024, Nature (Zhu et al. 2020), **b** Mechanism of ML-assisted biochip. Reproduced with permission. Copyright 2024, Wiley (Joshi et al. 2020), **c** ABB robotic arm, **d** RobotStudio platform; **e** Printed part. Reproduced with permission. Copyright 2024, Elsevier (Zhang et al. 2016a), **f** Ceramic printing. Reproduced with permission. Copyright 2024, Elsevier (Mostafavi and Bier 2016), **g** Thermosetting printing. Reproduced with permission. Copyright 2024, Elsevier (Urhal et al. 2019)

(Table 13.8). 3D printing applications with multi-axis robot-manipulated manufacturing methods are commonly employed for welding and pick-and-place duties in conventional additive manufacturing. It provides enhanced quality and consistency, increased productivity, increased safety for repetitive duties, and reduced labor costs. It provides enhanced quality and consistency, increased productivity, improved safety for repetitive duties, and reduced labor costs (Neto 2013; Neudecker et al. 2016; Brooks et al. 2017). An ABB robot was employed by Zhang et al. to regulate a filament-based extrusion printing head. Robot Studio, a computational platform developed by ABB, enabled the simulation, optimization, and visualization of process fabrication, as shown in Fig. 13.10c–e. The ABS parts were manufactured; however, the composite parts can be produced by extending the system to other materials, such as carbon fibres (Zhang et al. 2016a). A team at the University of Michigan (USA) employed a robotic extrusion system to fabricate a monolithic elastic net utilizing a thermoplastic elastomer (TPE). The Institute for Advanced Architecture of Catalonia (Spain) developed an elastic net for the ABB-Robotstudio project using an ABB 2400 L robotic arm as shown in Fig. 13.10g (Urhal et al. 2019). De Marzi et al. employed robot-assisted hybrid printing technology, demonstrating an optimized strength-to-weight ratio achievable through this printing method. A highly reactive resin is utilized as the ink, incorporating silica particles to modify its rheological characteristics. A strong correlation exists between the printed components and the digital model, with mechanical properties surpassing conventionally additive manufactured octets (Marzi et al. 2023). The computational 3D model approximates the prototype. However, the two remain primarily distinct due to emergent material features. The distinctions between virtual and material illustrate the evolving aesthetics intrinsic to material action. The emergent aesthetics of the prototype result from both the 3D stacking process and the variability of material extrusion along the way, as shown in Fig. 13.10f (Mostafavi and Bier 2016).

13.8 Applications and Case Studies

13.8.1 Fashion and Apparel Industry

Textile printing, especially 3D printing, provides a cost-efficient, sustainable, and easy way to develop prototypes and designs. This includes material selection and incorporation of materials to build a product that is not only functional but also fashionable, as shown in Fig. 13.11a–c (Chakraborty and Biswas 2020). Blending 3D printing techniques with fashion, the textile industry can produce a fabric that is not only fashionable but also has functional properties, as shown in Fig. 13.11d–g. Textiles made using 3D printing techniques show flexibility, and elasticity and ensure comfort (Chakraborty and Biswas 2020).

Iris van Herpen, a renowned Dutch fashion designer, has revolutionized the fashion industry by integrating 3D printing technology into her creations,

Table 13.8 Summary of automation, AI, and robotics in textile printing

Material/fabric/ software	Printing method	Technology	Benefits/application	Refs.
Polyethylene terephthalate (PET) Nanoparticle: Silver nanoparticle ink	Inkjet printing	Machine learning	• Single-cell studies • Tumour heterogeneity studies • In point-of-care cancer diagnostics	Joshi et al. (2020)
Porous paper microfiber Nanoparticle: Reduced graphene oxide (rGO)	Inkjet printing	Machine learning	• Non-contact patterning • Cost-effectiveness • Sensing response: High • Robust mechanical performance	Zhu et al. (2021)
CAD	3D printing	Machine learning AI	• Prevent error-trial • Speed up overall production	Rahmani Dabbagh et al. (2022)
Blended pellets of biopolyesters containing approx. 5 wt% of Phormium tenax fiber	3D printing	Robot-assisted	• Robust software to convert various types of shapes to robot printable paths	Brooks et al. (2017)
Alginate-poly (ethylene glycol) diacrylate hydrogel	3D printing	Robot-assisted	• Printing accuracy: Average • Dimensional error: 0.06 ± 0.14 mm • Focal cartilage defect restoration	Lipskas et al. (2019)

blending fashion, art, and technology (How 3D Printing is Revolutionizing the Fashion Industry, https://www.tomorrow.bio/post/how-3d-printing-is-revolutionizing-the-fashion-industry-2023-06-4727466168-3d-printing). Her collections include garments that are visually stunning and groundbreaking in their construction, such as the use of flexible filaments, translucent polymers, and lightweight materials that adapt to the body's movement. One of her most iconic pieces, the "Capriole" collection, features structures mimicking biological processes. In contrast, the "Voltage" collection showcases electric-like patterns created with the help of 3D printing and laser cutting (van Herpen, https://high.org/exhibition/iris-van-herpen-transforming-fashion/). Through her work, Van Herpen demonstrates how 3D printing can transform fashion into wearable sculptures, offering new dimensions in sustainability, customization, and design innovation. To explore the fusion of fashion and technology, The Museum of Fine Arts, Boston, in collaboration with Nervous System, created a 3D-printed dress inspired by natural components like petals, feathers, and scales. The dress features over 1,600 interlocking elements, which are rigid individually but collectively behave like a flexible textile. Each of the aspects of this

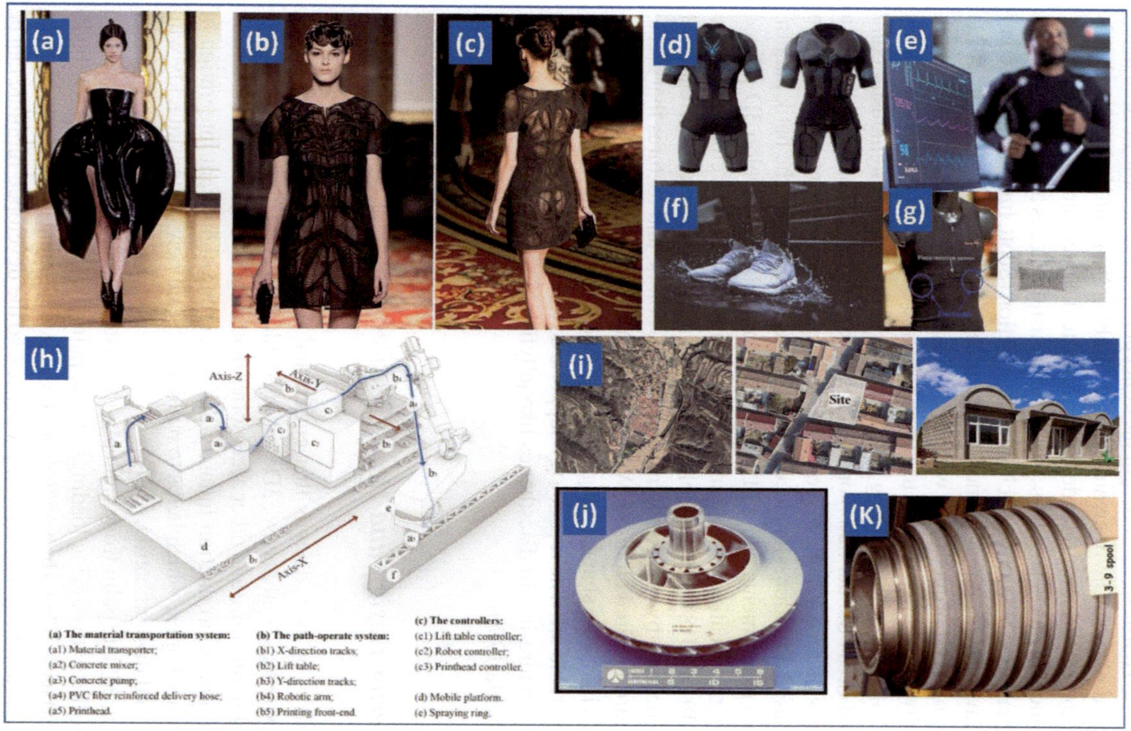

Fig. 13.11 **a** Dress in Paris Fashion Week AW-2012/13, **b**, **c** Outfit from voltage collection, titanium alloy based 3D printed components. Reproduced (adapted) with permission. Copyright 2024, Elsevier (Chakraborty and Biswas 2020), **d** Antelop suit equipped with electrodes, **e** Health monitoring textile, **f** Athletic footwear, **g** The e-textile electrodes for ECG (electrocardiographic), **h** 3D mobile robotic system, **i** Satellite image: Wujiazhuang village, The birds-eye view of the construction site, and 3D printing house. Reproduced with permission. Copyright 2024, Elsevier (Xu et al. 2022). **j** Fuel pump impeller for space shuttle, **k** Compressor spool for engine. Reproduced with permission. Copyright 2024, Elsevier (Alami et al. 2023)

dress is customizable in size, shape, and orientation, and the fit of the dress can be modified by 3D scanning (Nervous System 2016). Besides fashion items, 3D printing technology is now used to produce regular-wear clothes that will be easy to wear and made of comfortable and biodegradable material (Sculpteo 2024). For example, Julia Daviy, a designer, brought the concept of biodegradable fashion made out of 3D-printed plastic. Daviy creates lacy, wearable, 3D-printed clothing that is biodegradable and eco-friendly. Her designs eliminate the need for mass production, allowing customers to buy design files and 3D print custom pieces at home, offering a more ethical alternative to traditional clothing production (Scott, https://3dprint.com/212640/julia-daviy-3d-printing/). American designer Travis Fitch (Ma et al. 2021) created a multicolor 3D-printed dress with 30 flexible sections designed to move with the body. Usually, 3D-printed fashion uses monochrome colors, but this project shows the versatility of multicolor printing (Armstrong, https://3dprintingindustry.com/news/3d-printed-dress-debuts-new-york-fashion-week-95736/). Kuswanto et al., a group of Indonesian researchers, explored the application of 3D printing for fashion development based on Indonesian decorative patterns, specifically the batik pattern. They conducted experiments using PLA and ABS materials to develop 3D-printed textiles, focusing on creating flexible and durable designs suitable for clothing (Djoko Kuswanto and Iftira 2018). The technology of 3D printing also enables the creation of bionic fashion designs by replicating natural shapes, colors, and textures that create innovative, nature-inspired clothing. With 3D printing, abstract and realistic biological inspirations like wing shapes, pangolin shells, and snake morphologies are transformed into wearable art. The technology supports advanced color and texture replication, offering nature-inspired aesthetics like jellyfish hues or bird feather patterns (Fanglan and Kaifa 1790).

13.8.2 Interior Design and Home Textiles

The world thrives on new designs that enrich and decorate homes with creative, innovative, fresh ideas. In that case, 3D printing could be a good choice as it helps to build and develop furniture and other accessories that make the house more beautiful. The household goods production industry has undergone significant integration of 3D printing technologies, which helps to build and create new designs quickly with customization and diversification.

Noha et al. used date palm fronds and plastic waste from drinking water bottles to create a composite with varying loadings of date palm frond nanoparticles (DPFNP) at 10, 20, and 30 wt%. They showed that the developed filament components showed good thermoregulation and mechanical properties, so the developed furniture can be used for decoration purposes in the home (Elessawy et al. 2024). 3D printing techniques are used for the decoration of the home wall. The Picoroco (Dimension:48″ × 48″ × 6″) wall was designed by Ronald Rael, Virginia San Fratello, and Seong Koo Lee and has a random surface pattern using a series of interconnected pentagons, hexagons, and quadrilateral shapes whose terminus is a circle. This wall is also heatproof (IDJ_Volume 6_Issue 3_Pages 143–163.pdf). 3D printing methods can be an effective option for rapidly constructing dwellings. This method mitigates elevated manufacturing and labor expenses, enhances building adaptability, and tailors the residence to the owner's preferences. Xu et al. illustrated a design-to-printing workflow for the 3D printing of an entire farmhouse structure in Wujiazhuang (WJZ) hamlet, as shown in Fig. 13.11h–i. Design strategies are established by a sequence of interdisciplinary research initiatives, including printing system development, site planning, material design, and structural assessment. Secondly, construction methods that integrate the management of the printing process and strategies that optimize labor efficiency are utilized in 3D printing. The executed workflow achieves a 62.4% decrease in human resources relative to the traditional reinforced concrete (RC) method. In comparison, the utilization of 3D printing technology escalates from 18.3% in the regular 3DCP scenario to 76.5% in this project. The on-site building length of the WJZ farmhouse is reduced by 24.5% compared to the unplanned construction technique (Xu et al. 2022). Marcin et al. created a cement-glass composite brick (CGCB) with a printed internal scaffold from recycled Polyethylene Terephthalate Glycol (PET-G) to enhance residential structures' compressive and flexural strength. These bricks exhibited superior thermal properties compared to CGM, with an 8% increase in thermal diffusivity and a 10% reduction in specific heat (Małek et al. 2022).

13.8.3 Automotive and Aerospace Applications

Textile printing has applications in aerospace and automation, as shown in Fig. 13.11j–k. 3D printing facilitates the production of prototypes for complex designs based on CAD engineering concepts. This technique enables the development of lightweight designs with many applications. Swift iteration in design and prototyping facilitates the creation of functional end-use components, encompassing high-performance parts and customized elements, via 3D printing technology.

Dashboards, consoles, air vents, control levers, handles, body panels, bumpers, wheels, subframe components, gearboxes, engines, and pumps can be made using digital printing techniques (Dalpadulo et al. 2022). Diverse sensor types can be fabricated with flexibility and durability using 3D printing processes. Several nanoparticles, including copper, silver, carbon nanotubes (CNT), and indium tin oxide (ITO), serve as ink materials for creating flexible sensors in the automotive sector; nonetheless, most of these sensors do not operate effectively. Srinivasan and Muthuramalingam et al. analyzed the track fracture of the fabricated sensor, which developed a flexible printed sensor using silver ink. The sensing mechanism of the fabricated sensor track was improved, and the track with inclined track specimens was capable of producing superior performance and negligible fractures during the formability process (Srinivasan and Muthuramalingam 2021). Diatezo et al. have developed and optimized 3D-printed electronic coated fabric. The advanced heating technique was developed utilizing a 3D printer with a multilayer thin film applied to the fabric substrate. The developed coating effectively addressed the overheating issue and provided consistent temperature regulation (Diatezo et al. 2023). Maurya et al. made a self-powered strain sensor by 3D printing based on graphene. Using graphene ink-based materials, this sensor analyses the interactions between tires and roads at varying velocities. It is capable of wireless communication and self-power sensing. Cost-effective solutions for innovative tires are provided by these sensors, energy harvesting, and secure data transmission. The cost was approximately 2.7 cents for one sensor (Maurya et al. 2020).

NASA used 3D printing to fabricate various rocket engine components, such as chambers, injectors, nozzles, and bimetallic chambers. Then, they showed a significant hot-fire test and evaluated a full-scale additive manufacturing component, which will be transported on NASA's SLS (Space Launch System) RS-25 Pogo Z-Baffle. Using a 3D printing technique will reduce complexity from 127 to 4 welds (Balaji et al. 2022). The Airbus A320's nacelle brackets and the A350's hinge bracket, made with materials like Ti–6Al–4V and topology optimization, are examples of lightweight, optimized designs made possible by 3D printing. 3D printing fabricates spacecraft structures in initiatives like the Juno mission to Jupiter and the Sentinel satellites. Components comprise antenna brackets and star tracker camera brackets optimized through topology optimization utilizing 3D printing technology. Components of spaceship engines comprise impellers, turbine blades, and fuel nozzles. The LEAP fuel nozzle by GE was produced using 3D printing technology (Blakey-Milner et al. 2021).

13.9 Future Trends and Challenges

The textile printing sector is headed for significant advancements driven by digital printing technologies (3D printing, inkjet printing, 4D printing), market demands, and environmental considerations. As the sector progresses, future trends and difficulties will influence its direction.

Innovative technologies have the potential to transform textile printing. Nanoprinting enables the creation of highly exact designs at a microscopic size, facilitating enhanced functionalities in textiles, including embedded sensors and antibacterial qualities (Kuzubasoglu et al. 2021; Liu et al. 2019; Abdelrahman et al. 2020). Bio-based inks, obtained from natural and renewable materials, are increasingly recognized as sustainable substitutes for synthetic inks, minimizing environmental effects while maintaining rich color quality. Multifunctional fabrics that include thermochromic or light-sensitive pigments signify a new frontier. These advances can provide dynamic designs that adapt to temperature, light, or contact, presenting new possibilities for fashion and functional uses (Pinto et al. 2016; Ahmed et al. 2023; Yang et al. 2024; Liu et al. 2022). Although there was potential for these breakthroughs, obstacles remained in attaining their extensive implementation. The expense is a considerable obstacle, as advanced technologies frequently necessitate extensive research, development, and equipment investments. Scalability presents

Table 13.9 Summary of the applications of printing in home textiles, apparel industry, spacecraft design

Materials/fabric/fiber	Printing technique	Application	Benefits/ characteristics	Refs.
Unspun fiber	3D printing	Home textile	• Develop different aesthetics of soft and imprecise objects	Hudson (2014)
–	3D printing	Apparel industry	• Long, partially transparent dress	Chakraborty and Biswas (2020)
Space fabric	3D, 4D printing	Apparel industry	• Thermoregulation • Durable	Chakraborty and Biswas (2020)
Primary matrix: Thermoplastic Polyurethane (Ninjaflex) Continuous fiber: Stainless steel thread	3D printing	Apparel industry	• High degree of flexibility	Pattinson et al. (2019)
Poly (vinyl alcohol) (PVA) and boron nitride (BN)	3D printing	Apparel industry	• High mechanical strength (355 MPa) and favorable heat dispersion	Gao et al. (2017)
Hybrid inks (TEMPO@Ti_3C_2 MXene)	3D printing	Apparel industry	• Strain sensor	Cao et al. (2019)
PGS, CNTs	3D printing	Apparel industry	• Monitor: Motion • Efficiency: High	Chen et al. (2018)
Nylon	SLS printing	Apparel industry	• Petal dress has • Pieces: 1600 • Hinges: 2600 • Reconfigurable dress	Xiao and Kan (2022)
Additive manufacturing (AM) components	Additive manufacturing	Spacecraft, Structures, Design, Topology Optimization	• Designing for additive manufacturing: Lightweighting through topology optimization enables lunar spacecraft	Orme et al. (2017)

a further challenge; although some developing techniques demonstrate efficacy in controlled environments, their adaptation for industrial applications can be intricate and resource-demanding. Technical constraints, such as the resilience of nano-printed designs and the accessibility of bio-based inks, provide challenges. Resolving these difficulties necessitates cooperative endeavors by researchers, manufacturers, and policymakers to develop feasible solutions that harmonize creativity with practicality.

Environmental and ethical factors are becoming increasingly essential to the future of textile printing. Garbage management, waste reduction, and chemicals are critical concerns, especially in the fast fashion and textile industry (Huang et al. 2017; Wang et al. 2017; Zhang et al. 2016b). Progress in recycling technology and circular economy frameworks is crucial for alleviating this impact. Energy consumption is a significant issue, as energy-intensive processes, such as heat setting, may negate the environmental advantages of sustainable inks or materials. Ethical factors, including ensuring equitable labor standards and procuring sustainable raw resources, will significantly influence the industry's future.

13.10 Conclusion

The advancement of textile printing has been highlighted by notable technological developments, transitioning from conventional block and screen-printing methods to innovative digital processes such as 3D, inkjet, and hybrid printing. Recent developments in this field have significantly altered the landscape, facilitating enhanced customization, improved efficiency, and increased sustainability. Emerging technologies such as nano-printing, bio-based inks, and interactive textiles present significant potential for the future, providing innovative ways to integrate functionality with creativity. 3D printing techniques are employed to create objects possessing aesthetic and functional attributes, which are utilized in clothes and household textiles such as reconfigurable dresses and curtains and in automotive and aerospace engineering. Multifunctional fabrics are created by integrating nanoparticles with a colorant (dye or pigment) using inkjet and screen-printing techniques, exhibiting antibacterial and self-cleaning characteristics. Integrating thermochromic and photochromic printing with the screen-printing technique produces many functional textiles in camouflage, packaging, security detection devices, and sensing applications. However, expense, scalability, and ecological consequences must be resolved to implement these goals fully while using digital printing techniques. Cooperative efforts among stakeholders will be crucial in overcoming these obstacles and advancing the sector toward a more sustainable and ethical future.

Declaration The authors declare that they have no known competing financial interests or personal relationships that could have appeared to influence the work reported in this chapter. Generative AI and AI-assisted technologies have been used in the writing process to improve the readability and language of the work only. After using this tool, the authors carefully reviewed and revised the content to ensure accuracy and quality. The authors take full responsibility for the final version of this publication.

References

Abdelrahman MS et al (2020) Studies of polylactic acid and metal oxide nanoparticles-based composites for multifunctional textile prints. Coatings 10(1):58. https://doi.org/10.3390/coatings10010058

Ahmed H, Abdelrahman MS, Al-Balakocy NG, Wen Z, Khattab TA (2022) Preparation of photochromic and photoluminescent nonwoven fibrous mat from recycled polyester waste. J Polym Environ 30(12):5239–5251. https://doi.org/10.1007/s10924-022-02587-y

Ahmed E, Maamoun D, Abdelrahman MS, Hassan TM, Khattab TA (2023) Imparting cotton textiles glow-in-the-dark property along with other functional properties: photochromism, flame-retardant, water-repellency, and antimicrobial activity. Cellulose 30(6):4041–4055. https://doi.org/10.1007/s10570-023-05125-8

Ahn J et al (2019) Printed fabric heater based on Ag nanowire/carbon nanotube composites. Nanotechnology 30(45):455707. https://doi.org/10.1088/1361-6528/ab35eb

Akter N, Akter N, Pervin M, Repon MdR (2023) The influence of mixed thickeners on printing over lyocell knitted fabric. Heliyon 9(3):e14175. https://doi.org/10.1016/j.heliyon.2023.e14175

Alam F, Alsharif A, AlModaf FO, El-Atab N (2024) 3D-printed smartwatch fabricated via vat photopolymerization for UV and temperature sensing applications. ACS Omega 9(13):14830–14839. https://doi.org/10.1021/acsomega.3c07411

Alami AH et al (2023) Additive manufacturing in the aerospace and automotive industries: recent trends and role in achieving sustainable development goals. Ain Shams Eng J 14(11):102516. https://doi.org/10.1016/j.asej.2023.102516

Arbab AA et al (2019) Fabrication of conductive and printable nano carbon ink for wearable electronic and heating fabrics. J Colloid Interface Sci 539:95–106. https://doi.org/10.1016/j.jcis.2018.12.050

Armstrong K, 3D printed dress debuts at New York fashion week. 3D Printing Industry. [Online]. https://3dprintingindustry.com/news/3d-printed-dress-debuts-new-york-fashion-week-95736/

Balaji D et al (2022) Additive manufacturing for aerospace from inception to certification. J Nanomater 2022(1):7226852. https://doi.org/10.1155/2022/7226852

Behera M, Nayak J, Banerjee S, Chakrabortty S, Tripathy SK (2021) A review on the treatment of textile industry waste effluents towards the development of efficient mitigation strategy: an integrated system design approach. J Environ Chem Eng 9(4):105277. https://doi.org/10.1016/j.jece.2021.105277

Bihar E et al (2017) Fully printed electrodes on stretchable textiles for long-term electrophysiology. Adv Mater Technol 2(4):1600251. https://doi.org/10.1002/admt.201600251

Blakey-Milner B et al (2021) Metal additive manufacturing in aerospace: a review. Mater Des 209:110008. https://doi.org/10.1016/j.matdes.2021.110008

Boumegnane A, Nadi A, Cherkaoui O, Tahiri M (2022) Inkjet printing of silver conductive ink on textiles for wearable electronic applications. Mater Today Proc 58:1235–1241. https://doi.org/10.1016/j.matpr.2022.01.469

Broadbent AD (2001) Basic principles of textile coloration. Society of Dyers and Colourists, Bradford, West Yorkshire

Brooks BJ, Arif KM, Dirven S, Potgieter J (2017) Robot-assisted 3D printing of biopolymer thin shells. Int J Adv Manuf Technol 89(1–4):957–968. https://doi.org/10.1007/s00170-016-9134-y

Cabral I, Santiago D, Steffens F (2023) Chromic textiles: colour fastness properties and irreversible colour change behaviour of textiles screen printed with thermochromic, photochromic and hydrochromic colourants. Color Technol 139(2):200–208. https://doi.org/10.1111/cote.12660

Cao W, Ma C, Mao D, Zhang J, Ma M, Chen F (2019) MXene-reinforced cellulose nanofibril inks for 3D-printed smart fibres and textiles. Adv Funct Mater 29(51):1905898. https://doi.org/10.1002/adfm.201905898

Cao T, Yang Z, Zhang H, Wang Y (2024a) Inkjet printing quality improvement research progress: a review. Heliyon 10(10):e30163. https://doi.org/10.1016/j.heliyon.2024.e30163

Cao Y, Xu B, Li B, Fu H (2024b) Advanced design of soft robots with artificial intelligence. Nano-Micro Lett 16(1):214. https://doi.org/10.1007/s40820-024-01423-3

Chakraborty S, Biswas MC (2020) 3D printing technology of polymer-fiber composites in textile and fashion industry: a potential roadmap of concept to consumer. Compos Struct 248:112562. https://doi.org/10.1016/j.compstruct.2020.112562

Chatterjee K, Ghosh TK (2020) 3D printing of textiles: potential roadmap to printing with fibers. Adv Mater 32(4):1902086. https://doi.org/10.1002/adma.201902086

Chen S et al (2018) A single integrated 3D-printing process customizes elastic and sustainable triboelectric nanogenerators for wearable electronics. Adv Funct Mater 28(46):1805108. https://doi.org/10.1002/adfm.201805108

Court D, Torah R (2021) Development of a printed e-textile for the measurement of muscle activation via EMG for the purpose of gesture control. In: International conference on the challenges, opportunities, innovations and applications in electronic textiles, MDPI, p 8. https://doi.org/10.3390/proceedings2021068008

Cruz BDD et al (2024) Thermochromic poly(L-lactic acid) based materials and their printability on different substrates. Chem Eng J 494:152956. https://doi.org/10.1016/j.cej.2024.152956

Dalpadulo E, Petruccioli A, Gherardini F, Leali F (2022) A review of automotive spare-part reconstruction based on additive manufacturing. J Manuf Mater Process 6(6):133. https://doi.org/10.3390/jmmp6060133

De Marzi A, Vibrante M, Bottin M, Franchin G (2023) Development of robot assisted hybrid additive manufacturing technology for the freeform fabrication of lattice structures. Addit Manuf 66:103456. https://doi.org/10.1016/j.addma.2023.103456

Diatezo L, Le M-Q, Tonellato C, Puig L, Capsal J-F, Cottinet P-J (2023) Development and optimization of 3D-Printed flexible electronic coatings: a new generation of smart heating fabrics for automobile applications. Micromachines 14(4):762. https://doi.org/10.3390/mi14040762

Dip TM et al (2020) 3D printing technology for textiles and fashion. Text Prog 52(4):167–260. https://doi.org/10.1080/00405167.2021.1978223

Djoko Kuswanto OMH, Iftira NJ (2018) 3D printing for fashion development. In: 3D printing for fashion development. IEEE, p 6

Elessawy NA, El Shakhs A, Fahmy El-Saka M, Youssef ME, Youssef BAB, Malek Ali MA (2024) Sustainable and eco-friendly 3D printing filament fabricated from different recycled solid wastes and evaluate its impact on interior and furniture design. Res Eng 23:102428. https://doi.org/10.1016/j.rineng.2024.102428

Fanglan Z, Kaifa D (2021) Innovative application of 3D printing technology in Fashion design. J Phys Conf Ser 1790(1). https://doi.org/10.1088/1742-6596/1790/1/012030

Gao T et al (2017) Three-dimensional printed thermal regulation textiles. ACS Nano 11(11):11513–11520. https://doi.org/10.1021/acsnano.7b06295

Gao C, Xing T, Hou X, Zhang Y, Chen G (2021) Clean production of polyester fabric inkjet printing process without fabric pretreatment and soaping. J Clean Prod 282:124315. https://doi.org/10.1016/j.jclepro.2020.124315

Ge F, Yu W, Yin Y, Wang C (2021) Solar-driven thermochromic fabric based on photothermal conversion for light intensity monitoring. J. Mater. Chem. A 9(36):20565–20575. https://doi.org/10.1039/D1TA04073A

Hayta P, Oktav M, Ateş Duru Ö (2022) An ecological approach to printing industry: development of ecofriendly offset printing inks using vegetable oils and pine resin as renewable raw materials and evaluation of printability. Color Res Appl 47(1):164–171. https://doi.org/10.1002/col.22708

He P et al (2019) Screen-printing of a highly conductive graphene ink for flexible printed electronics. ACS Appl Mater Interfaces 11(35):32225–32234. https://doi.org/10.1021/acsami.9b04589

How 3D Printing is Revolutionizing the Fashion Industry. [Online]. https://www.tomorrow.bio/post/how-3d-printing-is-revolutionizing-the-fashion-industry-2023-06-4727466168-3d-printing

Hu G et al (2018) Functional inks and printing of two-dimensional materials. Chem Soc Rev 47(9):3265–3300. https://doi.org/10.1039/C8CS00084K

Huang T, Cui H, Yang D, Kong X, Lin J (2017) Continuous dyeing processes for zipper tape in supercritical carbon dioxide. J Clean Prod 158:95–100. https://doi.org/10.1016/j.jclepro.2017.04.167

Hudson SE (2014) Printing teddy bears: a technique for 3D printing of soft interactive objects. In: Proceedings of the SIGCHI conference on human factors in computing systems. ACM, Toronto, Ontario, Canada, pp 459–468. https://doi.org/10.1145/2556288.2557338

IDJ_Volume 6_Issue 3_Pages 143–163.pdf

Ibarhiam SF et al (2022) Novel nanocomposite film developed via screen-printing of viologen polymer for anti-counterfeiting applications: photochromism, thermochromism and vapochromic. React Funct Polym 172:105186. https://doi.org/10.1016/j.reactfunctpolym.2022.105186

Islam MR et al (2022) Fully printed and multifunctional graphene-based wearable e-textiles for personalized healthcare applications. iScience 25(3):103945. https://doi.org/10.1016/j.isci.2022.103945

Islam MR, Afroj S, Yin J, Novoselov KS, Chen J, Karim N (2024) Advances in printed electronic textiles. Adv Sci 11(6):2304140. https://doi.org/10.1002/advs.202304140

Jabli M et al (2020) Green synthesis of colloid metal oxide nanoparticles using Cynomorium coccineum: application for printing cotton and evaluation of the antimicrobial activities. Mater Chem Phys 249:123171. https://doi.org/10.1016/j.matchemphys.2020.123171

Jandyal A, Chaturvedi I, Wazir I, Raina A, Ul Haq MI (2022) 3D printing—a review of processes, materials and applications in industry 4.0. Sustain Oper Comput 3(October 2021):33–42. https://doi.org/10.1016/j.susoc.2021.09.004

Joshi K et al (2020) A machine learning-assisted nanoparticle-printed biochip for real-time single cancer cell analysis. Adv Biosyst 4(11):2000160. https://doi.org/10.1002/adbi.202000160

Kan CW, Yuen CWM (2012) Digital ink-jet printing on textiles. Res J Text Appar 16(2):1–24. https://doi.org/10.1108/RJTA-16-02-2012-B001

Kang T-K (2021) Inkjet printing of highly sensitive, transparent, flexible linear piezoresistive strain sensors. Coatings 11(1):51. https://doi.org/10.3390/coatings11010051

Kang DJ, Jüttke Y, González-García L, Escudero A, Haft M, Kraus T (2020) Reversible conductive inkjet printing of healable and recyclable electrodes on cardboard and paper. Small 16(25):2000928. https://doi.org/10.1002/smll.202000928

Khan Y, Thielens A, Muin S, Ting J, Baumbauer C, Arias AC (2020) A new frontier of printed electronics: flexible hybrid electronics. Adv Mater 32(15):1905279. https://doi.org/10.1002/adma.201905279

Khan S, Ali S, Khan A, Bermak A (2022) Developing pressure sensors from impregnated textile sandwiched in inkjet-printed electrodes. J Mater Sci Mater Electron 33(1):541–553. https://doi.org/10.1007/s10854-021-07325-z

Komazaki Y, Uemura S (2019) Stretchable, printable, and tunable PDMS-CaCl2 microcomposite for capacitive humidity sensors on textiles. Sens Actuators B Chem 297:126711. https://doi.org/10.1016/j.snb.2019.126711

Kulkarni MV, Apte SK, Naik SD, Ambekar JD, Kale BB (2013) Ink-jet printed conducting polyaniline based flexible humidity sensor. Sens Actuators B Chem 178:140–143. https://doi.org/10.1016/j.snb.2012.12.046

Kuzubasoglu BA, Sayar E, Bahadir SK (2021) Inkjet-printed CNT/PEDOT:PSS temperature sensor on a textile substrate for wearable intelligent systems. IEEE Sens J 21(12):13090–13097. https://doi.org/10.1109/JSEN.2021.3070073

La T et al (2018) Two-layered and stretchable e-textile patches for wearable healthcare electronics. Adv Healthc Mater 7(22):1801033. https://doi.org/10.1002/adhm.201801033

Lee JY, An J, Chua CK (2017) Fundamentals and applications of 3D printing for novel materials. Appl Mater Today 7:120–133. https://doi.org/10.1016/j.apmt.2017.02.004

Li BM, Kim I, Zhou Y, Mills AC, Flewwellin TJ, Jur JS (2019) Kirigami-inspired textile electronics: K.I.T.E. Adv Mater Technol 4(11):1900511. https://doi.org/10.1002/admt.201900511

Li L, Chu R, Yang Q, Li M, Xing T, Chen G (2022) Performance of washing-free printing of disperse dye inks: influence of water-borne polymers. Polymers 14(20):4277. https://doi.org/10.3390/polym14204277

Li Y et al (2023) High-performance inkjet printing ink: properties and application effects of vinyl sulfone reactive dye-based inks. J Mol Liq 369:120864. https://doi.org/10.1016/j.molliq.2022.120864

Li Z et al (2025) 3D printing of thermochromic multilayer flexible film for multilevel information encryption. J Colloid Interface Sci 678:776–788. https://doi.org/10.1016/j.jcis.2024.09.143

Liang J, Tian B, Li S, Jiang C, Wu W (2020) All-printed MnHCF-MnO_x-based high-performance flexible supercapacitors. Adv Energy Mater 10(12):2000022. https://doi.org/10.1002/aenm.202000022

Liang Y et al (2021) Construction of new surface on linen fabric by hydroxyethyl cellulose for improving inkjet printing performance of reactive dyes. Prog Org Coat 154:106179. https://doi.org/10.1016/j.porgcoat.2021.106179

Lipskas J, Deep K, Yao W (2019) Robotic-assisted 3D bio-printing for repairing bone and cartilage defects through a minimally invasive approach. Sci Rep 9(1):3746. https://doi.org/10.1038/s41598-019-38972-2

Liu K et al (2019) Carbon black from diesel soot for high-performance wearable pressure sensors. Adv Mater Technol 4(10):1900475. https://doi.org/10.1002/admt.201900475

Liu Y et al (2022) A three-dimensional-printed recyclable, flexible, and wearable device for visualized UV, temperature, and sweat pH sensing. ACS Omega 7(11):9834–9845. https://doi.org/10.1021/acsomega.2c00128

Ma Y, Yu Y, Li J, Liu S, Huang W, Zhao Q (2021) Stimuli-responsive photofunctional materials for green and security printing. InfoMat 3(1):82–100. https://doi.org/10.1002/inf2.12125

Madhan S et al (2024) Preparation of eco-friendly inks for primary & secondary packaging. In: Proceedings—the twelfth international symposium GRID 2024. University of Novi Sad, Faculty of Technical Sciences, Department of Graphic Engineering and Design, Novi Sad. https://doi.org/10.24867/GRID-2024-p85

Majumdar A, Textile Institute (eds) Process control in textile manufacturing. In: Woodhead publishing series in textiles, no. 131. Woodhead Pub, Oxford, UK, Philadelphia, PA

Małek M et al (2022) Cement-glass composite bricks (CGCB) with interior 3D printed PET-G scaffolding. J Build Eng 52:104429. https://doi.org/10.1016/j.jobe.2022.104429

Manivannan R, Ryu J, Son Y-A (2021) Ultrasonic assisted fabrication of dual function surface on PET and preparation of single component ink to attain efficient self-cleaning function via digital printing. J Mol Liq 324:114668. https://doi.org/10.1016/j.molliq.2020.114668

Mauchauffé R, Lee SJ, Han I, Kim SH, Moon SY (2019) Improved de-inking of inkjet-printed paper using environmentally friendly atmospheric pressure low temperature plasma for paper recycling. Sci Rep 9(1):14046. https://doi.org/10.1038/s41598-019-50495-4

Maurya D et al (2020) 3D printed graphene-based self-powered strain sensors for smart tires in autonomous vehicles. Nat Commun 11(1):5392. https://doi.org/10.1038/s41467-020-19088-y

Miles LWC (ed) (1981) Textile printing. Dyers Company Publications Trust, Bradford

Mostafavi S, Bier H (2016) Materially informed design to robotic production: a robotic 3D printing system for informed material deposition. In: D. Reinhardt, R. Saunders, and J. Burry, Eds., Robotic fabrication in architecture, art and design 2016. Springer International Publishing, Cham, pp 338–349. https://doi.org/10.1007/978-3-319-26378-6_27

Nechyporchuk O, Yu J, Nierstrasz VA, Bordes R (2017) Cellulose nanofibril-based coatings of woven cotton fabrics for improved inkjet printing with a potential in e-textile manufacturing. ACS Sustain Chem Eng 5(6):4793–4801. https://doi.org/10.1021/acssuschemeng.7b00200

Nervous System (2016) Kinematic petals dress

Neto P (2013) Off-line programming and simulation from CAD drawings: robot-assisted sheet metal bending. In: IECON 2013—39th annual conference of the IEEE industrial electronics society. IEEE, Vienna, Austria, pp 4235–4240. https://doi.org/10.1109/IECON.2013.6699815

Neudecker S et al (2016) A new robotic spray technology for generative manufacturing of complex concrete structures without formwork. Procedia CIRP 43:333–338. https://doi.org/10.1016/j.procir.2016.02.107

Ngo TD, Kashani A, Imbalzano G, Nguyen KTQ, Hui D (2018) Additive manufacturing (3D printing): a review of materials, methods, applications and challenges. Compos Part B Eng 143:172–196. https://doi.org/10.1016/j.compositesb.2018.02.012

Nigusse AB, Malengier B, Mengistie DA, Tseghai GB, Van Langenhove L (2020) Development of washable silver printed textile electrodes for long-term ECG monitoring. Sensors 20(21):6233. https://doi.org/10.3390/s20216233

Nourbakhsh S, Ashjaran A (2012) Laser treatment of cotton fabric for durable antibacterial properties of silver nanoparticles. Materials 5(7):1247–1257. https://doi.org/10.3390/ma5071247

Offsetdruckfarbe W (2008) (54) Waterless waterwashable offset printing ink and method of production thereof

Ohiri KA et al (2022) E-textile based modular sEMG suit for large area level of effort analysis. Sci Rep 12(1):9650. https://doi.org/10.1038/s41598-022-13701-4

Orme ME, Gschweitl M, Ferrari M, Madera I, Mouriaux F (2017) Designing for additive manufacturing: lightweighting through topology optimization enables lunar spacecraft. J Mech Des 139(10):100905. https://doi.org/10.1115/1.4037304

Özen İ, Schneider R, Buchmeiser MR, Wang X (2022) Revisiting the sublimation printability of cellulose-based textiles in light of ever-increasing sustainability issues. Color Technol 138(6):581–589. https://doi.org/10.1111/cote.12639

Ozguney AT (2007) The comparison of laser surface designing and pigment printing methods for the product quality. Opt Laser Technol 39(5):1054–1058. https://doi.org/10.1016/j.optlastec.2006.05.004

Paiva A, Vieira D, Cunha J, Carvalho H, Providência B (2019) Design of a smart garment for cycling. In: Machado J, Soares F, Veiga G (eds) Innovation, engineering and entrepreneurship, vol 505. Lecture notes in electrical engineering. Springer International Publishing, Cham, pp 229–235. https://doi.org/10.1007/978-3-319-91334-6_32

Pan K et al (2018) Sustainable production of highly conductive multilayer graphene ink for wireless connectivity and IoT applications. Nat Commun 9(1):5197. https://doi.org/10.1038/s41467-018-07632-w

Pattinson SW et al (2019) Additive manufacturing of biomechanically tailored meshes for compliant wearable and implantable devices. Adv Funct Mater 29(32):1901815. https://doi.org/10.1002/adfm.201901815

Pei E, Shen J, Watling J (2015) Direct 3D printing of polymers onto textiles: experimental studies and applications. Rapid Prototyp J 21(5):556–571. https://doi.org/10.1108/RPJ-09-2014-0126

Peng H, Mankoff J, Hudson SE, McCann J (2015) A layered fabric 3D printer for soft interactive objects. In: Proceedings of the 33rd annual ACM conference on human factors in computing systems. ACM, Seoul Republic of Korea, pp 1789–1798. https://doi.org/10.1145/2702123.2702327

Pinto TV et al (2016) Screen-printed photochromic textiles through new inks based on SiO_2 @naphthopyran nanoparticles. ACS Appl Mater Interfaces 8(42):28935–28945. https://doi.org/10.1021/acsami.6b06686

Quintero AV et al (2015) Capacitive strain sensors inkjet-printed on PET fibers for integration in industrial textile. Procedia Eng 120:279–282. https://doi.org/10.1016/j.proeng.2015.08.613

Rabbi MdB, Tania IS, Sani AA, Uddin MdZ (2024) Facile approach for preparing printed cotton fabric with antimicrobial activity by utilizing the functional characteristics of nano-silver. J Inorg Organomet Polym Mater 34(8):3675–3688. https://doi.org/10.1007/s10904-024-03047-x

Ragab M, Othman H, Hassabo A (2022) An overview of printing textile techniques. Egypt J Chem. https://doi.org/10.21608/ejchem.2022.131477.5793

Rahmani Dabbagh S, Ozcan O, Tasoglu S (2022) Machine learning-enabled optimization of extrusion-based 3D printing. Methods 206:27–40. https://doi.org/10.1016/j.ymeth.2022.08.002

Rožić M, Šegota N, Vukoje M, Kulčar R, Šegota S (2020) Description of thermochromic offset prints morphologies depending on printing substrate. Appl Sci 10(22):8095. https://doi.org/10.3390/app10228095

Scott C, Julia Daviy uses 3D printing to create beautiful biodegradable fashion. [Online]. https://3dprint.com/212640/julia-daviy-3d-printing/

Sculpteo, 3D printed clothes in 2024: what are the best projects? [Online]. https://www.sculpteo.com/en/3d-learning-hub/applications-of-3d-printing/3d-printed-clothes/

Seipel S, Yu J, Periyasamy AP, Viková M, Vik M, Nierstrasz VA (2018) Inkjet printing and UV-LED curing of photochromic dyes for functional and smart textile applications. RSC Adv 8(50):28395–28404. https://doi.org/10.1039/C8RA05856C

Shah MA, Lee D-G, Lee B-Y, Hur S (2021) Classifications and applications of inkjet printing technology: a review. IEEE Access 9:140079–140102. https://doi.org/10.1109/ACCESS.2021.3119219

Shahariar H, Kim I, Soewardiman H, Jur JS (2019) Inkjet printing of reactive silver ink on textiles. ACS Appl Mater Interfaces 11(6):6208–6216. https://doi.org/10.1021/acsami.8b18231

Siddiqua UH, Ali S, Tufail A, Gatasheh MK, Riaz L, Yasir MW (2023) Relationship between reactive group chemistry and printing properties of heterofunctional reactive dyes via screen printing. Sci Rep 13(1):7259. https://doi.org/10.1038/s41598-023-33819-3

Silva AF et al (2020) High resolution soft and stretchable circuits with PVA/liquid-metal mediated printing. Adv Mater Technol 5(9):2000343. https://doi.org/10.1002/admt.202000343

Soleimani-Gorgani A, Avinc O, Alborz R (2023) Sustainable antibacterial cotton fabrics with in situ formed silver nanoparticles by bio-inkjet printing. J Clean Prod 386:135796. https://doi.org/10.1016/j.jclepro.2022.135796

Song Y et al (2023) Fabrication of antireflective coatings on cotton surface using dye-loaded nanoparticles for eco-friendly textile inkjet printing. Prog Org Coat 182:107607. https://doi.org/10.1016/j.porgcoat.2023.107607

Spahiu T, Canaj E, Shehi E (2020) 3D printing for clothing production. J Eng Fibers Fabr 15:1558925020948216. https://doi.org/10.1177/1558925020948216

Srinivasan KP, Muthuramalingam T (2021) Design, fabrication, and crack analysis of silver track printed flexible sensor for automobile infotainment application. IEEE Sens J 21(12):13910–13915. https://doi.org/10.1109/JSEN.2021.3071582

Sun M et al (2020) Smart cotton fabric screen-printed with viologen polymer: photochromic, thermochromic and ammonia sensing. Cellulose 27(5):2939–2952. https://doi.org/10.1007/s10570-020-02992-3

Swetha TG, Harini MS, Pachiyappan KM, Divya Sathyam R, Sharmila M (2024) Review of textile printing: differentiation of block, screen & digital printing. EPRA Int J Multidiscip Res IJMR 122–125. https://doi.org/10.36713/epra18574

Tkalec M, Glogar M, Penava Ž, Forte Tavčer P, Kuščer D, Stojanoska I (2024) The complexity of colour/textile interaction in digital printing as an integral part of environmental design. Arts 13(1):29. https://doi.org/10.3390/arts13010029

Toral V et al (2020) Cost-effective printed electrodes based on emerging materials applied to biosignal acquisition. IEEE Access 8:127789–127800. https://doi.org/10.1109/ACCESS.2020.3008945

Tse S, Kan C (2020) Effect of laser treatment on pigment printing on denim fabric: low stress mechanical properties. Cellulose 27(17):10385–10405. https://doi.org/10.1007/s10570-020-03461-7

Uddin Rabbi MdB, Haque S, Bedoura S (2024) Advancements in synthesis, immobilization, characterization, and multifaceted applications of silver nanoparticles: a comprehensive review. Heliyon 10(24):e40931. https://doi.org/10.1016/j.heliyon.2024.e40931

Ugur Koseoglu A (2019) Innovations and analysis of textile digital printing technology. Int J Sci Technol Soc 7(2):38. https://doi.org/10.11648/j.ijsts.20190702.12

Urhal P, Weightman A, Diver C, Bartolo P (2019) Robot assisted additive manufacturing: a review. Robot Comput Integr Manuf 59:335–345. https://doi.org/10.1016/j.rcim.2019.05.005

van Herpen I: Transforming Fashion. [Online]. https://high.org/exhibition/iris-van-herpen-transforming-fashion/
Wang L, Yan K, Hu C (2017) Cleaner production of inkjet printed cotton fabrics using a urea-free ecosteam process. J Clean Prod 143:1215–1220. https://doi.org/10.1016/j.jclepro.2016.11.182
Wang L, Zhu Q, Yu H, Xie G, Yu Z (2022) Agricultural crops of sarsaparilla root as source of natural dye for inkjet printing and bio-functional finishing of cotton fabric. Ind Crops Prod 187:115490. https://doi.org/10.1016/j.indcrop.2022.115490
Wang L et al (2023a) Regulating the alkyl chain length of quaternary ammonium salt to enhance the inkjet printing performance on cationic cotton fabric with reactive dye ink. ACS Appl Mater Interfaces 15(15):19750–19760. https://doi.org/10.1021/acsami.3c02304
Wang M et al (2023b) Green fabrication of inkjet printed antibacterial wool fabric with natural gardenia yellow dye. Ind Crops Prod 206:117700. https://doi.org/10.1016/j.indcrop.2023.117700
Wang M et al (2023c) Simple surface low temperature grafting for antibacterial and anti-felting inkjet printing wool fabrics. Prog Org Coat 183:107723. https://doi.org/10.1016/j.porgcoat.2023.107723
Xiao Y-Q, Kan C-W (2022) Review on development and application of 3D-printing technology in textile and fashion design. Coatings 12(2):267. https://doi.org/10.3390/coatings12020267
Xu W et al (2022) Toward automated construction: the design-to-printing workflow for a robotic in-situ 3D printed house. Case Stud Constr Mater 17:e01442. https://doi.org/10.1016/j.cscm.2022.e01442
Yang Y, Li M, Fu S (2021) Screen-printed photochromic textiles with high fastness prepared by self-adhesive polymer latex particles. Prog Org Coat 158:106348. https://doi.org/10.1016/j.porgcoat.2021.106348
Yang L et al (2024) Reversible dual-responsive color-changing fabric based on thermochromic microcapsules for textile fashion and intelligent monitoring. Dyes Pigments 231:112397. https://doi.org/10.1016/j.dyepig.2024.112397
Zhang GQ, Spaak A, Martinez C, Lasko DT, Zhang B, Fuhlbrigge TA (2016a) Robotic additive manufacturing process simulation—towards design and analysis with building parameter in consideration. In: 2016 IEEE international conference on automation science and engineering (CASE). IEEE, Fort Worth, TX, USA, pp 609–613. https://doi.org/10.1109/COASE.2016.7743457
Zhang H, Gao A, Song X, Hou A (2016b) Cleaner production applied to urea-free printing of cotton fabrics using polyethylene glycol polymers as alternative additives. J Clean Prod 124:126–131. https://doi.org/10.1016/j.jclepro.2016.02.090
Zhang Q et al (2021a) Highly sensitive resistance-type flexible pressure sensor for cuffless blood-pressure monitoring by using neural network techniques. Compos Part B Eng 226:109365. https://doi.org/10.1016/j.compositesb.2021.109365
Zhang W-W et al (2021b) A novel multicolor viologen-derived Zn-organic coordination polymer for environment friendly ink free erasable printing. J Solid State Chem 304:122597. https://doi.org/10.1016/j.jssc.2021.122597
Zhao H et al (2022) Insights into coloration enhancement of mercerized cotton fabric on reactive dye digital inkjet printing. RSC Adv 12(17):10386–10394. https://doi.org/10.1039/D2RA01053D
Zhu Z, Ng DWH, Park HS, McAlpine MC (2020) 3D-printed multifunctional materials enabled by artificial-intelligence-assisted fabrication technologies. Nat Rev Mater 6(1):27–47. https://doi.org/10.1038/s41578-020-00235-2
Zhu J et al (2021) Machine learning-enabled textile-based graphene gas sensing with energy harvesting-assisted IoT application. Nano Energy 86:106035. https://doi.org/10.1016/j.nanoen.2021.106035

MIX
Papier aus verantwortungsvollen Quellen
Paper from responsible sources
FSC® C105338

If you have any concerns about our products,
you can contact us on
ProductSafety@springernature.com

In case Publisher is established outside the EU,
the EU authorized representative is:
Springer Nature Customer Service Center GmbH
Europaplatz 3, 69115 Heidelberg, Germany

Printed by Libri Plureos GmbH
in Hamburg, Germany